INTRODUCTION TO CIRCLE PACKING

The Theory of Discrete Analytic Functions

KENNETH STEPHENSON

University of Tennessee

CAMBRIDGE
UNIVERSITY PRESS

CAMBRIDGE UNIVERSITY PRESS
Cambridge, New York, Melbourne, Madrid, Cape Town, Singapore, São Paulo

Cambridge University Press
40 West 20th Street, New York, NY 10011-4211, USA

www.cambridge.org
Information on this title: www.cambridge.org/9780521823562

First published 2005

Printed in the United States of America

A catalog record for this publication is available from the British Library.

Library of Congress Cataloging in Publication Data

Stephenson, Kenneth, 1945–
Introduction to circle packing : the theory of discrete analytic functions /
Kenneth Stephenson.
p. cm.
Includes bibliographical references and index.
ISBN 0-521-82356-0 (hardback : alk. paper)
1. Circle packing. 2. Discrete geometry. 3. Analytic functions. I. Title.
QA640.7.S74 2005
516′.11 – dc22 2004054523

ISBN-13 978-0-521-82356-2 hardback
ISBN-10 0-521-82356-0 hardback

INTRODUCTION TO CIRCLE PACKING

The topic of "circle packing" was born of the computer age but takes its inspiration and themes from core areas of classical mathematics. A circle packing is a configuration of circles having a specified pattern of tangencies, as introduced by William Thurston in 1985. This book lays out their study, from first definitions to the latest theory, computations, and applications. The topic can be enjoyed for the visual appeal of the packing images – over 200 in the book – and the elegance of circle geometry, for the clean line of theory, for the deep connections to classical topics, or for the emerging applications. Circle packing has an experimental and visual character that is unique in pure mathematics, and the book exploits that character to carry the reader from the very beginnings to links with complex analysis and Riemann surfaces. There are intriguing, often very accessible, open problems throughout the book and seven Appendices on subtopics of independent interest. This book lays the foundation for a topic with wide appeal and a bright future.

To my lovely wife, Dolores,
and our two wonderful children, Laura and Jon

Contents

Preface

The circle is arguably the most studied object in all of mathematics, so it was a surprise at a conference in 1985 to hear William Thurston introducing a new topic called circle packing. And if encountering a new idea is one of the pleasures of mathematics, then seeing it attach to your favorite topic is a true joy. When Thurston conjectured a connection between circle packing and the venerable Riemann Mapping Theorem of 1851, I was hooked.

Now, nearly 20 years later, one sees that this was no mere passing encounter for these topics. Circle packing has opened a discrete world that both parallels and approximates the classical world of conformal geometry – a "quantum" complex analysis that is classical in the limit. In this book, I introduce circle packing as a portal into the beauties of conformal geometry, while I use the classical theory as a roadmap for developing circle packing.

Circle packings are configurations of circles with specified patterns of tangency. They should not be confused with sphere packings; here, it is the pattern of tangencies that is central – the connection between combinatorics and geometry. We will study the existence, uniqueness, computation, manipulation, display, and application of circle packings from the ground up. There are no formal prerequisites for this study; indeed, I shamelessly exploit the visual nature of circle packing and our native intuition about circles so that even the novice mathematician can penetrate deeply into the subject. At the same time, I have an obligation to circle packing itself as a new field, so the book is mathematically rigorous.

To balance access with rigor, I have structured the book in four parts. For most readers, Part I will be the first encounter with circle packings, so it is a broad overview: from a circle packing managerie to the function-theory paradigm. We become more formal in Part II with a proof of the fundamental existence and uniqueness result for maximal packings from first principles. Because the key roles are played by surprisingly elementary geometric arguments, this can serve even the non-mathematician as an exemplar of a robust and self-contained mathematical theory. In the classroom, Part II serves well as a one semester course for advanced undergraduate or beginning graduate students.

I define a discrete analytic function theory based on circle packings in Part III. At its core, analyticity is a profoundly geometric property, and this comes out in the discrete setting in a very compelling way. The amazing integrity of the analogies is confirmed in Part IV, when we prove that under refinement, the objects of the discrete theory converge to their classical counterparts. We prove Thurston's 1985 conjecture on the approximation of conformal maps (the Rodin/Sullivan Theorem) from first principles and then extend it broadly to other functions and to conformal structures. The circle packing methods described here have made the 150-year-old Riemann Mapping Theorem computable in many situations for the very first time. I demonstrate a number of applications, the most surprising of which involves "cortical brain mapping." Material in Parts III and IV could augment the traditional complex analysis sequence or serve as a basis for advanced topics courses.

The book also provides a wealth of material for individual or group projects from the undergraduate to the research level. I promote an intuitive and hands-on approach throughout, posing many open questions and experimental opportunities; see in particular, the several independent topics in the appendixes. I have provided "practica" on computational and software issues for those willing to get their hands dirty, and one can always download and run my software package, `CirclePack`. The book closes with a full circle packing bibliography.

People are drawn to mathematics for any number of reasons, from the clarity in elementary geometry, the challenge of richly layered theory and open questions, the discipline of computation, to the need for results in other areas. Read with an open mind and you can find all of these in circle packing – and I have not even mentioned the pure aesthetic pleasure of the pictures, which can sustain us all through the rough patches. I hope you enjoy circle packing.

I have many people to thank for their advice, encouragement, and patience over the years of this book's writing. Thanks go to my circle packing collaborators and friends, from whom I've learned so much: Dov Aharonov, Phil Bowers, Charles Collins, and Alan Beardon; also to my former students Tomasz Dubejko and Brock Williams, great circle packers both. Special thanks to Alan Beardon and Fred Gehring for their unfailing encouragement and sage advice over the years; to Oded Schramm, Jim Cannon, and Bill Floyd for many enjoyable and insightful conversations; and to William Thurston for the audacious notion of circle packing. To my many friends in classical function theory: I'm still one of you!

I began writing during a sabbatical at the University of Cambridge; my thanks to the department and, particularly, to Alan Beardon, Keith Carne, and my part-III class. Of course, this project could never have succeeded without the continued support and encouragement of wonderful colleagues and staff here at the University of Tennessee. Thanks to the circle packing class who helped me hone my notes: James Ashe, Matt Cathey, Denise Halverson, and Jason Howard. Finally, I acknowledge a debt of gratitude to the National Science Foundation for its financial support over the years and, likewise, to the Tennessee Science Alliance.

Part I

An Overview of Circle Packing

Circle packing has arrived so recently on the mathematical scene as to be totally new to many readers. Therefore, I am devoting Part I to an informal and largely visual tour of the topic, introducing only the most basic terminology and notation, but giving the reader a glimpse of how the full story will unfold. We all understand circles, but the reader may be surprised at how deeply they can carry us into the heart of conformal geometry.

Chapter 1 begins with a visit to a "menagerie" of circle packings, a wide-ranging collection that I hope you enjoy as much in the touring as I did in the collecting. The Menagerie suggests our first theme: Given a specified combinatoric pattern, what can one say about the existence, uniqueness, and variety of circle packings having those combinatorics? In Chapter 2 we get a view of the landscape beyond existence and uniqueness, where the central theme of the book – the emergence of fundamental parallels with analytic functions – plays out. I hope to set a style here that will carry on throughout the book, namely, one that engages the reader's native intuition not about static pictures, but about packing dynamics: How does a packing react if we change its boundary radii? If we introduce branching? How are the combinatorics and geometry feeding off one another? Deep classical themes will bubble to the surface with a surprising ease and clarity if one only remains open to the possibilities.

At the end of Part I is the first of four practica inviting the reader into the experimental side of the topic. Circle packings exist both in theory and in *fact*. The book requires nothing more than mental experiments, but those with an adventurous spirit may wish to grapple with my software package `CirclePack` or even do their own coding.

1
A Circle Packing Menagerie

1.1. First Views

In the belief that images speak louder than words, I will begin with a preliminary ramble through a menagerie of circle packings. Look for the common features and the differences in preparation for the guided tour to follow.

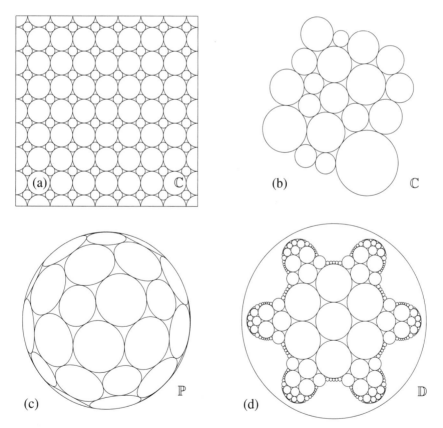

Figure 1.1. Collection 1.

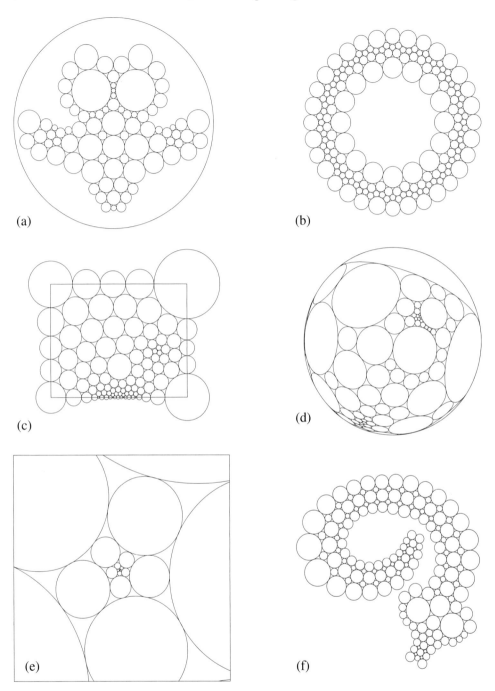

(a)

(b)

(c)

(d)

(e)

(f)

Figure 1.2. Collection 2.

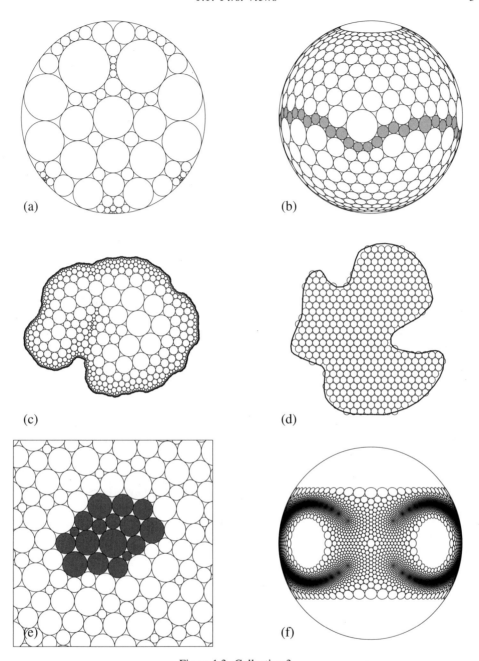

Figure 1.3. Collection 3.

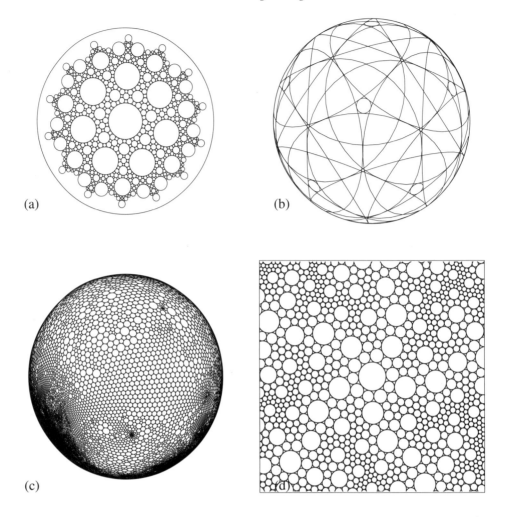

(a)

(b)

(c)

(d)

Figure 1.4. Collection 4.

The examples we have seen so far have been relatively tame. Before we start our guided tour, let us turn up the heat a notch with the more involved examples in Figure 1.4. Some of these reflect the complexity of applications; others contain internal symmetries, the more subtle of which may not be immediately evident. They represent topics we will touch upon in Parts III and IV of the book.

1.2. A Guided Tour

We start with some basics that you may already have deduced. First, there are three geometric settings for our packings, *euclidean*, *spherical*, and *hyperbolic*. In Figure 1.1

we see the familiar *euclidean plane* $\mathbb{R}^2$ in (a) and (b), the sphere in (c), and the interior of a disc in (d). Throughout the book, we treat $\mathbb{R}^2$ as the complex plane $\mathbb{C}$ and make use of complex arithmetic. The sphere will be the *Riemann sphere*, represented as the ordinary unit sphere in $\mathbb{R}^3$ and denoted by $\mathbb{P}$ (for *complex projective space*). The outer circle enclosing the packing of Figure 1.1(d) is not part of the packing; rather it is the boundary of the unit disc $\mathbb{D}$ in the plane. Here, however, $\mathbb{D}$ represents the Poincaré disc, a standard model of the *hyperbolic plane*. The geometries of $\mathbb{P}$, $\mathbb{C}$, and $\mathbb{D}$ will be of central importance in our work and we will have more to say about their distinct personalities shortly.

Next, observe that each packing involves an underlying pattern of tangencies. The hierarchy of structures is indicated in Fig. 1.5. All our tangencies are *external*, each circle lying outside the disc bounded by the other. In fact, however, tangencies do not occur in isolated pairs; rather the fundamental units of the patterns are mutually tangent triples of circles (*triples*, for short), with each triple forming a (curvilinear) triangular *interstice*. Triples are as important to the rigidity associated with circle packings as cross-bracing is to the rigidity of a bookcase. In turn, the triples of a pattern are linked together through shared pairs of circles to form the next level of structure, the *flower*, consisting of a central circle and some number of *petal* circles, the chain of successively tangent neighbors. The number of petals defines the *degree* of the central circle. The condition that every

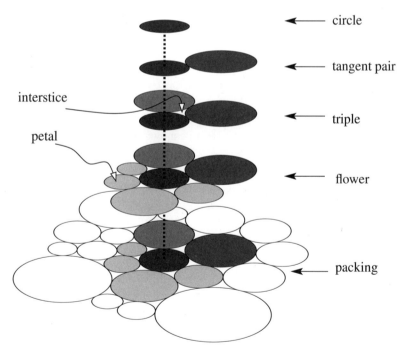

Figure 1.5. A hierarchy of circle packing structure.

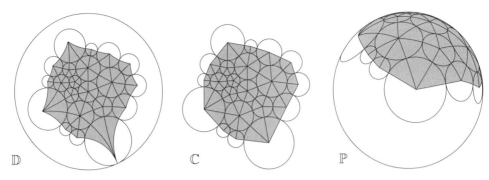

Figure 1.6. Carriers in the three geometries.

circle have such a flower is a *local planarity* condition that we will enforce on all our packings. Last, of course, the overall pattern of a packing consists of interlinked flowers. There will be some mild additional restrictions when we get to the function theory of Part III.

Within the context of the pattern of tangencies, further variability resides with the radii of the circles. Indeed, each triple may be thought of as a discrete piece of "geometry," its shape determined by the three radii involved. But in sewing these pieces together, even the first step, formation of a flower from several triples, involves a compatibility condition on the ensemble. And then, of course, the flowers must further cooperate to form the overall pattern. So although one might say that the combinatorics provide a master plan, it is the patient cooperation of the individual circles that allows that plan to be realized geometrically as a circle packing.

Various auxiliary structures in a circle packing may highlight, or even suggest, important properties. Foremost among the structures is the *carrier*, obtained by connecting the centers of tangent circles with geodesic segments to form *edges* and *faces*. The carrier is a geometric realization of a packing's abstract combinatorics. The packing of Figure 1.5 has been transplanted to Figure 1.6, where it is shown with its carrier in the hyperbolic, euclidean, and spherical settings.

Identifying various portions of a packing or carrier, appropriately grouping faces or circles, even coloring for visual effect, can make all the difference in understanding a particular packing. I have reproduced the packings of Figure 1.4 in Figure 1.7 with various decorations. The shaded faces in Figure 1.7(a) reproduce one of the most famous illustrations from 19th-century function theory, a fundamental domain for the *Klein surface*. The packing in Figure 1.7(b) is a *branched* packing and the dark circles are among the 12 of degree 5 at which the branching occurs. (Incidentally, this packing has the same combinatorics as Figure 1.1(c).) The packing in Figure 1.7(c) represents a flattened cortical surface from a magnetic resonance imaging (MRI) scan of a human cerebellum; colors are essential in visually distinguishing various cortical lobes. By drawing only certain edges in Figure 1.7(d) we reveal a hidden tiling, and even more selective darkened edges show an emerging fractal curve. We will use decorations like these to highlight features of packings throughout the book.

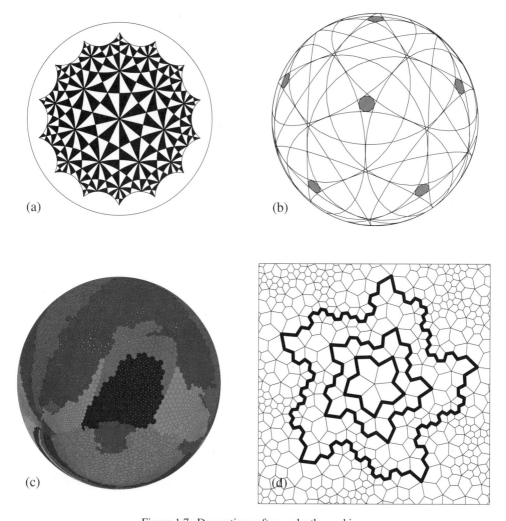

Figure 1.7. Decorations often make the packing.

Putting aside these decorations, there are two central sources of flexibility in circle packing: *combinatoric* and *geometric*. The clash and cooperation between these two provides the speciation seen in our Menagerie. Let us pin down some of that variability.

1.2.1. Combinatorial Flexibility

Combinatorial flexibility has to do with the pattern of tangencies. Circle packings may be finite or infinite; some circles will be *interior* to their patterns, fully surrounded by their flowers, while others are *boundary* circles, ones at the edge with flowers that do not close up. Combinatorics may be conceived abstractly, derived from some natural configuration, as with lattices, or obtained from existing packings. Thus Figure 1.3(d) was

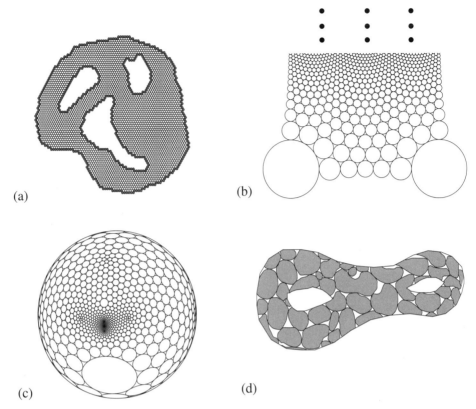

(a)

(b)

(c)

(d)

Figure 1.8. Illustrations of topological variety.

cookie-cut from an infinite packing, while Figure 1.3(b) resulted from an abstract gluing of two patterns, the line of circles showing the weld. Topological variety is illustrated in Figure 1.8. Packings can have more than one boundary component and even infinite packings can have boundary circles. Finite packings might have no boundary, as in the packing of the sphere in (c) or the packing of a compact genus 2 surface suggested by (d). (Note with regard to the latter that we will need to clarify the sense in which these are "circles" – what metric is in play?)

Certain of our patterns have highly regular combinatorics. Archetypes are the constant-degree complexes $K^{[n]}$ in which every circle has n neighbors. Figure 1.9 shows us a packing of $K^{[5]}$ in the sphere and a packing of $K^{[7]}$ in the disc. And everyone will recognize in the background the familiar "penny" packing, the *regular hexagonal packing* of the plane for $K^{[6]}$ (also denoted H). Each of these packings has circles of constant radius in its respective geometry. Since we are talking combinatorics, however, you might note that Figure 1.2(e) is also hexagonal, though quite distinct globally from a penny packing.

The *ball-bearing* motif of Figure 1.1(a) has a nice rectangular feel, but of course the large circles alone would leave quadrangular *versus* triangular interstices, so we have to add in the smaller ball-bearing circles to meet our basic triples requirement.

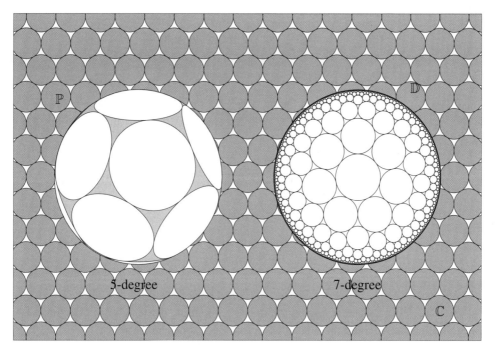

Figure 1.9. Constant degree packings: 5-, 6-, and 7-degree.

Combinatorial symmetries will be common in our work, but they can be very subtle. In Figure 1.3(e), for instance, one can with careful scrutiny and the aid of the shaded circles confirm that there is an underlying double translational symmetry; we will see later that this shaded fragment should be treated as a circle packing lying on a torus.

Several examples, such as Figure 1.3(c), were chosen to emphasize that no combinatorial regularity is required a priori. In point of fact, there are few combinatoric restrictions on our circle packings; beyond the requirement that the circles form triples and flowers, we have natural connectivity and orientability conditions – all will be described in detail at the appropriate time. We ultimately require only *sensible* combinatorics.

Keep in mind through all these examples that the pattern of tangencies of a packing is a combinatoric and not a geometric notion. It is the radii which imbue the pattern with size and shape, so let us move to the discussion of that source of variability.

1.2.2. Geometric Flexibility

Check out Figures 1.2(a) and 1.3(a). See any similarity? They're clearly distinct packings, but when they are shown side by side in Figure 1.10, one can – with a magnifying class and great patience – verify that they have the same underlying combinatorics. One could number the circles of P, say, and then transfer that numbering to Q in such a way that

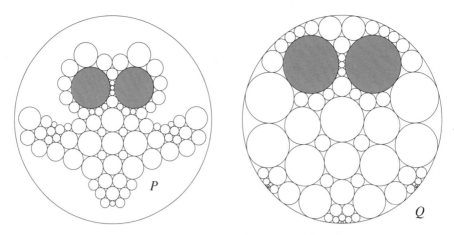

Figure 1.10. Packings sharing the same combinatorics.

two circles of Q are tangent if and only if the corresponding circles of P are tangent. (The "eyes" are shaded here and in some later pictures as helpful landmarks.)

It is clear that the radii are providing the geometric flexibility on display here. As our topic unfolds, you will find that *making adjustments in radii is the principal method for manipulating packings.* This brings up an important point that I want to emphasize right at the outset. Parts of the circle packing literature assume (often without comment) that the circles of a packing have mutually disjoint interiors. As will become abundantly clear, *this is **not** to be assumed here. Neighboring* circles will naturally have disjoint interiors, being externally tangent, but *non*-neighbors may well overlap or form *extraneous* tangencies.

Packings whose circles have mutually disjoint interiors, termed *univalent*, play a central role in both theory and practice (and have clear advantages for pictures), but they are in a sense atypical. Figure 1.11 illustrates. Figure 1.11(a) is univalent, while (b) and (c) depict the two basic ways in which univalence fails, one global and one local. The packing of (b) satisfies what is known as *local univalence*; that is, the faces formed by the flower of any interior circle are nonoverlapping. Nonetheless, this packing is globally nonunivalent because different parts of the packing happen to end up covering the same region of the plane. On the other hand, in Figure 1.11(c) (admittedly difficult to interpret right now), local univalence fails at a *branch circle.* Zooming in on this special flower, one can see that its seven petal circles wrap twice (rather than once) around the shaded center circle.

Another issue is the ambient metric; radii of a packing depend on the geometry in use. As we saw in Figure 1.6, a packing in the disc lives as well in the plane or the sphere, so its radii and carrier may be given in the hyperbolic metric of $\mathbb{D}$, the familiar euclidean metric of $\mathbb{C}$, or the spherical metric appropriate to $\mathbb{P}$. In a sense, it is the "same"

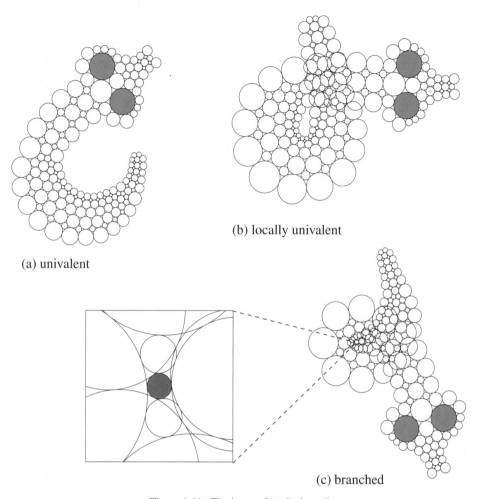

(b) locally univalent

(a) univalent

(c) branched

Figure 1.11. The issue of "univalence".

packing – the choice of geometry may be a matter of convenience, computational efficiency, or even aesthetics.

Within a single geometry, there is also the matter of normalization. Each of our standard spaces comes equipped with a family of *conformal automorphisms*, which coincidentally are the homeomorphisms mapping circles to circles! We routinely designate two circles of our packing for normalizations: one, called the α-circle, is centered at the origin, while the second is centered on the positive imaginary axis (best for recognizing symmetries). A packing that is unique up to a conformal automorphism will be termed *essentially unique*. It is important to note that the richness of the set of conformal automorphisms depends on which of the three geometries we happen to be working in and has a marked impact on the computations.

1.2.3. Tour Summary

This, then, completes the visit to our Menagerie. The term *reference collection* might be appropriate, since we come back to these images throughout the book. It is an impressive and pretty collection, don't you think? However, it may well leave the reader unsettled. While displaying the broad genera and species, one hardly feels these to be living things, subject to change and action. How are such packings created? What tools do we have for manipulating them? We need to move to their native environments and see circle packings in the wild, so to speak. This book is really about the dynamics of circle packings – free, not caged.

2

Circle Packings in the Wild

Have we simply been fortunate to find a few ossified specimens of circle packings, some rare accidents of nature? No. In point of fact, there are huge herds of these things, many of which have never been observed! We will learn how to create and manipulate new species with great facility. How do we proceed? What do we control and how do these packings respond? The dynamic issues are what this book is really about, and I want to give the reader an overview of the wilder landscape we are entering and a feel for some of the key geometric issues.

At this juncture we need a modicum of notation and terminology – what I call the "bookkeeping" aspect of circle packing. I intend to keep this to a minimum here so that we can continue rather informally. Let yourself go with the visual flow; we will be getting to the specifics soon enough.

2.1. Basic Bookkeeping

Just to be perfectly clear, our task is no longer mere classification and examination of given packings. Now we are to start with desired combinatorics and to *construct, manipulate*, and *study* packings with those combinatorics. The essential bookkeeping ingredients involve methods for encoding the combinatorics, for storing circle radii and centers, and of course for verifying that the data are consistent with a concrete circle packing.

To encode the combinatorics of a packing P one basically needs lists of circles, required tangencies, and triples. We structure this abstractly as vertices, edges, and faces, which together will be called the *complex* for P, generally written as K. Each vertex represents a circle, each edge between vertices represents a tangency between two circles, and each face represents a triple. The easiest way to capture the restrictions we need is to declare that K *must represent a triangulation of a surface*.

Of course, K is an abstract combinatorial object, with no metric and no geometry. The metric data will reside in a *label* R attached to K which associates a positive number (a putative radius) to each circle (i.e., to each vertex of K). Circle *centers* turn out to be secondary data. We will prove early on that they are essentially determined by the radii and combinatorics: you can lay down one circle and a neighbor, and the placements of the rest are then forced on you by the tangency relationships.

15

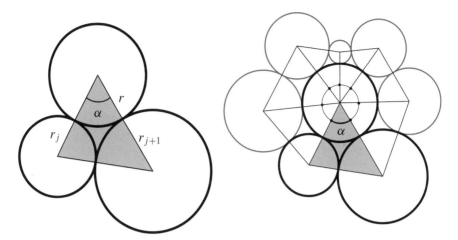

Figure 2.1. The Law of Cosines and flower angle sums.

The "label" vector R associated with K has been given this rather neutral name because its entries serve as the main parameters in our work. It is evident that a generic label – say, some random collection of positive values – is very unlikely to represent radii of circles that actually fit together in the pattern encoded by K. Indeed, that is the first surprise I warned you about: that there exist any such labels! What is a criterion for the "good" labels? A necessary condition involves local compatibility through the *angle sum*. The notion is quite simple: Suppose v is a vertex of K with neighboring vertices $v_1, \ldots, v_k$ in counterclockwise order. You are given associated radii $\{r; r_1, \ldots, r_k\}$ and asked if circles with these radii form a closed k-flower. To check, we would all probably do the same thing: lay down a circle of radius r, place one of radius r_1 tangent to it, and then proceed (counterclockwise) with successive petals, laying each tangent to the center and the previous petal. As we place the last circle, radius r_k, the answer becomes clear: if it is tangent to the first, then we have our k-flower; otherwise, the petals either reach too far around or leave a gap – these radii have failed the compatibility check.

Simple trigonometry makes this criterion explicit. In euclidean geometry, for each triple $\langle r, r_j, r_{j+1} \rangle$ of radii, the Law of Cosines gives the angle α in a corresponding triple of circles, as on the left in Figure 2.1:

$$\alpha = \arccos \left(\frac{(r + r_j)^2 + (r + r_{j+1})^2 - (r_j + r_{j+1})^2}{2(r + r_j)(r + r_{j+1})} \right).$$

If we add these individual angles over the k triples involved (and don't forget the last one, $\langle r, r_k, r_1 \rangle$), we get the *angle sum* for this label at v, denoted $\theta_R(v)$. The radii are compatible if and only if $\theta_R(v) = 2\pi n$ for some integer $n \geq 1$, and in that case we say that R satisfies a *packing condition* at v. The integer n is the number of times the petals wrap around the center, so it must be 1 for local univalence.

This rather elementary notion of angle sum is the key to all of our later work. If a label *R* satisfies the packing condition at each interior vertex *v* of *K* (flowers of boundary circles are not required to close up), then we say that *R* is a *packing label* for *K*. Is this, then, enough to warrant describing *R* as a label of "radii"? In many cases – for example, if *K* is simply connected – the answer is yes, but in others there can be global obstructions to realizing a packing. This local-to-global tension provides one of the themes that we will see played out as we get deeper into the topic.

You may notice a rather casual use of the term "circle" throughout this book. In manipulating packings, one treats circles as objects to be moved, resized, and shuffled around. Indeed, we work largely with the radii alone and encounter concrete circles only as we lay out a packing.

2.2. The Storyline

With the bookkeeping in place, I can now provide an overview of the book's storyline. The geometry of circle packing is a juxtaposition of *rigidity* and *flexibility*. In Part II of the book, we will see the forces of rigidity in the form of extremal packings, where combinatorics determine everything. In Part III, we will marshal forces of variability in boundary conditions and branching, variability that we capture in the form of discrete analytic functions and discrete conformal structures. Then in Part IV we resolve the apparent conflict between these forces; circle packings are part of the landscape of classical analytic function theory – the archetype for *local* rigidity within *global* flexibility – and our discrete notions actually approximate their classical counterparts.

2.2.1. Part II: The Rigidity of Maximal Circle Packings

Our work begins with the recognition that one can generate circle packings in wild abundance. Part II of the book establishes this quite amazing fact:

Given any sensible combinatorics, there exists a circle packing with tangency pattern matching those combinatorics.

This result had its genesis in seminal ideas of William Thurston, E. M. Andreev, and, earlier, Paul Koebe. The sensible combinatorics turn out to be those of a triangulation of a surface. This means that one can generate them in wholly free and abstract terms – simply drawing on a surface, if one likes – and then watch the geometry emerge as these combinatorics are "circle packed." This highlights a crucial subtext: namely, that *the geometry in which the packing lives may be forced by the combinatorics*. In the constant-degree examples of Figure 1.9, for example, it is quite clear that the 5-degree packing must live in the sphere. But what does one make of the infinite 6- and 7-degree cases? Their combinatorics choose their geometries: one is by nature euclidean, the other hyperbolic.

The central existence result derives from circle packings with certain extremal proper-ties; we will be calling them *maximal* packings. Given the combinatorics, the maximal packing is univalent, *fills* (in an appropriate sense) the space in which it lives, and is essentially unique. This is where one sees most starkly the *rigidity* ascribed to packings, an issue that threads throughout our study.

In Part II we will introduce the formal definitions and terminology of circle packing and give a rigorous proof for the existence, completeness, and uniqueness of maximal packings from first principles. This proof is surprisingly accessible; the reader does not need a background in conformal geometry. I will caution, however, that we use some of the obscure properties of circles – for example, that distinct circles can intersect in at most two points!

2.2.2. Part III: The Variety of Discrete Analytic Functions

If the theory of circle packing were to stop with maximal packings, it would already be a rich topic, with connections to several areas of mathematics, a range of applications, and open questions that beckon.

However, maximal packings are only the tip of the iceberg. One can see the possibil-ities even in the grandfather of rigidity results, Dennis Sullivan's proof that the regular hexagonal packing – the penny packing – is, up to similarity, the only *univalent* packing with hexagonal combinatorics. Remove the univalence condition and one finds an infi-nite family of spirals, like that of Figure 1.2(e), which share these combinatorics. Allow branching and one finds a full family of what we will term "polynomial" packings, as we will do in Section 14.2.

Since Part III concerns the *flexibility* of circle packings, it might be good to illustrate by fixing a relatively elementary combinatorial pattern (that is, a complex K) and showing what we can do with it. For convenience, I will work in the euclidean setting with our friend Owl – its complex is simple, with one line of symmetry and "eyes" that are easy to pick out in almost any repacking. Figure 2.2 displays some moderate contortions; the boundary edge-path, like the eyes, is provided for visual reference.

While the number of circles in Owl and their tangency relationships are fixed, it is clear that their sizes and the global aspect of the configuration can be manipulated endlessly. Broadly speaking, there are two sources of flexibility once you have fixed the combinatorics: *boundary conditions* and *branching*. Boundary conditions are in play for Figures 2.2(a)–(d). We will be establishing the following fundamental result, which applies in both the euclidean and the hyperbolic settings:

Given any assignment of positive numbers to the boundary vertices of K, there exists an essentially unique locally univalent circle packing for K whose boundary circles have these numbers as their radii.

In other words, we can solve *boundary value problems*. Owl was manipulated for (a), (b), and (c) in Figure 2.2 by merely adjusting boundary radii and repacking for interior

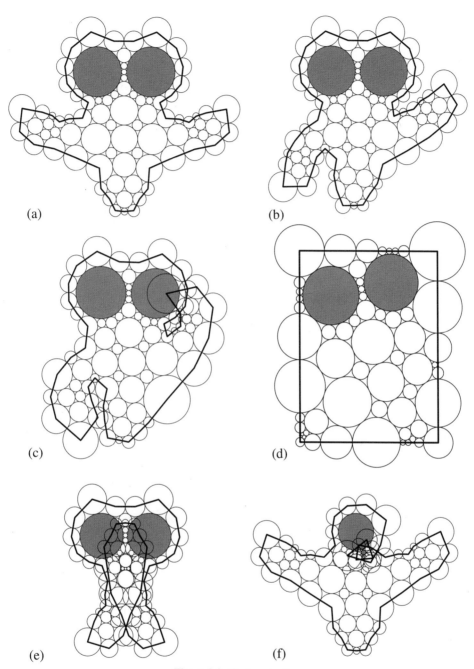

Figure 2.2. Owl contortions.

radii. ("Repack" is shorthand for "compute a packing label.") Packing (d) illustrates the parallel result for prescribed boundary *angle sums*; all were set to π with the exception of the four corner vertices, set to $\pi/2$. We will find that maximality itself is a form of boundary condition, and that even infinite complexes may be subject to more subtle "boundary" conditions at infinity – at what is known as the ideal boundary of the complex.

The second source of flexibility is branching. Typically, the branch vertices for a packing are specified in advance of the packing computations – they represent prescriptions for the interior local behavior. Figure 2.2(e), for instance, was obtained by specifying a single branch point at Owl's center; the boundary radii are those of (a), but the branching has necessitated recomputation of the interior radii. Packing (f) is obtained similarly, but now with a double branch point at one of Owl's eyes. We will be establishing necessary and sufficient (purely combinatorial) conditions for branch structures, allowing us to prove existence of essentially unique packings satisfying both prescribed boundary values and prescribed branching.

I hope the images convince you that this packing is a dynamic business. But don't forget the undercurrent of rigidity. If a packing contains n circles, q of them on the boundary and p interior, then we have q degrees of freedom (assuming a fixed branch structure) in the n-dimensional parameter space represented by the labels. The lost p degrees of freedom reflect, of course, the packing constraints at the p interior circles, and this represents a rigidity imposed by the combinatorics. You will find that you can control many of a packing's features, but that nature controls the rest. You may grab this poor owl, mash it, twist it, and stretch it – but the owl fights back in interesting ways that we want to understand.

In view of the tremendous flexibility we have now observed, even with very simple combinatorics, how can one hope to get a handle on circle packing behavior? We will use a time-honored method; namely, we will represent the variety of packings for a given complex K in terms of *mappings*. With fixed combinatorics (and that is the name of the game here), it seems only natural to treat the canonical maximal packing $\mathcal{P}_K$ as a common domain. Every other circle packing P for K may be regarded as an image of $\mathcal{P}_K$ under some map f. With luck, the properties of P translate to properties of f and *vice versa*. We embark on this course with our most important definition:

> A **discrete analytic function** is defined to be a map $f : Q \longrightarrow P$ between circle packings that preserves tangency and orientation.

This turns out to be a very fruitful approach; the local rigidity manifests itself in ways that are strongly reminiscent of the rigidity in the classical setting for conformal and analytic mappings. For the author this represents the central unifying theme, indeed the central source of inspiration and intuition, for the whole topic. In writing this book, I was torn between the urge to use the classical road map on the one hand, and the desire to sit back and let you, the reader, enjoy the process of discovery (or recognition) on the other. I largely settled on the latter for Part II (though I typically gave results their classical

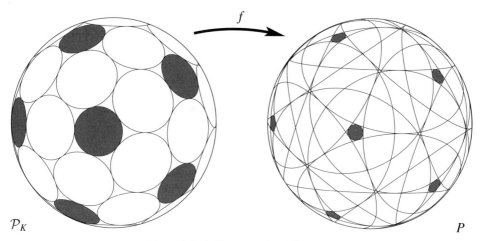

Figure 2.3. A discrete rational function.

names). However, in Part III we jump whole hog into the classical landscape, where we find that the parallels enrich both theories.

Let me give the reader a quick pictorial tour of discrete function theory. Note that the underlying complex K will change from one example to the next, but within a *particular* example, the packings will share the same complex K and the domain will be its maximal packing $\mathcal{P}_K$.

The Sphere

Our first example is based on a complex K of 42 vertices triangulating the sphere. Its packings live in spherical geometry, so the associated functions f mapping $\mathbb{P}$ to itself are *discrete rational* functions. The particular image packing in Figure 2.3 involves 12 branch circles (shaded), and a careful look at P will show that the five neighbors of each branch circle wrap twice around it. Note as a result of the branching, the image packing is highly nonunivalent – in fact, by the *Riemann–Hurwitz formula*, it covers the sphere seven times! Not an easy picture to interpret, is it? The very existence of this branched packing is a rather happy accident of symmetry (see Appendix H). The fact is that you will see many subtle and not-so-subtle differences in the geometries as we go along, and the sphere is the toughest. Very basic questions regarding existence and uniqueness for branched packings in $\mathbb{P}$ remain largely unresolved.

The Disc

We have a much larger repertoire of functions on the unit disc. The flexibility in packings on display in Figure 2.4, for instance, is now encoded in three discrete analytic functions, with the maximal packing in $\mathbb{D}$ as common domain (you just have to trust me that the circles are all there).

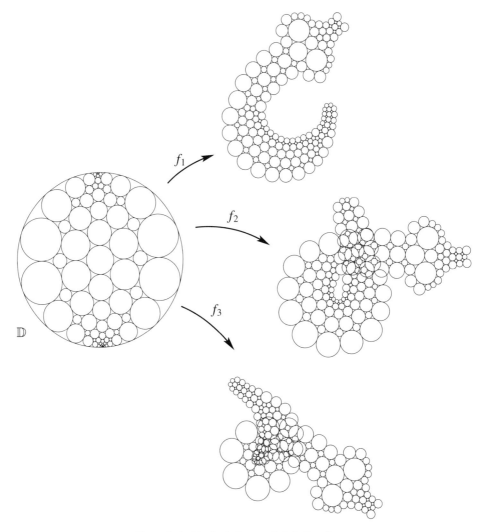

Figure 2.4. Manipulations and their functions.

Of special importance in our theoretical work will be *self-maps* of the disc; that is, maps where both domain and image packings lie in $\mathbb{D}$. Figure 2.5 is such a self-map, and if you could compare the circles of P with their counterparts in $\mathcal{P}_K$ you would observe the following properties:

Every circle of P is closer to the origin (than the corresponding circle of $\mathcal{P}_K$), every interior circle has smaller hyperbolic radius, and every pair of interior circles is closer together in the hyperbolic metric.

For the associated discrete analytic function $f : \mathcal{P}_K \longrightarrow P$, this result is the *hyperbolic contraction principle* or, in function theory terms, the discrete version of the all-important

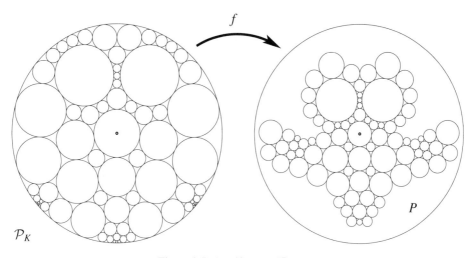

Figure 2.5. A self-map of $\mathbb{D}$.

Schwarz–Pick Lemma. If the term "analytic" has conjured up derivatives, here is an opening. The ratio between the radius of an image circle $f(C)$ and the radius of C itself represents the local stretching/shrinking of the map f, and this ratio, defined as the *ratio function* $f^{\#}$, plays the role of $|f'|$. In Figure 2.5, for instance, the circles at the origin correspond (i.e., $f(0) = 0$), and the results above tell us that $f^{\#}(0) \leq 1$, which is the discrete version of the classical *Schwarz Lemma*.

Figure 2.6 illustrates another self-map of $\mathbb{D}$. This is a *discrete Blaschke product*, representing a family of functions whose classical counterparts play important roles in many parts of function theory. The image packing has an extremal aspect like the maximal packing in that its boundary circles are all internally tangent to the unit circle – these are called *horocycles* and represent circles of infinite hyperbolic radius. I have included a second pair of discs to highlight the boundary horocycles and three interior circles. What do you make of this picture? On close inspection the boundary horocycles march around the unit circle once in the domain but four times in the range. This number $4 \, (= 3 + 1)$ is no accident, for the function harbors three branch circles (shaded). Their tiny (!) image circles result from hyperbolic contraction. And what is to be made of the obvious (to our euclidean eyes) growth in horocycle size between domain and range? This reflects *angular derivatives*, which exceed one if we are to believe the classical model.

We will see a variety of additional functions on the disc, from discrete Riemann mappings to discrete disc algebra functions. It is our richest setting – richest in examples, theory, and open questions.

The Plane

On reaching the plane, we find new issues entering the mix. The complexes are termed *parabolic*, since their maximal packings fill $\mathbb{C}$ (rather than $\mathbb{D}$), and of course, they

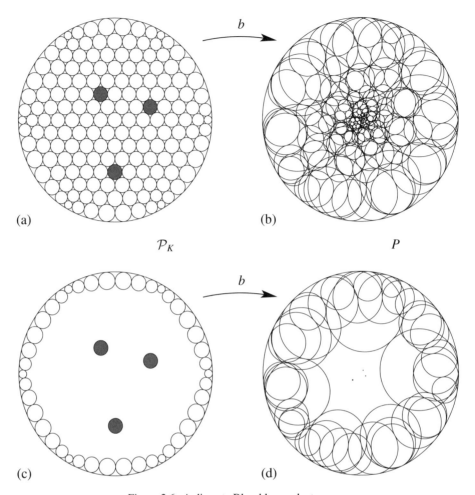

Figure 2.6. A discrete Blaschke product.

necessarily involve infinitely many circles. *Value distribution* is a key theme, and among our first results will be the following:

> *There can be no circle packing for a parabolic complex K that lives in the hyperbolic plane.*

Discrete analytic functions with domain $\mathcal{P}_K$ are naturally termed *entire*, so this result is the *Discrete Liouville Theorem*: there exist no bounded discrete entire functions.

Discrete entire functions are a challenge to create, both in theory and in practice. Nature has provided at least one accessible example to get us started, the *discrete exponential function* of Figure 2.7. Let us take a minute with this.

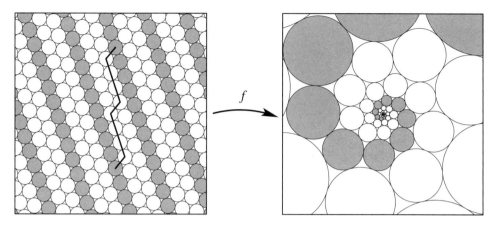

Figure 2.7. A discrete exponential function.

The domain is a regular hexagonal packing and the range is a Doyle spiral of a type we have seen in the Menagerie. The classical exponential, $\exp : z \mapsto e^z$, is nonvanishing and has period $2\pi i$. Likewise, this discrete version never takes the value zero, and its periodicity is evident: each chain of shaded circles on the left maps to the chain of shaded circles spiraling in to zero (and out to infinity) on the right. In particular, each circle on the right is actually the image of infinitely many circles in the domain; the discrete exponential is infinite-to-one. These spirals come in many forms and have an intriguing internal geometry that we revisit in Appendix C. In addition to exponentials, we will establish the existence of a full range of *discrete polynomials* and a *discrete sine* function. The key challenge in the parabolic setting is to find further examples.

Moving beyond the three standard spaces, we will initiate the study of discrete analytic functions on Riemann surfaces, largely through failed experiments, but with some notable successes as well. We also define and investigate discrete conformal structures, and I lure the reader into some *random walks* on circle packings. Part III covers a lot of ground and offers many challenges.

2.2.3. Part IV: Resolution – Discrete Approximates Classical

Again, I would have to say that even if this were the end of the road, if our circle packing maps were simply clever mimics of classical analytic functions, this would be an impressively rich topic. But, in fact, the discrete objects we create also *approximate*: our *discrete* analytic functions approximate their classical counterparts, the *discrete* conformal structures imposed via circles approximate classical conformal structures, and *discrete* conformal maps approximate actual conformal maps.

To me, approximation is the acid test of integrity for the parallels which we are proposing. Since approximation was the initial impetus for the topic and is the foundation of emerging applications, let me wrap up our overview by looking at the past and the future.

Thurston's Conjecture

Thurston defined the notion of circle packing in his *Notes*, but he brought it to the attention of complex analysts only with his 1985 talk in Purdue. The classical situation he addressed is, perhaps, familiar. Given a simply connected domain Ω in the plane, there exists an essentially unique mapping $F : \mathbb{D} \longrightarrow \Omega$ which preserves angles between intersecting curves – a conformal map. This implies that F is *analytic* and brings the impressive machinery of analytic function theory to bear on its study.

The reader may appreciate that we have put ourselves in position to construct discrete analogues of F, that is, *discrete conformal maps*. Look to the top panel in Figure 2.8. Lay down a regular hexagonal packing of circles in the plane, say each of radius $1/n$. Use $\partial\Omega$ like a cookie-cutter to cut out a packing P_n. Let K_n be its complex and compute its maximal packing $\mathcal{P}_{K_n}$ in $\mathbb{D}$. Thurston conjectured, and Rodin and Sullivan proved, that as $n \longrightarrow \infty$ the discrete conformal maps $f_n : \mathcal{P}_{K_n} \longrightarrow P_n$ (appropriately normalized) converge uniformly on compact subsets of $\mathbb{D}$ to the classical conformal mapping F.

In Part IV of the book we prove and extend the Rodin–Sullivan Theorem. Moreover, we show that it applies much more broadly, that discrete analytic functions approximate their classical models in a wide variety of other settings. It is a well-known adage in describing the geometry of analytic functions that they "map infinitesimal circles to infinitesimal circles." We are simply using real circles! As the packings become finer – more and smaller circles – the ensemble behavior converges to that of the illusive "infinitesimal" circles. It all becomes perfectly natural; but of course we need proof, not just intuition.

Conformal Structures

Part IV concludes by mixing the foundations we laid in Part III and the expanded approximation results so that we can discuss emerging applications of circle packing. We are certainly in no position to go into detail now, but I would like to end with a picture that might whet your appetite.

There is a classical genus 3 surface, called *Picard's surface*, which is associated with the solution of the equation $y^3 = x^4 - 1$. The key information regarding this surface is encoded in what is known as its "conformal structure." That in turn can be studied by means of an equivalent piecewise flat structure obtained by triangulating the surface in a carefully chosen way and identifying each triangle with a unit-sided equilateral triangle.

By the time we reach this material in the book, the reader's first impulse on being confronted with such a triangulation will be to "circle pack" it, to find and display the associated maximal circle packing. Nature has been kind, and we will see that this impulse

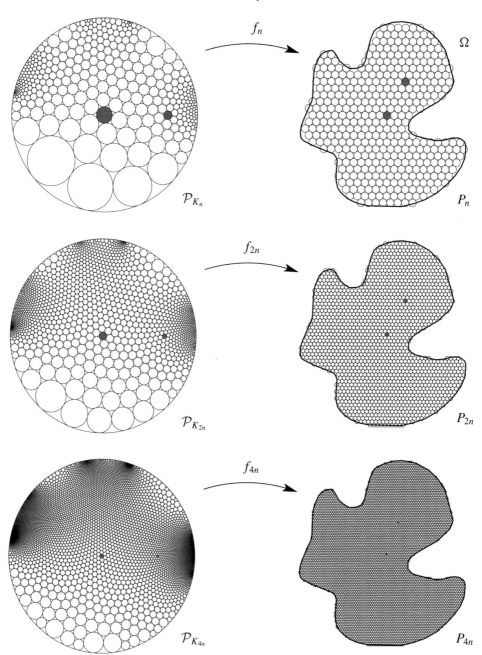

Figure 2.8. A sequence of finite Riemann mappings.

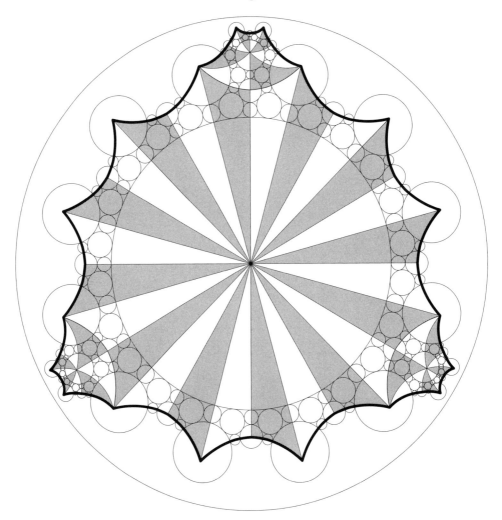

Figure 2.9. The coarse stage packing of the Picard surface.

is the right one: the conformal structures given by circle packings of this triangulation and its refinements actually converge to the conformal structure of the Picard surface itself. So let me leave you with Figure 2.9, which shows the triangulation of a fundamental domain of the Picard surface as embedded via its circle packing in the hyperbolic plane. Stay tuned.

This concludes our preview of the book's themes. Circle packing enters increasingly sophisticated topics as the book progresses, but readers who hold tightly to their native intuitions about circles can get carried along without all the formalities. Stick with it, and I hope you enjoy the ride.

P.S. And do not forget the Appendixes for some topics that stand on their own.

Practicum I

Leaving the auditorium after Thurston's Purdue talk, spring 1985, I did not realize that I was hooked. Exciting talk, full of ideas totally new to me – an undefinable pull, but hardly my cup of tea, right? Summer, normal routine. Fall, something is wrong – I need to see more "packings." Now, how did he get those pictures?

So it began for me, programming from scratch to understand the pictures. The effort was a revelation. It is good that the Zen of programming is not about working code – my early attempts all failed. It is about the discipline of *algorithmic thinking*, about process, trial and error, observation, about being one with the circles. My encounter with this parallel world of computation and experimentation has fundamentally altered my view of mathematics itself, and I am writing "Practicum" sections at transition points in the book in order to share the view. I encourage the reader to program at least the circle packing algorithm, if not more. And do not worry if your code never quite works – it is the thought that counts!

Let us get started with Thurston's basic packing algorithm, which I've formulated here as "meta-code." We take as given a prescribed pattern, that is, an indexed list $\{c_v\}$ of the circles and, for each c_v, a list of its neighbors in order. We keep track of putative radii in a vector R.

Repack meta-code:

1. Initialize R: set boundary radii to assigned values, set interior radii to arbitrary values.

2. For each interior circle c_v:

    a. Compute the interior angle sum $\theta(v)$ using the Law of Cosines.

    b. Adjust radius $R(v)$ for c_v to decrease the difference $|2\pi - \theta(v)|$:

        i. If $\theta(v) < 2\pi$, decrease $R(v)$.

        ii. If $\theta(v) > 2\pi$, increase $R(v)$.

3. If $|2\pi - \theta(v)| < \epsilon$ for all interior circles c_v, then you are **done**; otherwise, repeat step 2.

There are no exact values in computers; this algorithm leads to a succession of adjustments to R which, according to upcoming theory, converges to the *packing* label – the radii that are the true goal. In practice, of course, we do not reach that limit, but R eventually passes the test in Step 3, at which point we deem our algorithm to have succeeded.

Of course, this code provides no actual *configuration* – we know the circle radii and pattern of tangencies, but we need *centers* to actually display the packing. This involves a *layout* process that I liken to the growth of an ice crystal: nucleation consists of placing an initial pair of tangent circles, and then additional circles accrete to the growing configuration as they find appropriate attachment sites. The meta-code is on the next page.

You can look to Figure I.1 to test this or some strategy of your own device. It indexes the 19 circles; numbers #1 – #7 are interior. Set the radii for #8 – #19 to any values you like and run the packing algorithm to compute R for the seven interior circles. For the layout, take circle #7 as the α-circle: you have two degrees of freedom in placing it in the plane. Locate circle #8 tangent to it – its tangency point is one further degree of freedom. Now run the accretion process. Assuming the list L of Step 2 is in index order, for example, you will lay out successive circles in this order:

Circle order: 6 1 2 3 4 5 9 10 11 12 13 14 15 16 17 18 19.

Layout meta-code:

1. Center a circle at the origin; place a neighboring circle tangent to it and centered on the positive x-axis.

2. Create list L of the indices for all remaining circles.

3. Pass once through list L and for each index $v \in L$:

   a. Inspect indices u, w whose circles are neighbors of v and neighbors of one another.

   b. If c_u and c_w are both in place, place c_v in the the unique location determined by the Law of Cosines and the orientation of the triple $\langle v, u, w \rangle$.

   c. If c_v has been placed, remove v from L.

4. If L is empty, you are **done**; otherwise, repeat step 3.

And don't forget the payoff – actually *seeing* the circle packing on your computer screen for visual confirmation that everything is working.

I cannot guarantee that you will learn anything profound by actually programming these two processes. I will not argue whether *C++*, or *Maple*, or *Mathematica* is the best means. I have to say, however, that in my experience, *thinking* about these things is not the same as *doing* them.

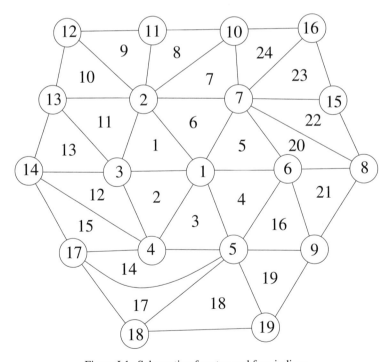

Figure I.1. Schematic of vertex and face indices.

As a matter of fact, you will probably decide quite quickly that my meta-code would benefit from a little more organization. Figure I.1 gives the indices of the 24 faces of the pattern – these are the triples that need to be consulted when placing circles. Instead of focusing on the circles, perhaps you should think of this as a process for placing faces: once circles #7 and #8 are down, for instance, placing face 20 is the equivalent of placing circle #6. Here is the face order corresponding to the circle order above:

Face-order: 20 5 6 1 2 3 16 7 8 9 10 13 22 23 15 17 18.

A minor change? We will see when we pick up this layout theme again in Practicum II.

Notes I

With hindsight, one can find isolated instances of circle packings reaching back to the ancient Greeks and beyond. As a topic, however, circle packing began with William Thurston's famous "Notes," *The Geometry and Topology of 3-Manifolds*. His aim was construction of hyperbolic 3-manifolds. The sphere $\mathbb{P}$ is the boundary of the hyperbolic ball $\mathbb{H}^3 = \{(x, y, z) \in \mathbb{R}^3 : x^2 + y^2 + z^2 < 1\}$. A circle on the sphere is the boundary of a geodesic plane in $\mathbb{H}^3$, and circles overlapping with some angle ϕ define planes intersecting with dihedral angle ϕ. Thurston constructed polyhedra in $\mathbb{H}^3$ with prescribed dihedral angles by building patterns of circles on $\mathbb{P}$ with specified angles of overlap.

Thurston's existence and uniqueness statement (see Morgan, 1984) is Theorem E.2 of Appendix E. See (Bowers and Stephenson, 1996; Marden and Rodin, 1990) for proofs. Thurston attributes the result to Andreev (1970–1970c) whose work concerned groups of automorphisms of $\mathbb{H}^3$ generated by reflections. Reiner Kühnau later drew attention to a proof for packings of *tangent* circles, the topic of this book, by Paul Koebe (Koebe, 1936). Tangency packings were also discussed in Thurston's "Notes" and subsequently used by Robert Brooks (1985) in classifying Schottky surfaces; see Appendix D. Circle packing became widely known only with Thurston's 1985 talk. His conjecture was proven by Rodin and Sullivan (1987).

In his 1985 talk, Thurston described an iterative algorithm for computing circle radii; refined versions of that algorithm, coded in `CirclePack` (Stephenson 1992–2004; Collins and Stephenson 2000) underlie all the packings in the book. Alternate variational and fixed point algorithms were developed by Colin de Verdière (1989, 1991), Brägger (1992), Rivin (1994), and others, culminating in recent methods of Bobenko and Springborn (2003).

A full circle packing bibliography is included at the end of the book; explicit citations will appear in *Notes* sections like this one after each part of the book and at the ends of the appendixes.

Note. There exists an active classical geometric topic known as "sphere packing," which concerns *density* and *coverage* by spheres of fixed sizes in two and higher dimensions; see (Conway and Sloane, 1999) for an overview. Our "circle packing" is *not* two-dimensional sphere packing. Images may appear, at first glance, to be related, but no significant links between these topics have been discovered (i.e., please do not ask me questions about sphere packing).

Part II

Rigidity: Maximal Packings

In this Part we give a complete and essentially self-contained proof of the fundamental existence result in the theory of circle packing: *Given any triangulation of a surface, there exists an essentially unique extremal circle packing with the pattern of that triangulation.*

There is much hidden within this statement; we will need to clarify the combinatorics and the meaning of "unique" and "extremal." Of course there must also be a geometric space in which the claimed circle packing lives. This is perhaps the theorem's deepest feature, that the combinatorics force a particular geometry on what appears at first to be a purely combinatorial setting. With circle packing one sees the noneuclidean geometries and geometries on general surfaces arising in completely natural ways. Indeed, the beauty of this topic, to the author's mind, lies in the constant interaction between the combinatorics and the geometry.

We begin with the obligatory background on topology, combinatorics, and geometry, and then the statement of the fundamental result in Chapter 4, followed by the all-important "bookkeeping" conventions. The proof itself takes several chapters, as it breaks naturally into cases based on combinatorics. Each case requires certain geometric tools, and I follow a "just-enough, just-in-time" philosophy: you get the level of result you need, when you need it. You will find, nonetheless, that two notions emerge as the linchpins of the theory: "monotonicity" and "winding numbers." It is the truly elementary nature of these that makes the proof accessible – their beauty and power are really quite stunning.

3

Preliminaries: Topology, Combinatorics, and Geometry

This chapter provides a summary of the broader background material that we rely on throughout the book; details can be found in the standard references cited in *Notes II*. I know that this phase of a book, coming right at the beginning, can really throw off the novice reader. Let me share something with you. Mathematicians *write* linearly – I am *compelled* to put this stuff here! On the other hand, if I were *reading* this, I would glance briefly, then skip ahead to the main show, returning only when I found myself in a bind. I recommend that you do the same, throughout the book – use a cloaking device in the messy spots, sneak through to the action. And be comforted by the fact that the situations of real interest to us here are very concrete and accessible.

Having said that, let me keep you a moment for a tabletop experiment. Arrange seven pennies in a flower, one in the center and a chain of six petals. Easy enough – they fit together perfectly on a flat table. Try the same thing with just six pennies, however. Laid flat around the center, the five petals obviously leave a gap. Lift the outer edges of the petals slightly, however, and you can close up the chain – the result looks like a spherical shell, the archetype of positive curvature. Now try eight pennies. It takes a little thought (and dexterity), but if you are able to tilt some petals one way, some the other, you can again complete the chain – this flower, like a floppy hat with too long a brim, shows the characteristic saddle shape of negative curvature. Of course, in practice we will have the luxury of changing the *radii* of our circles. Nonetheless, these penny flowers reflect the nub of the combinatoric/geometric thread that runs through all of circle packing. You might tuck that away for now as we fill in the obligatory preliminaries.

3.1. Surfaces and Their Triangulations

I will assume the reader is familiar with basic metric topology and elementary notions associated with surfaces. By a *surface* we mean a connected topological 2-manifold, that is, a connected Hausdorff space S in which each point has a neighborhood that is homeomorphic to an open subset of the plane.

For every point $p \in S$ there exist an (open) neighborhood U and a homeomorphism ψ mapping U to an open subset $\psi(U)$ of $\mathbb{R}^2$. The pair (U, ψ) is typically called a *(coordinate) chart* for the surface and $z = \psi(p)$ is known as a *local coordinate* for

U. A collection $\mathcal{A} = \{(U_j, \psi_j) : j \in J\}$ of charts forms an *atlas* for S if the sets U_j cover S. For every pair of indices $i, j \in J$ with $U_i \cap U_j$ nonempty the *transition* map $\psi_j \circ \psi_i^{-1} : \psi_i(U_i \cap U_j) \longrightarrow \psi_j(U_i \cap U_j)$ is continuous and tells one how to convert between different local coordinates. By convention, *positive* orientation in $\mathbb{R}^2$ is taken to mean *counterclockwise*. If there exists an atlas $\mathcal{A}$ for S whose transition maps are orientation-preserving, then one can consistently lift the orientation in $\mathbb{R}^2$ to S, and S is said to be *oriented*. *All surfaces in this book are assumed to be so oriented.*

3.1.1. Classification

The theory of surfaces is well established and quite complete. The coarsest classification distinguishes *compact* from noncompact or *open* surfaces. Finer structure is reflected in the *fundamental group*, $\pi_1(S)$, associated with equivalence classes of closed curves in the surface; we will discuss this at greater length when the need arises. Among the distinctions we should point out now, the first involves connectivity. A *simply connected* surface is one in which every closed curve is homotopic to a constant curve – loosely speaking, every closed curve in the surface can be shrunk continuously to a point while remaining in the surface. Other surfaces are called *multiply connected*. As examples, a *topological disc* – that is, a topological space homeomorphic to an open disc in $\mathbb{R}^2$ – and a *topological sphere* are simply connected, while a topological annulus, such as $A = \{(x, y) : 1 < \sqrt{x^2 + y^2} < 2\} \subset \mathbb{R}^2$, and a torus (the surface of a "donut") are multiply connected.

These latter two examples illustrate another distinction. A surface has a *hole* if it is obtained from a larger surface by removing a topological disc. (Annulus A actually has two holes; it is a sphere with two closed discs removed, called *2-connected*.) On the other hand, a torus has no holes in this sense; the "hole" in a torus is, mathematically speaking, a *handle*. Handles are reflected in the structure of $\pi_1(S)$, and the number of handles is called the *genus* of the surface, genus(S). A topological sphere has genus 0, a torus has genus 1, and the surface suggested in Fig. 1.8(d) has genus 2. A notable accomplishment in the theory of surfaces is this classification theorem: *Compact oriented surfaces S_1 and S_2 are topologically equivalent (i.e., homeomorphic) if and only if* genus(S_1) = genus(S_2). Noncompact surfaces represent greater variety, but even there, genus is important in classification. For example, it is known that *every noncompact oriented surface S of genus 0 is planar;* that is, S can be embedded in $\mathbb{R}^2$.

3.1.2. Bordered Surfaces

We will need a natural generalization of surfaces to accommodate boundaries. Thus a *bordered surface* S is a connected Hausdorff space such that each point p has a neighborhood U that is homeomorphic either to an open set in the plane or to the intersection of such a set with the closed half-plane $\{(x, y) : y \geq 0\}$. Points having a neighborhood of the former type are called *interior* points; the others are called *boundary* points. The *boundary* of S, denoted ∂S, decomposes into connected components, each homeomorphic to a line or a circle.

Our term "surface" will include both those with and those without boundary, as the context will generally make the meaning clear. When we need to emphasize the existence of a boundary we will, as is customary, use the term *bordered surface*. The connectivity and genus of a bordered surface S are just those of Int(S), which is an open surface. In this regard, we observe a standard convention: the term *compact surface* always refers to a surface *without* boundary.

3.1.3. Triangulations

We discretize the notion of surface by means of triangulations. A *topological triangle* is a topological closed disc with three distinguished boundary points, the *vertices*, breaking the boundary into three boundary arcs, the *edges*. Note that there is nothing about edges being "straight"; we are talking topology here, not geometry.

Definition 3.1. *A* **triangulation** T *of a surface* S *is a locally finite decomposition of* S *into a collection of topological closed triangles,* $T = \{t_j\}$, *so that any two either are disjoint, intersect in a single vertex, or intersect in a single complete edge. (Locally finite means that every point of* S *has a neighborhood that intersects at most finitely many triangles of* T; *the collection* T *itself may be finite or countably infinite.)*

The *combinatorics* of a triangulation refer to the abstract relationships among its parts. For bookkeeping we rely on elementary notions of *simplicial complexes*. Given T, each triangle (*face*) is a 2-simplex; each *edge* is a 1-simplex; each *vertex* is a 0-simplex. The triangulation itself is then a *(simplicial) 2-complex* and has a canonical realization (some would denote this as $|T|$) as a locally compact Hausdorff space. We will have occasion to introduce simplicial maps, barycentric coordinates, and other simplicial machinery later, but all of it relates to very concrete situations.

In our setting, of course, the realization is just the surface S. However, we would like the flexibility to begin with abstract combinatorics rather than with surfaces.

Lemma 3.2. *The following are necessary and sufficient conditions for a 2-complex L to represent a triangulation T of an oriented surface S: (i) L is connected. (ii) Every edge of L belongs to either one or two faces (the former are called boundary edges, the latter, interior edges). (iii) Every vertex of L belongs to at most finitely many faces, and these form an ordered chain in which each face shares an edge from v with the next. (iv) Every vertex of L belongs either to no boundary edge or to exactly two boundary edges (the former are called interior vertices, the latter, boundary vertices). (v) Any two faces are either disjoint, share a single vertex, or share a single edge. (vi) An order $\langle u, v, w \rangle$ may be assigned to the vertices in every face of L in such a way that any pair of faces intersecting in an edge will induce opposite orientations on that edge.*

Henceforth, all the 2-complexes L we consider in this book will satisfy the conditions of this lemma. Topologically, the surface S is just the realization and we say that "L triangulates S." Note that when L has boundary edges, then S must necessarily be treated as a bordered surface.

Verification of this lemma is left to the reader, but the key steps involve certain neighbor relationships which should be defined now for future use. Vertices u and v of L are called *neighbors* if $\langle u, v \rangle$ is an edge in L; write $u \sim v$. One can show that the neighbors of a vertex v can be put in an ordered list, $v_1, \ldots, v_k$, so that $v_j \sim v_{j+1}$, $j = 1, \ldots, k-1$. This is a local *planarity* condition. $\{v; v_1, \ldots, v_k\}$ will be termed the *(combinatorial) flower* of v and the integer k is the *degree* of v, $\deg(v)$. If v is a boundary vertex, then the first and last vertices, v_1 and v_k, are necessarily boundary vertices as well, while if v is interior, then $v_k \sim v_1$, so the neighbors form a closed list. The order assigned to the vertices in faces of L induces a consistent orientation on all flowers $\{v; v_1, \ldots, v_k\}$, so that every face $\langle v, v_j, v_{j+1} \rangle$ has the assigned order. The boundary vertices and edges are denoted ∂L.

When L triangulates S, there is an opportunity for combinatorial analogues of the various properties and characteristics of S. We will see this throughout the book, but for now let us just note some instances. If S is compact, then L is necessarily finite (i.e., has finite numbers of vertices, edges, and faces) and without boundary. If S is open and without boundary, then L is necessarily infinite. For bordered surfaces, cardinality can go either way. All combinations of finite/infinite, bordered/unbordered are represented by circle packings in our Menagerie, where the carriers represent the triangulations.

A *path* γ in a surface S is a continuous map $\gamma : [0, 1] \longrightarrow S$; it is *closed* if $\gamma[1] = \gamma[0]$. The natural analogue in L would seem to be an *edge-path*, a sequence $\gamma = \{e_1, \ldots, e_m\}$ of oriented edges, each e_j ending at the vertex shared with e_{j+1}. We will certainly have uses for edge-paths. However, we will also find it practical to mimic paths in L with chains of faces. Thus define a *chain* in L as a sequence $\Gamma = \{f_0, f_1, \ldots, f_n\}$ of faces, where each f_j shares an edge with its successor f_{j+1}; the chain Γ is *closed* if $f_n = f_0$.

In a surface, paths can be modified via homotopies. In L, chains can be modified by *discrete homotopies*, which involve a finite succession of simple "moves" called *local modifications*. Consider a chain $\Gamma = \{f_0, f_2, \ldots, f_n\}$. A subchain $\gamma = \{f_j, \ldots, f_k\}$ of Γ will be called *local at vertex v* if its faces belong to the star of v (i.e., have v as a common vertex). A new chain Γ' is obtained from Γ by a *local modification* if γ is replaced in Γ by any other local subchain at v having the same first and last faces as γ. We say that chains Γ_1 and Γ_2 are *homotopic* if one can be obtained from the other by a finite succession of local modifications. (Since the first and last faces never change, these are *fixed-endpoint* homotopies.) Among the local modifications are elementary ones which carry out these pattern simplifications:

$$\{\ldots, f, f, \ldots\} \longrightarrow \{\ldots, f, \ldots\} \tag{1}$$

$$\{\ldots, f, g, f, \ldots\} \longrightarrow \{\ldots, f, \ldots\}. \tag{2}$$

However, a more typical (and interesting) local modification is one that reverses the direction in which a chain passes around some vertex.

The collection of closed chains, say starting and ending at some base face f_0, becomes a group when homotopic chains are identified. The group operation is *concatenation*: namely, $\Gamma_1 \cdot \Gamma_2$ represents Γ_1 followed by Γ_2. The *null* chain is the trivial chain $\Gamma_0 = \{f_0\}$, the group identity. A closed chain Γ is *null homotopic* if it is homotopic to Γ_0; in other words, a finite succession of local modifications reduces it to the trivial chain.

The group of closed chains, modulo homotopy, is the *fundamental group* of L, denoted $\pi_1(L)$, and is isomorphic to the familiar topological fundamental group $\pi_1(S)$. In particular, *a 2-complex L is simply connected if and only if every closed chain in L is null homotopic.*

A famous classical bit of combinatorial data reflecting connectivity and boundary components is provided by the Euler characteristic.

Definition 3.3. Suppose L is a finite simplicial 2-complex. Let V, E, and F denote the cardinalities of 0, 1, and 2-simplices of L, respectively. Then the **Euler characteristic** of L is the number

$$\chi(L) = V - E + F.$$

If L triangulates a surface S, then one defines the *Euler characteristic* of S by $\chi(S) = \chi(L)$.

In the case of surfaces, it is known that $\chi(S)$ is a topological invariant related to the genus g of S and the number m of components of ∂S by $\chi(S) = 2 - 2g - m$. Thus we can define *genus* g for a finite 2-complex L which triangulates a surface with m boundary components by

$$\chi(L) = 2 - 2g - m \qquad \Longleftrightarrow \qquad g = \frac{1}{2}(2 - m - \chi(L)).$$

All these rather abstract conditions have very practical consequences when it comes to organizing our bookkeeping efforts. A few preliminary observations: A finite 2-complex with no boundary and genus 0 triangulates a sphere. A finite 2-complex with no boundary and genus 1 triangulates a torus. A finite complex triangulates a topological closed disc if and only if $\chi(L) = 1$.

3.2. The Classical Geometries

We work with the three standard 2-dimensional geometries, namely, spherical, euclidean, and hyperbolic. Those studying geometry for its own sake have the advantage of starting at any of several levels of abstraction, but our goals are again best served by a very concrete approach. Our spaces will be modeled on the unit sphere in space, the euclidean plane, and the unit disc, respectively:

$$\mathbb{P} = \{(x, y, z) : x^2 + y^2 + z^2 = 1\} \subset \mathbb{R}^3 \qquad \text{(Riemann sphere)}$$
$$\mathbb{R}^2 = \mathbb{C} = \{z = x + iy : x, y \in \mathbb{R}\} \qquad \text{(euclidean plane)}$$
$$\mathbb{D} = \{(x, y) : \sqrt{x^2 + y^2} < 1\} = \{|z| < 1\} \qquad \text{(hyperbolic plane)}$$

We will derive great benefits from these particular models. In the first place, they are conveniently nested; the disc is a subset of the plane, which in turn is identified as a subset

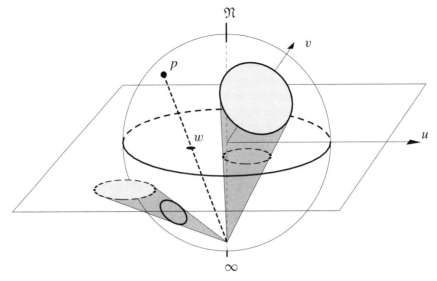

Figure 3.1. Stereographic projection.

of the sphere via stereographic projection. In addition, by identifying the euclidean plane with $\mathbb{C}$ we get the added convenience of an arithmetic that we can use in all three settings.

Recall that complex numbers have the form $z = x + iy$, where $i^2 = -1$. I will assume the reader is familiar with the operations of addition and multiplication of complex numbers and with the numerous algebraic properties that are summarized by saying that $\mathbb{C}$ is an *algebraically closed field*. The complex numbers $z = x + iy$ are identified with the points $(x, y) \in \mathbb{R}^2$, so $\mathbb{C}$ inherits the geometry of the euclidean plane. We routinely mix the algebraic and geometric properties of $\mathbb{C}$ as circumstances require. Complex addition is simply vector addition. Geometry for multiplication comes with the *polar* representation, $z = re^{i\theta}$, where $r = |z|$ is the *modulus* (or *absolute value*) of z, and $\theta = \arg(z)$ is the *argument* of z, the angle which the line from the origin to z makes with the positive real axis. Multiplying by z corresponds to scaling by factor r and rotating by angle θ.

The unit disc $\mathbb{D}$ is, of course, a subset of $\mathbb{C}$. The containment $\mathbb{C} \subset \mathbb{P}$ is a bit more complicated. The plane can be identified with the sphere punctured at a point in a natural way *via* stereographic projection. Let me explain this book's convention, which has certain advantages over the standard mathematical version.

Stereographic projection is illustrated in Figure 3.1. Given a line through the *South* pole, the point $w = u + iv$ where it punctures $\mathbb{C}$ and the point $p = (x, y, z)$ where it punctures $\mathbb{P}$ are identified. That identification is termed *stereographic projection*, whether going from the sphere to the plane or vice versa. Using similarity of triangles and recalling that $x^2 + y^2 + z^2 = 1$, one can confirm the following relationships:

$$u = \frac{x}{1+z}, \qquad v = \frac{y}{1+z},$$

$$z = \frac{1-(u^2+v^2)}{1+(u^2+v^2)}, \qquad x = u(1+z), \qquad y = v(1+z). \tag{3}$$

Note that the origin is identified with the north pole $\mathfrak{N} = (0, 0, 1)$, the unit circle $\mathbb{T}$ with the equator, the disc $\mathbb{D}$ with the "northern hemisphere," and $\mathbb{C}$ with everything but the south pole $\mathfrak{S} = (0, 0, -1)$, which by convention is denoted *infinity*, ∞. All three geometries can be pictured simultaneously on $\mathbb{P}$, but in practice we use whichever representation is most natural or convenient.

We will talk about circles in $\mathbb{C}$ and $\mathbb{P}$ shortly, but they are certainly familiar objects to the reader, so we can make an important observation now. Namely, *stereographic projection identifies circles of $\mathbb{C}$ with circles of $\mathbb{P}$*. In particular, because straight lines in $\mathbb{C}$ correspond with circles through ∞ in $\mathbb{P}$, straight lines are often included among the "circles" in $\mathbb{C}$, and we follow that convention: lines are "circles of infinite radius with centers at infinity."

3.2.1. Common Features

We will learn the personalities and idiosyncrasies of these geometries in turn, but before specializing we might note their considerable commonality.

Let $\mathbb{G}$ for now denote any one of these spaces. In each instance we are furnished with a *metric* (in the sense of Riemann), which allows us to measure the distances between points and thereby to define various standard geometric quantities. This involves nothing particularly sophisticated here since each of our models lives within familiar 2- or 3-space. It simply means that we have an *element of arclength*, ds, which we can integrate along sufficiently smooth curves to define distances. Thus a rectifiable curve $\gamma \subset \mathbb{G}$ has length defined by

$$\text{Length}(\gamma) = \int_\gamma ds.$$

From this we define the *distance* $d(a, b)$ between points $a, b \in \mathbb{G}$ as the infimum of the lengths of rectifiable curves from a to b,

$$d(a, b) = \inf \left\{ \int_\gamma ds : \gamma \text{ from } a \text{ to } b \right\}.$$

There is a companion *area element*, ds^2, so that the area of a Borel set B is given by

$$\text{Area}(B) = \iint_B ds^2.$$

Finally, the *angle* between two smooth curves that intersect at a point p can be defined using ds and the Law of Sines. Since our surfaces, and hence the curves, are in $\mathbb{R}^2$ or $\mathbb{R}^3$, this angle is simply the angle between the tangent vectors to γ_1 and γ_2 at p. Note that these are *signed* angles; throughout the book we use the standard convention that counterclockwise angles are positive.

We have in each setting the classical objects of geometry: A *geodesic*, or "straight line", is a curve which locally minimizes distances. Our spaces $\mathbb{G}$ are *geodesic spaces* in that any two points are connected by a geodesic; moreover, that geodesic is unique except in the case of antipodal points on the sphere. A *circle* C is the set of points a given *radius* $r > 0$ from a fixed *center*. The (open) *disc* bounded by C is the component of the complement of C containing the center. *Triangle* refers in a geometric setting to a *geodesic triangle*, that is, one having geodesic edges; likewise for quadrilaterals and other polygonal curves. We will also see that each geometry demands an appropriate *trigonometry*.

Central to any geometry is the notion of *congruence*. This involves the so-called "rigid motions," one-to-one transformations of $\mathbb{G}$ onto itself that preserve distances, areas, angles, and (by explicit convention) orientation. Also termed *isometries*, these clearly form a mathematical group under composition and we will denote it by Isom($\mathbb{G}$). An *a priori* less rigid set of transformations are those that preserve angles (including orientation), termed *(conformal) automorphisms*. They form the group denoted Aut($\mathbb{G}$) that contains Isom($\mathbb{G}$) as a subgroup. Mathematicians showed convincingly at the close of the 19th century that the transformations are key to a geometry, and we will certainly see their importance in later developments.

One of the geometric threads most evident in our work will be *curvature*. In its more concrete incarnations, curvature has to do with shape: in a landscape, for example, a round hilltop or a bowl-shaped valley has positive curvature, a soccer field has zero curvature (i.e., it is *flat*), while a saddle-shaped mountain pass has negative curvature. We are speaking here of *gaussian curvature k*, which is *intrinsic* (i.e., depends only on the surface metric, not on how that surface resides in some larger setting). Its geometric influences are more subtle (as in the "penny" experiment at the beginning of the chapter) but totally pervasive. Our length elements will all be represented in the form $ds = \rho(z)|dz|$ for smooth density functions $\rho(z)$. In these cases gaussian curvature can be computed directly:

$$k(z) = \frac{-\Delta \log(\rho(z))}{\rho(z)^2}, \quad \text{where } \Delta = \frac{\partial^2}{\partial x^2} + \frac{\partial^2}{\partial y^2}. \tag{4}$$

Here Δ is the well-known *Laplace operator* and log denotes the natural logarithm.

We turn now to the individual characters of the geometries, beginning with that most familiar of settings, the euclidean plane.

3.2.2. The Euclidean Plane

Our model is the well-known cartesian plane $\mathbb{R}^2$. However, we work instead with the complex plane $\mathbb{C}$, since that adds arithmetic to the geometry. Distance is defined by $d(z, w) = |z - w|$ and the elements of arclength and area are

$$ds = |dz| = \sqrt{dx^2 + dy^2}, \qquad ds^2 = dz\,d\bar{z} = dx\,dy.$$

We list a few basic facts for later comparison. (Here and in the later cases, t denotes a triangle with edge lengths a, b, c and opposite angles α, β, γ; C denotes a circle of radius $r > 0$; and D denotes the disc bounded by C.)

Elements:	$ds = \sqrt{dx^2 + dy^2}, \ ds^2 = dx \, dy$
Law of Cosines, $\mathbb{C}$:	$c^2 = a^2 + b^2 - 2ab \cos(\gamma)$
Circumference:	$\text{Length}(C) = 2\pi r$
Disc area:	$\text{Area}(D) = \pi r^2$
Triangle angles:	$\alpha + \beta + \gamma = \pi$
Aut($\mathbb{C}$) :	$\{\sigma : \sigma(z) = az + b, \ a, b \in \mathbb{C}, a \neq 0\}$
Isom($\mathbb{C}$) :	$\{\sigma : \sigma(z) = az + b, \ a, b \in \mathbb{C}, |a| = 1\}$
Curvature:	0 (constant, "flat")

Note that an automorphism $\sigma : z \longrightarrow az + b$ is *complex linear*; it involves rotation by $\arg(a)$, dilation by $|a|$, and then translation by b. In particular, circles are mapped to circles. Pure translations (i.e., $a = 1$) have no fixed point, but every other nontrivial automorphism has a single fixed point z (i.e., $\sigma(z) = z$).

3.2.3. The Sphere

The unit sphere is known as the *Riemann sphere* or the *complex projective line* (hence the notation $\mathbb{P}$). In computations $\mathbb{P}$ can be identified with the plane *via* stereographic projection, in which case the arclength and area elements take the form

$$ds = \frac{2|dz|}{1 + |z|^2}, \qquad ds^2 = \frac{4dxdy}{(1 + |z|^2)^2}$$

In $\mathbb{P}$ itself, ds turns out to be the restriction of the ambient euclidean metric of $\mathbb{R}^3$; the length of a curve $\gamma \subset \mathbb{P}$ is simply its length in $\mathbb{R}^3$. The geodesic between points $a, b \in \mathbb{P}$ is the shorter arc of the great circle containing them and is unique unless a and b are antipodal. Because $\mathbb{P}$ is a unit sphere, $d(a, b)$ is seen to be the *radian measure* of the angle η which a and b subtend at the origin. This is illustrated in Figure 3.2 and is a particularly handy fact to keep in mind. One consequence is that no two points are separated by more than π, so circle radii r satisfy $r < \pi$. One can deduce, also, that a circle C in $\mathbb{P}$ is the intersection $C = \Gamma \cap \mathbb{P}$ for some plane $\Gamma \subset \mathbb{R}^3$. C is a great circle if and only if Γ goes through the origin. Note that the complement of C has two simply connected components; the "disc" bounded by C refers to the component containing the circle center.

Spherical trigonometry takes over in formulas and we see something new: angles and area are now related.

Elements:	$ds = \dfrac{2|dz|}{1 + |z|^2}, \ ds^2 = \dfrac{4dxdy}{(1 + |z|^2)^2}$
Law of Cosines, $\mathbb{P}$:	$\cos c = \cos a \cos b - \sin a \sin b \cos \gamma$
Circumference:	$\text{Length}(C) = 2\pi \sin r$
Disc area:	$\text{Area}(D) = 2\pi(1 - \cos r)$

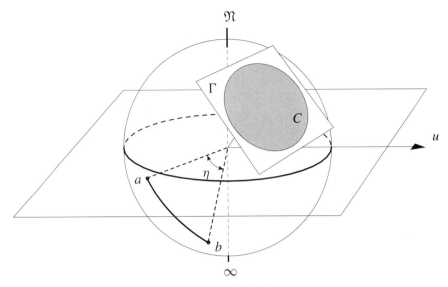

Figure 3.2. A bit of spherical geometry.

Triangle angles: $\alpha + \beta + \gamma = \pi + \mathrm{Area}(t)$

Aut($\mathbb{P}$) : $\{\sigma : \sigma(z) = \dfrac{az+b}{cz+d},\ a, b, c, d \in \mathbb{C}, ad - bc = 1\}$

Isom($\mathbb{P}$) : $\cong \mathrm{SO}(3, \mathbb{R})$

Curvature: 1 (constant, "positive")

The isometries are the rotations about axes through the origin. These are restrictions to $\mathbb{P}$ of linear transformations of $\mathbb{R}^3$ which preserve lengths and orientation, so they are represented as real 3×3 matrices O with $O \cdot O^{\mathrm{tr}} = I$ and $\det(O) = +1$; they from the *special orthogonal group*, $\mathrm{SO}(3, \mathbb{R})$.

The group of automorphisms of $\mathbb{P}$ is far, far richer. It consists of the *Möbius* or *fractional linear* transformations of $\mathbb{P}$. Using complex arithmetic, each $\sigma \in \mathrm{Aut}(\mathbb{P})$ may be described by $\sigma : z \mapsto \frac{az+b}{cz+d}$, for complex numbers a, b, c, d satisfying $ad - bc = 1$. This means the following: given $p \in \mathbb{P}$, let z be its (stereographic) projection to $\mathbb{C}$, apply the formula to z to obtain a complex number w, and then project w to $q \in \mathbb{P}$; thus $q = \sigma(p)$. The arithmetic can be extended to accommodate ∞: if $p = \infty$, then $\sigma(p) \in \mathbb{P}$ is (the projection of) a/c, while if $p \in \mathbb{P}$ is $-d/c$, then $\sigma(p) = \infty$.

The automorphisms of $\mathbb{P}$ can be identified with 2×2 matrices *via*

$$\sigma(z) = \frac{az+b}{cz+d} \qquad \longleftrightarrow \qquad m = \begin{pmatrix} a & b \\ c & d \end{pmatrix}.$$

These matrices are invertible, since $ad - bc = 1$, and (quite remarkably!) multiplication of matrices corresponds with composition of the associated automorphisms. Therefore,

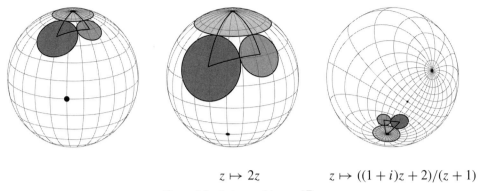

$$z \mapsto 2z \qquad\qquad z \mapsto ((1+i)z+2)/(z+1)$$

Figure 3.3. Automorphisms of $\mathbb{P}$.

Aut($\mathbb{P}$) is isomorphic to the much-studied Lie group PSL(2, $\mathbb{C}$). Figure 3.3 illustrates two automorphisms and their effects on a triple of tangent circles. A grid of latitude and longitude lines and dots at 0, 1, and ∞ are shown for visual reference. Nontrivial automorphisms have either one or two fixed points.

You may notice that the longitude and latitude markings on Fig. 3.3 seem to be circles in every case. There is no *a priori* reason to expect "angle-preserving" homeomorphisms of $\mathbb{P}$ to map circles to circles, but that is in fact the case. *Möbius transformations of $\mathbb{P}$ map circles to circles.* More about this shortly.

Before we leave the sphere, note that the geometry of $\mathbb{P}$ is said to be "noneuclidean" because it fails the parallel postulate of Euclid in a spectacular way: straight lines are great circles, so *every* pair of straight lines will intersect – there are no parallel lines!

3.2.4. The Hyperbolic Plane

Hyperbolic geometry is perhaps the least familiar of our three, so I want to say a little more about how it works. It is a rich and beautiful setting – one that we will be using extensively – and I hope you come to enjoy it as much as I do.

We use the Poincarè disc model: the open unit disc $\mathbb{D}$ with these arclength and area elements:

$$ds = \frac{2|dz|}{1-|z|^2}, \qquad ds^2 = \frac{4\,dx\,dy}{(1-|z|^2)^2}, \qquad |z| < 1.$$

Note that the density $\rho(z) = 2/(1-|z|^2)$ goes to infinity as z moves toward the boundary of the disc, so small euclidean distances near the boundary can represent large hyperbolic distances. A direct computation can be made of the length of the real segment $[0, x]$ for $0 < x < 1$, and a fairly easy comparison shows that to be less than the length of any competing curve from 0 to x. Therefore,

$$d(0, x) = \int_0^x \frac{2dx}{1 - x^2} = \log \frac{1 + x}{1 - x}. \tag{5}$$

From here, the properties of geodesics and circles can best be developed from an understanding of the automorphisms of $\mathbb{D}$. These are "conformal" automorphisms; since they preserve angles and orientation, they are necessarily analytic self-maps of $\mathbb{D}$ and can be shown to have the following form:

$$\phi \in \text{Aut}(\mathbb{D}) \quad \Longleftrightarrow \quad \phi(z) = \lambda \frac{z - z_0}{1 - \overline{z}_0 z}, \quad |\lambda| = 1, \quad z_0 \in \mathbb{D}. \tag{6}$$

A calculation will show that the element ds is invariant under these maps. In particular, if γ is a path from a to b, then $\sigma = \phi(\gamma)$ is a path from $\phi(a)$ to $\phi(b)$ and Length$(\sigma) =$ Length(γ). It is an easy step, to show that ϕ is an isometry. In particular, Isom$(\mathbb{D}) \equiv$ Aut$(\mathbb{D})$, so there are no similarities in $\mathbb{D}$ that are not congruences.

We may now observe from (6) that ϕ is the restriction to $\mathbb{D}$ of a Möbius transformation, and therefore preserves circles on $\mathbb{P}$. Equation (5) implies that a hyperbolic circle C centered at the origin of hyperbolic radius r is also a euclidean circle centered at the origin of euclidean radius $(e^r + 1)/(e^r - 1)$. To obtain the hyperbolic circle of radius r at some point a, choose ϕ to map 0 to a (this can be done explicitly using (6)) and observe that the image $\phi(C)$ is the desired circle. Since $\phi(C)$ is again a euclidean circle, we have the following result: *A hyperbolic circle in $\mathbb{D}$ is also a euclidean circle.*

Now consider geodesics. By Eq. (5), those through the origin are euclidean diameters of $\mathbb{D}$. If a and b are arbitrary points of $\mathbb{D}$, choose ϕ to map, say, a to the origin. If γ is the diameter through 0 and $\phi(b)$, then $\phi^{-1}(\gamma)$ is the geodesic through a and b. Since γ lies on a euclidean straight line orthogonal to the unit circle, its image under ϕ^{-1} lies on a euclidean circle (or straight line) orthogonal to the unit circle. We may draw the following conclusion: *The hyperbolic geodesics of $\mathbb{D}$ are arcs of euclidean circles that are orthogonal to the unit circle.* Figure 3.4(a) illustrates geodesics, circles, and triangles and the effects of automorphisms of $\mathbb{D}$.

There are special considerations that apply in the hyperbolic plane, for we plan to allow circles of *infinite* hyperbolic radius. These are represented by *horocycles*, euclidean circles internally tangent to the unit circle. The point of tangency is defined to be the circle center. The justification is suggested in Figure 3.4(b): hyperbolic circles passing through a fixed point z and having centers that move along a geodesic toward a boundary point $e^{i\theta}$ will converge as point sets to a horocycle with tangency point $e^{i\theta}$.

One consequence of treating horocycles as circles is that geodesics meeting at a circle center intersect only as they reach the unit circle, which is often called the "ideal boundary" in this model. Geodesics which meet there meet with zero angle. The triangle in the Figure 3.4(c) has two *ideal* vertices. There is a point of consistency to note here: when two horocycles are tangent to one another, then the geodesic between their (ideal) centers goes through that point of tangency. This is precisely what we see in each of the geometrics

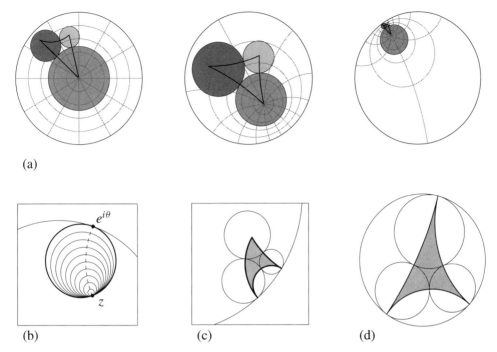

(a)

(b)   (c)   (d)

Figure 3.4. Hyperbolic matters.

with geodesic segments between ordinary circle centers. The triangle of Figure 3.4(d) has three ideal vertices and is called an *ideal* triangle. We will see shortly that its area is π, the maximal area for any hyperbolic triangle.

Our hyperbolic formulas naturally enough involve hyperbolic trigonometry. Again we find that areas and angles of triangles are linked:

Elements: $\qquad ds = \dfrac{2|dz|}{1 - |z|^2}, \quad ds^2 = \dfrac{4dxdy}{(1 + |z|^2)^2}$

Law of Cosines, $\mathbb{D}$: $\quad \cosh c = \cosh a \cosh b - \sinh a \sinh b \cos \gamma$

Circumference: $\qquad \text{Length}(C) = 2\pi \sinh r$

Disc area: $\qquad \text{Area}(D) = 4\pi \sinh^2 \left(\dfrac{1}{2} r \right)$

Triangle angles: $\qquad \alpha + \beta + \gamma = \pi - \text{Area}(t)$

$\text{Aut}(\mathbb{D}) = \text{Isom}(\mathbb{D})$: $\quad \left\{ \phi : \phi(z) = \lambda \dfrac{z - z_0}{1 - \bar{z}_0 z}, \ |\lambda| = 1, z_0 \in \mathbb{D} \right\}$

Curvature: $\qquad -1$ (constant, "negative")

Finally in our visit to hyperbolic geometry, note that Euclid's parallel postulate fails in quite a different way than in $\mathbb{P}$. Defining parallel lines as geodesics that do not intersect, we have the following: *Given a line l and a point p not on l, there exist infinitely many distinct lines through p that are parallel to l.*

3.3. Circles, Automorphisms, Curvature

Let us count the ways that nature has been kind to us. Our geometries are not only nested – the disc in the plane, the plane (stereoscopically) in the sphere – but *circles are circles are circles.* One can move among our model geometries without any ambiguity about circles. You do have to be more cautious about radii, centers, and geodesics, however; these depend on the geometry, as Fig. 3.4 shows in the hyperbolic setting.

If we look at automorphisms, our luck continues to hold: they are all Möbius transformations. Aut($\mathbb{C}$) and Aut($\mathbb{D}$) are the subgroups of Aut($\mathbb{P}$) consisting of the maps which fix $\mathbb{C}$ and $\mathbb{D}$, respectively. Every Möbius transformation is known to be a composition of (an even number of) reflections in circles (and lines); and every such reflection carries circles to circles. This underlies the fact that the automorphisms in each geometry map circles to circles. (The converse also holds, by the way; in each of our geometries, the only orientation preserving self-homeomorphisms which map circles to circles are the conformal automorphisms.)

Now, since automorphisms map circles to circles, *automorphisms carry circle packings to circle packings!* One has to be cautious again, however, when it comes to centers, radii, and geodesics, and hence to packing carriers. Automorphisms of $\mathbb{D}$ are isometries, so they preserve everything. In $\mathbb{C}$, an automorphism can change radii, but circle centers, geodesics, and hence carriers are preserved. In $\mathbb{P}$, nothing is safe: the image of a packing is indeed a packing, but (generically) the radii will be changed, the centers will not be mapped to centers, and geodesics will not be mapped to geodesics. That is just life on the sphere.

Back in the plus column, each of the groups Aut($\mathbb{G}$) ($\mathbb{G} = \mathbb{P}$, $\mathbb{C}$, or $\mathbb{D}$) is *doubly transitive*: given points p_1, $p_2 \in \mathbb{G}$ and directions β_1, β_2, there is an automorphism $\sigma \in$ Aut($\mathbb{G}$) which maps p_1 to p_2, with lines in direction β_1 at p_1 carried to lines in direction β_2 at p_2. This makes automorphisms valuable for routine normalizations and for manipulations, like those illustrated in Figure 3.5 for $\mathbb{D}$ and $\mathbb{P}$.

(I mentioned earlier that our version of stereographic projection was nonstandard. The two images on the left in Figure 3.5 suggest why. These show a packing in $\mathbb{D}$ and its projection to $\mathbb{P}$; note that the snake coils in the same direction in both. The typical mathematical projection puts infinity at the *north* pole; it would reverse the snake's orientation (as seen from *outside* the sphere). This plays havoc in applications – imagine a flat map of the United States projecting New York to the *left* coast!)

Let us move to metric considerations. The densities for our spaces are $\rho(z) = 2/1 + |z|^2$, $\rho(z) = 1$, and $\rho(z) = 2/1 - |z|^2$. Recalling (4) and $|z|^2 = x^2 + y^2$, direct computations lead to *constant curvatures* 1, 0, and -1, for $\mathbb{P}$, $\mathbb{C}$, and $\mathbb{D}$, respectively. The trichotomy among spaces with positive, zero, and negative curvatures is arguably the most powerful single distinction in all of geometry – so basic that it is the model for similar notions in topics from group theory to combinatorics to differential equations. Our three spaces are the archetypes for all of this, and were we studying the geometries

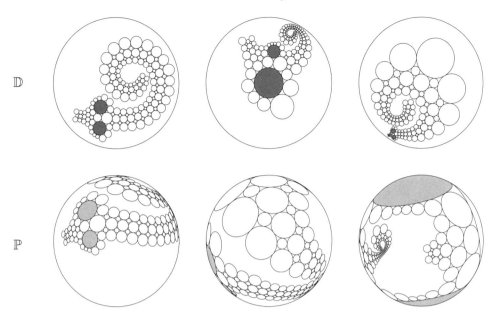

Figure 3.5. Automorphisms of circle packings in $\mathbb{D}$ and in $\mathbb{P}$.

for their own sakes, we might well talk at length about curvature's connections with the automorphisms, length/area comparisons, harmonic functions, Gauss–Bonnet, and so forth. As it happens, the reader will experience many of these effects first-hand. For now let me just mention one comparison.

Consider a circle C of radius r. The *circumference* of C is $2\pi r$ in the flat geometry of a soccer field. (This is in fact the very definition of π.) On the other hand, if C is centered on a hilltop, its circumference will be smaller than $2\pi r$, while if it is centered at the saddle point of a mountain pass, its circumference will exceed $2\pi r$. Similar effects occur with our metrics, and if we plot circumference versus radius from the formulas of the previous sections we get the starkly different behaviors of Figure 3.6. (You might think back to the tabletop experiment with pennies.)

These graphs all come in to $r = 0$ with slope 2π, reflecting their common *locally euclidean* nature. The hyperbolic distance between two points near $z_0 \in \mathbb{D}$ is roughly $2/(1 - |z_0|^2)$ times the euclidean distance; the spherical distance between two points near $w_0 \in \mathbb{P}$ is roughly $2/(1 + |w_0|^2)$ times the euclidean distance. In other words, when you zoom into small regions, the noneuclidean objects – the circles and geodesics – will *look* euclidean.

3.4. Riemann Surfaces

When we come to working with general surfaces, the "conformal geometries" associated with Riemann surfaces are the ones appropriate to our work. As it happens, each of these looks locally like one of our exemplars, $\mathbb{P}$, $\mathbb{C}$, or $\mathbb{D}$. We discuss Riemann surfaces at

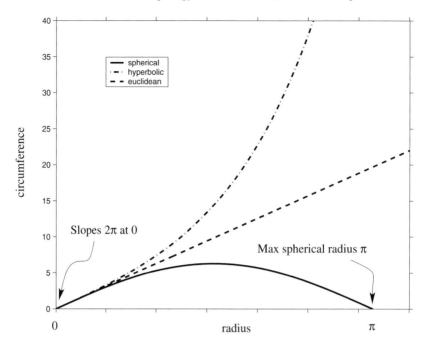

Figure 3.6. Circumference versus radius in the three geometries.

greater length in Chap. 9, but we need a few key properties now in order to state our main theorem.

If a topological surface S has an atlas $\mathcal{A} = \{(U_j, \phi_j)\}$ with the added property that its transition maps $\phi_j \circ \phi_i^{-1}$ are *analytic*, then we say that $\mathcal{A}$ defines a *conformal structure* for S. (See Appendix A for background on analytic functions.) A surface S with a conformal structure is called a *Riemann surface* (and we typically switch our notation from S to $\mathcal{S}$). A topological sphere supports an essentially unique conformal structure; a topological open disc supports precisely two, that of $\mathbb{C}$ and that of $\mathbb{D}$; every other surface supports uncountably many distinct conformal structures.

In essence, a conformal structure adds a strong layer of geometric rigidity to a surface on top of its topology. Every Riemann surface $\mathcal{S}$ supports an essentially unique complete metric of constant curvature 1, 0, or -1, which in this book will be termed the *intrinsic metric*. Every point of $\mathcal{S}$ has a neighborhood U which is isometrically isomorphic to an open set in one of $\mathbb{P}$, $\mathbb{C}$, or $\mathbb{D}$, respectively. As a geometric space, $\mathcal{S}$ is then classified as *spherical*, *euclidean*, or *hyperbolic*. The intrinsic metric is, unless stated otherwise, the metric used in defining circles on $\mathcal{S}$. It also permits a consistent notion of "angle" between intersecting curves. Riemann surfaces $\mathcal{S}_1$ and $\mathcal{S}_2$ are *conformally equivalent* if there exists a homeomorphism $F : \mathcal{S}_1 \longrightarrow \mathcal{S}_2$ which preserves these angles in magnitude and orientation. This concludes our preliminaries, we are now ready for the main show.

4

Statement of the Fundamental Result

A circle packing is a configuration of circles with a specified pattern of tangencies. A central theme in our study is the connection between combinatorics and geometry, so a description of the underlying pattern is an essential preliminary. Legitimate patterns are remarkably general, *local planarity* being the key requirement. However, we will concentrate exclusively on *triangulations*. They enjoy a rigidity missing in more general patterns, and we can use the well-established terminology of *simplicial complexes*.

Definition 4.1. The prescribed pattern for a circle packing will be encoded as an abstract simplicial 2-complex (or **complex** for short) which is (simplicially equivalent to) a triangulation of an oriented topological surface. The notation K will be used both for the complex and for its realization as a topological surface.

Topology and combinatorics are not enough; for circles we need a notion of distance. Assume $\mathcal{G}$ is an oriented surface with a metric.

Definition 4.2. A collection $P = \{c_v\}$ of circles in $\mathcal{G}$ is said to be a **circle packing** for a complex K (or K is the **complex** of P) if (1) P has a circle c_v associated with each vertex v of K, (2) two circles c_u, c_v are (externally) tangent whenever $\langle u, v \rangle$ is an edge of K, and (3) three circles c_u, c_v, c_w form a positively oriented triple in $\mathcal{G}$ whenever $\langle u, v, w \rangle$ forms a positively oriented face of K.

The fundamental existence and uniqueness result for circle packings has a rather involved history which I related in Notes I. It was proven for the sphere first by Koebe in 1936 and then independently by E. M. Andreev and by William Thurston; the extension stated here is due to Beardon and Stephenson.

Discrete Uniformization Theorem 4.3. *Let K be a complex that triangulates a topological surface S. Then there exist a Riemann surface $\mathcal{S}_K$ homeomorphic to S and a circle packing P for K in the associated intrinsic spherical, euclidean, or hyperbolic metric on $\mathcal{S}_K$ such that P is univalent and fills $\mathcal{S}_K$. The Riemann surface $\mathcal{S}_K$ is unique up to conformal equivalence and P is unique up to conformal automorphisms of $\mathcal{S}_K$.*

The packing P guaranteed by the Discrete Uniformization Theorem will be denoted $\mathcal{P}_K$ and referred to as the *maximal circle packing* for K. The complex K itself is termed *spherical*, *parabolic*, or *hyperbolic* as the geometry of the Riemann surface $\mathcal{S}_K$ is *spherical*, *euclidean*, or *hyperbolic*, respectively. Some notes are in order:

- When K is simply connected, then $\mathcal{S}_K$ is one of the sphere, the plane, or the hyperbolic plane with its standard metric and $\mathcal{P}_K$ is unique up to conformal automorphisms of $\mathcal{S}_K$.

- $\mathcal{S}_K$ is homeomorphic to S and hence to K (treated as a surface); one can therefore say that "$\mathcal{P}_K$ has endowed K with a conformal structure."

- The condition that $\mathcal{P}_K$ "fills" $\mathcal{S}_K$ will be evident in examples and will be defined more carefully later. Basically, it means that the packing has a carrier that is metrically "complete" in the geometry of $\mathcal{S}_K$ and it reflects the extremal nature of the packing.

- The packing P of the Discrete Uniformization Theorem is only *essentially unique*, meaning unique up to automorphisms of $\mathcal{S}_K$; although we will call it "the" maximal packing, there is in practice some normalization involved in choosing $\mathcal{P}_K$.

Several maximal packings are included in the Menagerie of Part I. Archetypes for the simply connected setting are the constant-degree examples in Figure 1.9 – these hark back to the table top experiment of the previous chapter. See also Figures 1.1(a) and (c), Figure 1.2(d), Figures 1.3(b) and (e), Figures 1.4(c) and (d), Figure 1.8(c), and the domain in Figure 2.4.

Non-simply-connected packings are more difficult to illustrate, but in fact once you understand how to read the pictures, you will be well on your way to grasping the underlying theory itself. In our Menagerie, Figure 1.8(d) merely suggests a genus 2 example, while actual packings are given for the genus 3 Klein surface in Figure 1.4(a) (and Figure 1.7(a)), for the genus 3 Picard surface in Figure 2.9, and for a torus in Figure 1.3(e) (shaded circles only). I am afraid that other multiply connected examples and their interpretations will have to wait.

The Plan: The proof of the Discrete Uniformization Theorem and the geometric tools we accumulate along the way occupy the remainder of this part of the book. We emphasize that we are making a transition here from the rather informal and intuitive material presented so far. We proceed now with precise definitions and rigorous mathematical proofs – nothing is proven "by picture." I will not apologize, however, for trying to engage the reader's intuition with images whenever it may be helpful or for leaving some routine details for the reader to sort out.

We begin in the next chapter by describing our bookkeeping methods, including the Monodromy Theorem for laying out circle packings. We then proceed to prove the existence and uniqueness of packings in successive settings depending on the properties of K. The first and central case is that of the combinatorial closed disc in which we find that $\mathcal{P}_K$ packs the hyperbolic plane $\mathbb{D}$. It is then an easy step to prove the result when K triangulates a sphere, the historical setting of the theorem. We move next to combinatorial open discs, where K is infinite and simply connected. A fundamental and quite fascinating dichotomy emerges here between packings in the hyperbolic and euclidean settings. Finally, we handle the case of multiply connected complexes in Chap. 9. This will require some additional background for those not familiar with surface theory. However, by this time, the insights from circle packing will essentially carry the case themselves – indeed, we will see the classical arguments appearing in discrete and very accessible forms.

5

Bookkeeping and Monodromy

Our bookkeeping requirements are surprisingly modest. We have already introduced the complexes K for encoding the prescribed combinatorics. We use radii chosen from an appropriate one of our three geometries as our principal parameters and rely on angle sums for measuring their suitability. With this information in hand other quantities of interest can be computed as necessary. In particular, the circle centers turn out to be a secondary matter. We establish the Monodromy Theorem for simply connected complexes, which says that once you have computed radii for K that satisfy local packing conditions, then the corresponding circle packing can be laid out in a straightforward fashion. The key computational effort therefore involves finding the radii. There is a common point of misunderstanding when first encountering circle packing, so let me be blunt:

> **Warning:** *There are no actual circles and there is **no circle configuration** until the correct radii have been found.*

With this warning in mind, let us see what notions and terminology we will need to manage our circle packing enterprise.

5.1. Bookkeeping

Complexes: Tangency patterns for circle packings are encoded as *abstract simplicial 2-complexes*, which we shorten to *complex*. Certain classifications will be important for us. We say that K is a *combinatorial closed disc* if it triangulates a topological closed disc; that is, K is finite, is simply connected, and has nonempty boundary. Likewise, a *combinatorial open disc* is infinite, simply connected, and without boundary. The complex K is *compact* if it is finite and without boundary, and hence triangulates a compact surface; if in addition K is simply connected, it is evidently a *combinatorial sphere*. With the important exception of combinatorial closed discs, we will leave most consideration of complexes with boundary as exercises for the reader.

The vertices of K will be assumed at various times to be indexed, $\{v_1, v_2, \ldots\}$. In the normal course of events, however, this is far too cumbersome and we refer to vertices by generic names or use a local, temporary numbering scheme. Context should make notations and meaning clear.

Labels: Labels, which are putative radii, serve as our principal packing parameters. A radius r must lie in $(0, \pi)$ in spherical geometry or in $(0, \infty)$ in euclidean and hyperbolic geometry; in hyperbolic geometry, infinite radii are permitted for boundary circles.

Definition 5.1. Given a complex K with vertices $\{v_1, v_2, \ldots\}$, a **label** R for K in a particular geometry – spherical, euclidean, or hyperbolic – consists of a set $R = \{r_1, r_2, \ldots\}$ of real numbers that qualify as radii. We refer to $K(R)$ as a **labeled complex** (in the particular geometry). The label associated with vertex $v = v_j$ may be denoted by r_j, $R(v_j)$, or $R(v)$, depending on circumstances.

In case a label R consists of the radii associated with a circle packing P for K, we write $P \longleftrightarrow K(R)$. In other words, this means that for each vertex $v \in K$, $R(v) = \mathrm{radius}(c_v)$, where c_v is the circle of P associated with v.

Recall that we are using the neutral term "label" because in general the values in R will *not* constitute radii for any actual packing having the combinatorics of K – were you to use them as radii, the circles would not fit together. We call the labels "radii" only when they deserve it.

Metrics: A label R for K, whether associated with a circle packing or not, endows $K(R)$ with a metric d_R. Every face of $K(R)$ is isometrically isomorphic to some triangle in the geometry of R, so a (rectifiable) path lying in a single face will have a well-defined *length*. For a general path in K, one can define the length by summing up the lengths in successive faces through which it passes. One defines the distance $d_R(a, b)$ between any two points a, b of K as the infimum of the lengths of paths from a to b. This is a standard definition for what is known as a *path metric*. There is nothing particularly difficult about it, but one must exercise caution. For example, the shortest path between two points of the *same* face f might well pass *outside* of f, even though f is isometrically isomorphic to a convex triangle.

Angle Sums: The compatibility conditions on labels are expressed in terms of angle sums. Here is how the formulas arise. (There is one side condition for spherical circles, which we discuss momentarily.)

Given three (legal) labels x, y, z in any one of our geometries $\mathbb{G}$, there exists a positively oriented triple of mutually tangent circles $\langle c_x, c_y, c_z \rangle$ in $\mathbb{G}$ having these radii. This is easily verified by simply building the triple. Connecting the centers with geodesics forms a triangle T; the triple and T are unique up to isometries of $\mathbb{G}$. If $\alpha = \alpha(x; y, z)$ denotes the angle of T at the center of c_x, then the law of cosines appropriate to the geometry allows us to compute α from the numbers x, y, z. Figure 5.1 illustrates typical triangles in the three geometries and the explicit formulas are recorded for later reference in the table of Figure 5.2.

Please note this side condition in spherical geometry: circles of spherical radii x, y, z can form a triple in $\mathbb{P}$ if and only if $x + y + z \leq \pi$. In the case of equality, the three

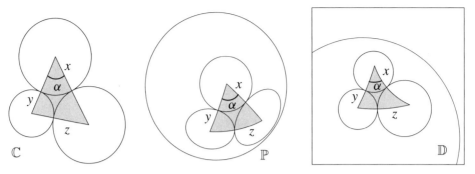

Figure 5.1. Typical triples and their triangles.

centers lie on a common great circle, which is the boundary of T (the interior of T is determined, as always, by orientation).

With triples in hand, we may now focus on the local geometry at a vertex v of the labeled complex $K(R)$. This local geometry is determined entirely by the labels for v and its immediate neighbors.

Definition 5.2. Let R be a label for K. The **angle sum map** $\theta_R : K^{(0)} \longrightarrow [0, \infty)$ assigns to each vertex $v \in K$ the sum of the angles at v in the faces of K to which v belongs:

$$\theta_R(v) = \sum_{\langle v, u, w \rangle} \alpha(R(v); R(u), R(w)).$$

Suppose that $F_v = \{v; v_1, \ldots, v_k\}$ is the (combinatorial) flower for v in K. Vertex v belongs to m faces, where $m = k$ if v is interior and $m = k - 1$ if v is boundary. In either case, the associated labels $\{r; r_1, \ldots, r_k\}$ taken from R allow us to compute the m angles at v. An alternative to the notation $\theta_R(v)$ if we want to focus only on these local data is $\theta(r; r_1, \ldots, r_k)$. This formula is also recorded in the table of Figure 5.2.

Packing Conditions: Let v be an interior vertex of a labeled complex $K(R)$ and let $F_v = \{v; v_1, \ldots, v_k\}$ be its combinatorial flower. There exists a corresponding *geometric flower* of circles in the geometric space of the label R if and only if the angle sum $\theta_R(v)$ is a positive integral multiple of 2π. The forward implication is self-evident, and the reverse implication is easily proven with a construction – we leave this to the reader.

Definition 5.3. A label R is termed a **packing label** for K if for every interior vertex v of K there exists a nonnegative integer β_v such that $\theta_R(v) = 2\pi(\beta_v + 1)$. If $\beta_v > 0$ we say that R has a **branch point** at v and β_v is the **branch order**. A label R having at least one branch point is a **branched** packing label; otherwise the label is **unbranched**.

Triangle Angle: In a triangle T formed by circles of radii x, y, z, the angle at the circle of radius x is given by the following, depending on geometry:

Euclidean:

$$\alpha(x; y, z) = \arccos\left(\frac{(x + y)^2 + (x + z)^2 - (y + z)^2}{2(x + y)(x + z)}\right)$$

Spherical: (side condition: $x + y + z \leq \pi$)

$$\alpha(x; y, z) = \arccos\left(\frac{\cos(y + z) - \cos(x + y)\cos(x + z)}{\sin(x + y)\sin(x + z)}\right)$$

Hyperbolic:

$$\alpha(x; y, z)$$

$$= \begin{cases} \arccos\left(\dfrac{\cosh(x + y)\cosh(x + z) - \cosh(y + z)}{\sinh(x + y)\sinh(x + z)}\right), & x, y, z \in (0, \infty) \\[2ex] \arccos\left(\dfrac{\cosh(x + y) - e^{(y-x)}}{\sinh(x + y)}\right), & x, y \in (0, \infty), z = \infty \\[2ex] \arccos\left(1 - 2e^{2x}\right), & x \in (0, \infty), y = z = \infty \\[2ex] 0 & x = \infty. \end{cases}$$

Angle Sums: In a flower having central label r and petal labels $\{r_1, \ldots, r_k\}$, the angle sum is given by the following summation formula, where $m = k$ and $r_{k+1} = r_1$ if the flower is closed, and $m = k - 1$ otherwise:

$$\theta(r; r_1, \ldots, r_k) = \sum_{j=1}^{m} \alpha(r; r_j, r_{j+1}).$$

Figure 5.2. Angle formulas.

(The packings in this part of the book will typically be unbranched, but we include branching along the way when it can be accommodated without additional effort, as in the Monodromy Theorem below.)

Circle Packings: The last bookkeeping detail concerns the circle centers and the actual circle packing configuration itself. As we have observed earlier, most of our effort will be directed toward existence and uniqueness of packing labels. In the simply connected case, the packing then comes essentially for free. We handle this laying out process and prove the Monodromy Theorem in the next section. In the multiply connected cases we will see later how the circles *force* the geometry to accommodate them.

Carriers: With every circle packing P comes its *carrier*, the *geometric complex* formed by connecting the centers of neighboring circles by geodesic segments. Our notation is carr(P). We will systematically treat the carrier as a geometric realization of the abstract complex K. Each vertex v is identified with the center of the corresponding circle, each edge $\langle v, u \rangle$ with the geodesic between the centers, and each face with the corresponding triangle of the carrier. If P is univalent, then carr(P) provides an embedding of K in $\mathbb{G}$, if P is locally univalent, it provides an immersion, and if P is branched, it provides a branched immersion.

5.2. The Monodromy Theorem

Let us assume for a moment that you have found a packing label for K. Where is the actual *packing P?* You might picture your task this way: you have a collection $\{c_v\}$ of circles, one for each vertex of K, each with its radius $R(v)$ – all you need to do is determine where to put them!

Were these cardboard discs, most of us would proceed in a rather obvious way, putting one disc in place and successively adding others to the pattern as their positions become clear. The process is just that straightforward when K is simply connected – we merely need a more rigorous description.

Theorem 5.4. *Let K be a simply connected complex with label R. Let $\mathbb{G}$ be one of $\mathbb{P}, \mathbb{C}$, or $\mathbb{D}$, depending on the geometry of R. Then there exists a circle packing P in $\mathbb{G}$ with $P \longleftrightarrow K(R)$ if and only if R is a packing label. The circle packing P is unique up to isometries of $\mathbb{G}$.*

Proof. Choose some vertex of K, say the designated α-vertex, v_1, and place its circle c_1 at some arbitrary but definite location in $\mathbb{G}$. Choose $v_2 \sim v_1$ and center its circle c_2 to be tangent to c_1. Thereafter, follow this routine: Look through the remaining unplaced circles to find one, c_v, having two contiguous neighbors, c_u and c_w, that are already in place and tangent to one another. These neighbors clearly determine a unique location for c_v. The orientation enters here since there are initially two possibilities, and c_v must be placed so that the triple $\langle c_v, c_u, c_w \rangle$ has the orientation of $\langle v, u, w \rangle$ in K. Continue placing circles until you are finished.

How do we organize this and what do we have to prove? The fact that K is connected ensures that every circle can eventually be placed. However, we need to show that the placement will be well-defined – the circle locations should be independent of the order in which the circles have been placed and unique after the first two circles are down. Then we need to show that the resulting configuration is a packing for K – each circle is tangent to all its neighbors, not just those used in placing it.

The face structure of K can serve as an organizing principle. If the triple of circles for a face f are in place, they determine a geometric triangle T associated with f. Any neighboring face f' has two of its circles already in place, and using the law of cosines and orientation one can place its third circle, giving a triangle T' for f'. In other words, we may focus on placing *triangles* rather than circles.

If we write $\mathcal{F}$ for the collection of faces of K, then here is the natural procedure: Designate a "base" face $f_0 = \langle v_1, v_2, v_3 \rangle$, where c_1 and c_2 are our starter circles. Placing c_3 then determines the "base" triangle T_0. Suppose $\Gamma = \{f_0, f_1, \ldots, f_n\}$ is a *chain* of (not necessarily distinct) faces from $\mathcal{F}$ (recall that this means that each face shares an edge with its predecessor). Since T_0 has already been placed, there is a unique location for the triangle T_1 associated with f_1. Proceeding inductively, we place triangles $T_2, T_3, \ldots$ associated with faces of Γ until we have placed a triangle T_n for f_n. We will say that the location of T_n was obtained from that of T_0 by a *development* along Γ.

We use developments along chains to define the locations of triangles for all faces. In particular, given any $f \in \mathcal{F}$, there exists a chain $\Gamma = \{f_0, f_1, \ldots, f_n\}$ with $f = f_n$ and we obtain the location of $T_f = T_n$ by a development along Γ.

The main work enters in proving that the location of T_f is independent of Γ, since there are many possible chains from f_0 to f. This is where you may recognize a version of "monodromy." A moment's reflection will convince you that two chains Γ_1 and Γ_2 from f_0 to f will end with the same triangle for f if and only if traveling out along Γ_1 and back along Γ_2 places T_0 back at its original location. That is, for our whole effort it suffices to consider closed chains $\Gamma = \{f_0, f_1, \ldots, f_n\}$ (that is, with $f_n = f_0$) and to show that the triangle T_n obtained by a development along Γ is identical to T_0. Proceed using homotopies. The local modifications defined abstractly in Sect. 3.1.3 are now quite concrete; Figure 5.3 illustrates a local modification in which the direction around a vertex v is reversed. Because the angle sum $\theta_R(v)$ is equal to an integral multiple of 2π – finally, the packing hypothesis on R enters! – the local chains $\gamma = \{f_1, f_2, f_3, f_4\}$ and $\gamma' = \{f_1, f_7, f_6, f_5, f_4\}$, if started at T_1, will lay the triangle for f_4 in precisely the same spot. This means that if γ were a subchain of a closed chain Γ, then replacing γ by γ' would not affect the rest of the development along Γ. This is true for all local modifications,

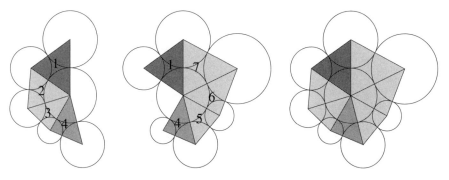

Figure 5.3. Two local chains at v.

and we conclude that *developments along homotopic closed chains will place their final triangles at identical locations.*

There is one chain where it is perfectly clear that T_0 ends up back where it started, namely, the *null* chain $\Gamma_0 = \{f_0\}$. Since K is simply connected, *every* closed chain is null homotopic, so by the fact noted above we conclude that *every* closed chain places T_0 where it started. In other words, the configuration P comes out the same with our construction procedure regardless of the order in which the faces are laid – it is well-defined. Is it a packing? Given any vertex v, once you have a chain placing one of its faces, the chain can be extended to proceed through its whole flower. In other words, v will have all the tangencies prescribed by K. Therefore, P is a circle packing for K and $P \longleftrightarrow K(R)$.

Uniqueness is easy: the locations of all the circles of P follow once you place those of the first triangle T_0. Given any two locations for T_0, there is a rigid motion mapping one to the other, and that rigid motion identifies the resulting circle packings. $\qquad\qquad\square$

5.2.1. Two Examples

The proof that P exists – that it is well-defined regardless of the order in which its faces are placed – is really quite subtle.

On the left in Figure 5.4 is a euclidean circle packing $P \longleftrightarrow K(R)$ for the complex K in Fig. I.1 of Practicum I. I have indicated the indices for two chains of faces running from face 20 to face 17, namely,

$$\Gamma_1 = \{20, 5, 4, 3, 14, 17\} \qquad \text{and} \qquad \Gamma_2 = \{20, 5, 6, 1, 2, 3, 14, 17\}.$$

As expected, using radii from the packing label R, developments from 20 along Γ_1 and Γ_2 place face 17 at the same location.

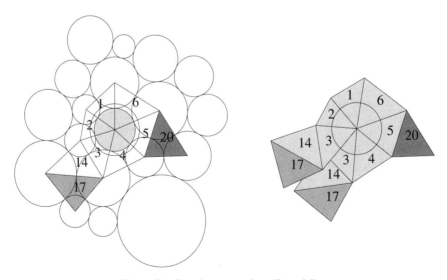

Figure 5.4. Developments along Γ_1 and Γ_2.

Suppose, however, that we increase the radius of the shaded circle, c_1, by, say, 20%. That single change gives a new label R' that is no longer a packing label. On the right in Figure 5.4 I have laid down face #20 and then developed along first Γ_1 and then Γ_2. These chains differ only in the way that they pass around c_1 – they are local modifications of one another. As you see, they place #17 in different spots. There is nothing special about #17; any face will have multiple locations depending on how you chain to it from #20. Where does face #17 really belong? Well . . . nowhere! When you have a label that is not a packing label, there is **no consistent way** to lay out faces – there is **no** circle packing!

A second example addresses the theorem's simple connectivity hypothesis. If K is not simply connected, the key monodromy argument still holds. Namely, *for a packing label, developments along homotopic chains will place faces in identical locations.* However, absent simple connectivity, there exist chains that are *not* null homotopic. Figure 5.5 shows that the conclusion of the theorem can fail. The complex K on the left is hexagonal, so the euclidean label $R \equiv 1$ is a packing label. However, on the right we see what happens when you try to lay the corresponding circles out. The three dashed circles on the right end are what we will call "ghosts" of the circles at the left end – they represent the same three vertices of K, but the layout places them in different locations depending on which way you develop around the hole in K.

When K is not simply connected, the packing condition on R is a necessary but not sufficient condition for existence of a packing P.

We will return to such situations in Chap. 16.

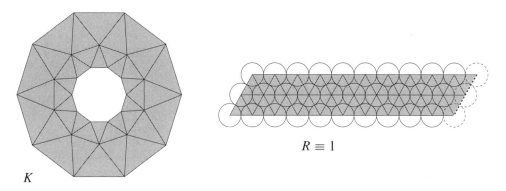

$R \equiv 1$

K

Figure 5.5. Development for a simple annulus.

6

Proof for Combinatorial Closed Discs

Proposition 6.1. *Let K be a combinatorial closed disc (that is, simply connected, finite, and with nonempty boundary). Then there exists an essentially unique univalent circle packing $\mathcal{P}_K \subset \mathbb{D}$ for K such that every boundary circle is a horocycle.*

This case turns out to be pivotal in the proof of the general Discrete Uniformization Theorem, yet when we use the appropriate geometry on the disc – namely, hyperbolic geometry – the proof is surprisingly elementary. We start with an example on which you can limber up your intuition, and then go on to geometric lemmas and the proof.

6.1. A Mind Game

To put us in the right frame of mind, consider the modest example of Figure 6.1. On the left is a packing P in the hyperbolic plane for a complex K having 19 vertices. On the right is the associated maximal packing $\mathcal{P}_K$; I hope you get some immediate sense of "maximality" – $\mathcal{P}_K$ seems to strain against the unit circle $\mathbb{T} = \partial \mathbb{D}$. The question is "How to go from P to $\mathcal{P}_K$?"

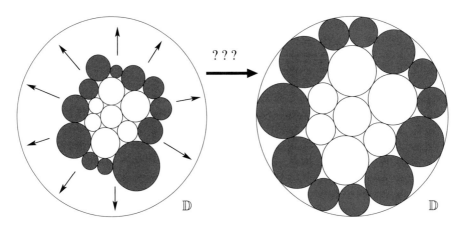

Figure 6.1. Setting for a mind game.

I suggest that you imagine (for the moment) these 19 circles as a community of sentient beings. They recognize two *prime directives*: Grow! and Maintain your community! The individuals fall into two categories. (1) The 12 shaded boundary circles are **Type A**; they are driven to *grow*. (2) The 7 interior circles are **Type B**; these are the peacemakers who strive to keep the community intact – that is, to maintain the agreed tangencies.

- At every opportunity a Type A circle will double its radius.
- At every opportunity a Type B circle will adjust its own radius to suit its immediate neighbors, meaning that it will adjust its radius so that its angle sum is 2π.

The circles are all community-minded, yet they recognize their differences. After some debate they agree to a reasonable strategy: they generate a roster of their names and repeatedly cycle through it, each circle acting its part in turn one after the other.

Your job: *Explain why $\mathcal{P}_K$ is the ultimate result.*

We will have to surrender our literary license in order to build a mathematical proof. But if you have thought through this mind game carefully enough you might already have sorted out the key intuitions. Take some time here, then proceed.

6.2. Monotonicities and Bounds

We start with fundamental behavior of small configurations of circles, namely, triples and flowers, and ask what happens when one circle decides to change size.

Lemma 6.2. (Monotonicity in Triangles). *Consider the triple of circles in $\mathbb{D}$ and the hyperbolic triangle T formed by their centers, as labeled in Figure 6.2. Assuming r_1 is finite, α is continuous and strictly decreasing in r_1, while area(T) is continuous and strictly increasing in r_1. If r_2 (resp. r_3) is also finite, then β (resp. γ) is continuous and strictly increasing in r_1.*

Proof. Figure 6.2 demonstrates these monotonicities when the original circle of radius r_1 is replaced by the larger (dashed) circle. It is tempting to rely on the picture or on the analogous results in the more familiar euclidean setting. However, with a judicious change of variables, the formal hyperbolic computations are quite straightforward.

Define auxiliary function g by

$$g(x, y, z) = \arccos\left[\frac{(xy + 1)(xz + 1) - 2x(yz + 1)}{(xy - 1)(xz - 1)}\right]. \tag{7}$$

Assuming that the hyperbolic radii r_1, r_2, r_3 of Figure 6.2 are finite, then according to formulae in the table of Figure 5.2, $\alpha = g(x, y, z)$, where $x = e^{2r_1}$, $y = e^{2r_2}$, $z = e^{2r_3}$.

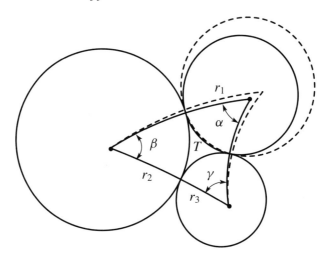

Figure 6.2. A typical hyperbolic triple.

Clearly g is continuous, and easy computations give the partial derivatives

$$\frac{\partial g}{\partial x}(x, y, z) = \frac{-(x^2 yz - 1)\sqrt{(y - 1)(z - 1)}}{(xy - 1)(xz - 1)\sqrt{x(x - 1)(xyz - 1)}} \tag{8}$$

$$\frac{\partial g}{\partial y}(x, y, z) = \frac{\sqrt{x(x - 1)(z - 1)}}{(xy - 1)\sqrt{(y - 1)(xyz - 1)}} \tag{9}$$

$$\frac{\partial g}{\partial z}(x, y, z) = \frac{\sqrt{x(x - 1)(y - 1)}}{(xz - 1)\sqrt{(z - 1)(xyz - 1)}}. \tag{10}$$

Noting that $x, y, z > 1$, the first of these is clearly negative, while the others are positive. Therefore, α is strictly decreasing in x, and hence in its own radius r_1, while it is strictly increasing in y and z, and hence in the radii r_2, r_3.

The effect of r_1 on area(T) is more subtle. In hyperbolic geometry, area$(T) = \pi - \alpha - \beta - \gamma$, in the notation of Figure 6.2. Since α decreases while β and γ increase with r_1, the effects on area involve cancellations. Though one could compute this, the result is geometrically clear in Figure 6.2. Since increasing the radius r_1 while keeping the other two circles fixed in place causes increases in angles β and γ, it is clear that the new triangle strictly contains the original, so area(T) goes up.

If one or both of the radii r_2, r_3 is infinite, then alternate formulae in the table apply, but the calculations are similar and are left for the reader. This completes our proof of triangle monotonicity. $\square$

Monotonicity associated with flowers flows immediately from that for triples, since a flower's angle sum is simply a sum of the central angles of its faces. Area of a flower, area(F), is defined as the sum of the areas of the faces forming it. For reference, the statement is this:

Lemma 6.3 (Monotonicity in Flowers). *Let $F = \{v; v_1, \ldots, v_n\}$ denote a combinatorial flower with central vertex v, let $\{r; r_1, \ldots, r_n\}$ be a hyperbolic label for F, and write $\theta(v)$ for the angle sum at v, a function of $r, r_1, \ldots, r_n$. Assuming $r < \infty$, $\theta(v)$ is continuous and strictly decreasing in r and is continuous and strictly increasing in petal radii that are finite. If $r = \infty$, then $\theta(v) \equiv 0$. Area(F) is continuous and strictly increasing in both r and the petal radii.*

Returning to the auxiliary function g and $x = e^{r_1}$, note that

$$\lim_{x \downarrow 1} g(x, y, z) = \pi, \qquad \lim_{x \uparrow \infty} g(x, y, z) = 0. \tag{11}$$

So the angle in a triple of circles can vary between π and 0 as one varies the corresponding radius r_1. This implies that for a flower F having n faces, the angle sum varies between $n\pi$ and 0 with the radius of the center circle. If the flower is closed, then $n \geq 3$. Together with monotonicity, we have the following lemma.

Lemma 6.4. *Let $F = \{v; v_1, \ldots, v_n\}$ denote a closed combinatorial flower and let $\{r_1, \ldots, r_n\}$ denote the hyperbolic labels for its petals. Then there exists a unique label r for the center v such that $\theta(v) = 2\pi$.*

We conclude the geometric preliminaries by proving a bound which nature enforces on the radii of interior circles in hyperbolic packings.

Lemma 6.5. *Let $F = \{v; v_1, \ldots, v_n\}$ denote a closed combinatorial flower, let $\{r; r_1, \ldots, r_n\}$ be a hyperbolic label for F, and let $\theta(v)$ denote the angle sum for F at v. Suppose $\theta(v) \geq 2\pi$. Then $r \leq -\log(\sin(\pi/n))$.*

Proof. It is easy to argue from monotonicity that the extremal situation occurs when the petal radii are all infinite and the central radius r is such that $\theta = 2\pi$, the situation pictured in Figure 6.3. Indeed, if θ were to exceed 2π, then r could be made larger and still satisfy the hypotheses. Likewise, if a petal were to have finite radius, then increasing it would increase θ, again permitting r to be increased.

If the central circle were any larger in Figure 6.3, the n petals, squeezed as they are against the unit circle, could not wrap all the way around it. The central angle α common to the faces in Figure 6.3 may be computed as $\alpha = \arccos(1 - 2e^{-2r})$ by letting y and z go to infinity in (7). Since $\alpha = \theta/n = 2\pi/n$, we have

$$\cos(2\pi/n) = 1 - 2e^{-2r} \implies \frac{1 - \cos(2\pi/n)}{2} = e^{-2r}$$

$$\implies \sin^2(\pi/n) = e^{-2r} \implies r = -\log(\sin(\pi/n)). \qquad \square$$

Observe that there is no analogous bound on radii in the euclidean setting, since scaling will make a flower as large as you wish without changing its angle sum. Along with the existence of circles having infinite radii, this bound is a crucial advantage for hyperbolic geometry.

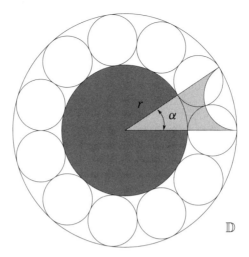

Figure 6.3. A maximal hyperbolic flower.

6.3. The Hyperbolic Proof

The first stage in the proof – namely, the situation described in our earlier Mind Game – may be separated out in this claim:

Claim. *The conclusion of Proposition 6.1 holds under the assumption that there exists some circle packing P for K in $\mathbb{D}$.*

Proof of Claim. The rather "algorithmic" and informal approach in our game is replaced here by what is known as a *Perron* method. Define the collection of labels

$$\Phi = \{R : \theta_R(v) \geq 2\pi \text{ for every interior vertex } v\}.$$

We show that the supremum $\widehat{R}$ of Φ, defined by $\widehat{R}(v) = \sup\{R(v) : R \in \Phi\}$, is our solution, that is, is the packing label whose circle packing $\mathcal{P}_K$ we seek.

The objects in a Perron class like Φ are often called "subsolutions," and as in all Perron-type arguments, one needs to show the following: (1) There exist subsolutions; that is, Φ is nonempty. (2) Φ is closed under taking maxima. (3) $\widehat{R}$ is nondegenerate. (4) $\widehat{R}$ is a solution (not just a subsolution).

(1) **Nonempty.** By hypothesis there is a circle packing P for K in $\mathbb{D}$; its radii form a label R_0 that is a member of Φ.

(2) **Closed under max.** Given $R_1, R_2 \in \Phi$, is $R = \max\{R_1, R_2\}$ in Φ? (Try confirming this yourself before you read on.) Let v be an interior vertex, and suppose, without loss of generality, that $R(v) = R_1(v)$. Comparing R to R_1, note that the labels for neighbors to v can only increase in R from their values in R_1. By the monotonicity

results for flowers, this implies $\theta_R(v) \geq \theta_{R_1}(v) \geq 2\pi$. This holds for all interior v, so R belongs to Φ.

(3) **Nondegenerate.** In our setting, this means that $\widehat{R}$ must be finite for interior vertices. Pick any interior v and let n be its degree. For each $R \in \Phi$, $\theta_R(v) \geq 2\pi$, so by Lemma 6.5, $R(v) \leq -\log(\sin(\pi/n))$. Consequently, $\widehat{R}(v) \leq -\log(\sin(\pi/n))$.

(4) **Solution.** Note first that given any label $R \in \Phi$ and any boundary vertex w, replacing $R(w)$ by ∞ can only increase angle sums at interior vertices, meaning that the modified label will be back in Φ. We conclude that $\widehat{R}(w) = \infty$ for all $w \in \partial K$.

The crucial questions, of course, one whether $\widehat{R}$ is itself in Φ and whether it is a *packing* label. Discreteness makes the proof here much easier than one might face in a classical continuous setting. For each integer $j > 0$ and interior vertex v, there exists $R_{j,v} \in \Phi$ such that $[\widehat{R}(v) - R_{j,v}(v)] < 1/j$. By (2), the label R_j defined by $R_j = \max_v\{R_{j,v}\}$ is back in Φ. This holds for all vertices v, and by the argument of the last paragraph, we may also assume R_j has all boundary labels infinite, so the sequence $\{R_j\}$ satisfies

$$R_j(v) \to \widehat{R}(v) \text{ as } j \to \infty \text{ for every vertex } v \in K. \tag{12}$$

Now, at an interior vertex v, $\theta_R(v)$ is continuous as a function of the radii in R, so (12) implies $\theta_{R_j}(v) \to \theta_{\widehat{R}}(v)$, and therefore $\theta_{\widehat{R}}(v) \geq 2\pi$. This holds for all interior v, so $\widehat{R} \in \Phi$. Moreover, suppose $\theta_{\widehat{R}}(v) > 2\pi$ at some particular interior v. Then again by continuity and monotonicity, a small increase can be made in the single label $\widehat{R}(v)$ without dropping this angle sum below 2π, and this increase can only add to the angle sums of neighbors. The result would be a label in Φ which exceeded $\widehat{R}$ at v, contradicting the definition of $\widehat{R}$. We conclude that $\theta_{\widehat{R}}(v) = 2\pi$ for every interior v, proving that $\widehat{R}$ is a packing label.

This completes the Perron argument, and we have found a promising packing label $\widehat{R}$. Let us refer to it henceforth as $\mathcal{R}_K$ and investigate the properties of associated circle packings; then we address uniqueness.

Since $\mathcal{R}_K$ is a packing label, Theorem 5.4 guarantees a circle packing $\mathcal{P}_K \longleftrightarrow K(\mathcal{R}_K)$ in $\mathbb{D}$, and the boundary circles, having infinite radii, will be horocycles. The only thing we need to establish is that $\mathcal{P}_K$ is univalent, and for that we use a standard topological result. Define a continuous map $\phi : K \longrightarrow \text{carr}(P)$ that initially maps each face of K one to one onto the corresponding hyperbolic triangle in $\text{carr}(P)$. The map ϕ is locally one-to-one; this is immediate at points interior to the faces and edges of K and at an interior vertex v the condition $\theta_R(v) = 2\pi$ means that the faces in the flower of v form a closed chain of triangles about $\phi(v)$ having mutually disjoint interiors. (There is an important point here that is easily overlooked: this condition at v does not itself guarantee that the *circles* of the flower of v have mutually disjoint interiors.)

We are now going to modify the definition of ϕ on any face having a boundary edge e as illustrated in Figure 6.4: namely, the image triangle is deformed continuously so that the boundary edge is mapped to the arc of the unit circle between the (ideal) centers of its endpoint horocycles. Precise details of the modification are unimportant as long as ϕ is unchanged on the other edges and carries the face continuously and one to one onto the

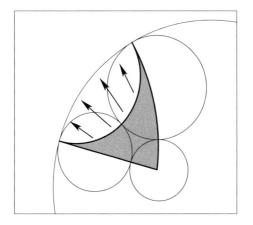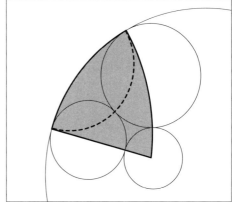

Figure 6.4. Pushing a boundary edge.

new image in $\overline{\mathbb{D}}$. (Note that some faces may need adjustment on more than one boundary edge.)

Continuing to use the name ϕ for the modified function, we now appeal to topology. K is a topological disc, and on any sequence of points $\{p_j\}$ that converge to its boundary, ∂K, the image sequence $\{\phi(p_j)\}$ converges to $\partial \mathbb{D}$. Thus $\phi : K \longrightarrow \overline{\mathbb{D}}$ is what is known as a *proper* map. Putting its properties together, ϕ is a proper, continuous, and locally one-to-one map between two topological discs. By the topological *argument principle*, ϕ is globally one-to-one. Since the circles of P are covered by the images of the faces of K, it is elementary to conclude that these circles have mutually disjoint interiors. Therefore, $\mathcal{P}_K \longleftrightarrow K(\mathcal{R}_K)$ is a univalent circle packing.

We next prove the uniqueness of $\mathcal{R}_K$, meaning that no other packing label can give a circle packing satisfying the conclusions of the proposition. (Of course, for $\mathcal{P}_K$ itself, this only implies uniqueness up to automorphisms of $\mathbb{D}$.) The proof is based on area considerations. In a labeled complex $K(R)$, we will write Area $(\cdot)$ for area in the metric d_R induced by R (see Section 5.1). Since each face of $K(R)$ is isometrically isomorphic to a hyperbolic triangle, its area is $\pi - \alpha - \beta - \gamma$. Adding areas over all faces gives

$$\text{Area}(K(R)) = F\pi - \Sigma,$$

where F is the number of faces of K and Σ is the sum of all angles of all the faces of $K(R)$. By grouping those angles according to whether they occur at interior or boundary vertices, we can further write $\Sigma = \Sigma_{\text{int}} + \Sigma_{\text{bdry}}$. Altogether, then, we have

$$\text{Area}(K(R)) = F\pi - \Sigma_{\text{int}} - \Sigma_{\text{bdry}}.$$

In the case of our label $\mathcal{R}_K$, we have $\Sigma_{\text{int}} = 2\pi N$, where N is the number of interior vertices of K, and $\Sigma_{\text{bdry}} = 0$, because all boundary circles are horocycles. Therefore,

$$\text{Area}(K(\mathcal{R}_K)) = (F - 2N)\pi.$$

Assume that R' is a second unbranched packing label for K having infinite boundary radii. $K(R')$ will then have precisely the same area. On the other hand, since R' would belong to Φ, the definition of $\mathcal{R}_K (= \widehat{R})$ would imply that $R' \leq \mathcal{R}_K$ (meaning this inequality holds at every vertex). Strict inequality at any vertex v would imply, by the last statement in Lemma 6.3, that $\text{Area}(K(\mathcal{R}_K)) > \text{Area}(K(R'))$. Since these are in fact equal, we conclude that $R'(v) = \mathcal{R}_K(v)$ for every interior v, and hence $R' \equiv \mathcal{R}_K$. This proves uniqueness of $\mathcal{R}_K$ and thereby the essential uniqueness of $\mathcal{P}_K$ and completes the proof of the claim. $\square$

Now continue the proof of Proposition 6.1. By the claim, it is enough to prove that a given K has *some* circle packing in $\mathbb{D}$. We proceed by induction on the number V of vertices of K. The minimal case is $V = 3$; K is a single triangle and $\mathcal{P}_K$ clearly consists of a triple of mutually tangent horocycles.

Assume that K is given with $V > 3$ and make the inductive hypothesis that the proposition holds for every triangulation having less than V vertices. Designate an arbitrary but fixed vertex $w \in \partial K$. There are two possibilities to consider:

Case 1. Suppose there exists an interior edge e from w whose other endpoint u is also a boundary vertex. In this circumstance we cut K into two pieces via the edge e. More formally, there exist complexes K_1 and K_2 having positively oriented boundary edges

$$e_1 = \langle w_1, u_1 \rangle \in K_1 \quad \text{and} \quad e_2 = \langle u_2, w_2 \rangle \in K_2 \tag{13}$$

such that

$$K = K_1 \bigcup_{e_1 \sim e_2} K_2.$$

That is, K is the disjoint union of K_1 and K_2 modulo the identifications

$$e_1 \longleftrightarrow e_2, \quad w_1 \longleftrightarrow w_2, \quad u_1 \longleftrightarrow u_2.$$

Since K triangulates a closed disc, the same is true of the pieces K_1 and K_2, and each of these has at least one less vertex than K.

Applying the induction hypothesis, there exist univalent circle packings P_1 and P_2 for K_1 and K_2, respectively, having all boundary circles as horocycles. It turns out that we can "paste" these together to get a circle packing for K. Consider P_1, for instance. The horocycles for w and u will be tangent, so normalization by an appropriate automorphism of $\mathbb{D}$ allows us to assume that the circles for $w = w_1$ and $u = u_1$ in P_1 are these two specific circles centered on the real axis:

$$c_w = \left\{ \left| z - \frac{1}{2} \right| = \frac{1}{2} \right\}, \quad c_u = \left\{ \left| z + \frac{1}{2} \right| = \frac{1}{2} \right\}.$$

Orientation considerations from (13) imply that the remaining circles of P_1 will lie above the real axis. Likewise with P_2, we can normalize so these same circles are the circles for w_2 and u_2, respectively. Now, however, (13) implies that the remaining circles of P_2 lie *below* the real axis. Simply superimposing these two normalized circle configurations results in a univalent hyperbolic circle packing for K. The claim now completes the induction step for Case 1.

Case 2. Suppose every interior edge from w ends at an interior vertex. If $F_w = \{w; v_1, \ldots, v_n\}$ denotes the combinatorial flower of w, then removing the open star of w from K leaves a reduced complex K' which triangulates a closed topological disc and has $V - 1$ vertices. The induction hypothesis implies existence of a univalent circle packing $P' = \mathcal{P}_{K'}$ whose boundary circles are all horocycles.

The petal vertices of flower F_w become boundary vertices of K', so their circles are horocycles in P'. If we now consider our situation on the *sphere*, we find that when we include the unit circle, the configuration $P'' = P' \cup \{\mathbb{T}\}$ qualifies as a univalent circle packing for K on the sphere. Here $\mathbb{T}$ is treated as bounding the *complement* of $\mathbb{D}$ so that it can serve as the circle for w; the circles for $v_1, \ldots, v_n$ are all externally tangent to it. (You might note that in fact *all* boundary circles of $\mathcal{P}_{K'}$ are tangent to $\mathbb{T}$ in P; those not neighboring w are *extraneous* tangencies as far as K is concerned.)

To wrap up this case, we need to move P'' into the hyperbolic plane; this can be done with an appropriate Möbius transformation. Choose a disc D which is disjoint from $\text{carr}(P'')$. There exists a Möbius transformation ψ which maps D onto the southern hemisphere of $\mathbb{P}$, implying that $P = \psi(P'')$ is a circle packing for K in the northern hemisphere. In other words P lies in the unit disc and we may apply the claim to complete the induction step for Case 2.

In both cases we have proven the existence of the packing $\mathcal{P}_K$, completing the induction; this concludes the proof of Proposition 6.1.   □

The geometry can be a bit confusing in this last result, so Figure 6.5 provides an example. K is a combinatorial closed disc of 19 vertices. A boundary vertex w and the three darkly shaded faces are removed, leaving a reduced complex K' of 18 vertices. The induction hypothesis gives P' of (b); the former neighbors of w are the four shaded horocycles. P' is stereographically projected to the sphere in (c), and the equator is included as the circle c_w in (d). Note that in P'', c_w is tangent to its four original neighbors, as indicated by the four edges. An appropriate ψ maps D (displayed in (b) but out of sight in (c) and (d)) onto the southern hemisphere and carries P'' to the packing $P = \psi(P'')$ in $\mathbb{D}$. Edges in (e) again indicate the tangencies of c_w with its original four neighbors, so this is a packing for K. This is, in fact, the starter packing which the claim converts into the maximal packing $\mathcal{P}_K$ in (f).

The following lemma is a direct corollary of the proof and justifies the name "maximal" packing.

Lemma 6.6. *If K is a combinatorial closed disc and R is a hyperbolic packing label for K, then $R \leq \mathcal{R}_K$.*

We will shortly use a form of "combinatoric" monotonicity that is an immediate consequence of this. I leave this as a homework problem, but it is not optional – we will need this shortly.

Homework: *If L and K are combinatorial closed discs with $L \subseteq K$ as simplices, then for any vertex $w \in L$, $\mathcal{R}_K(w) \leq \mathcal{R}_L(w)$.*

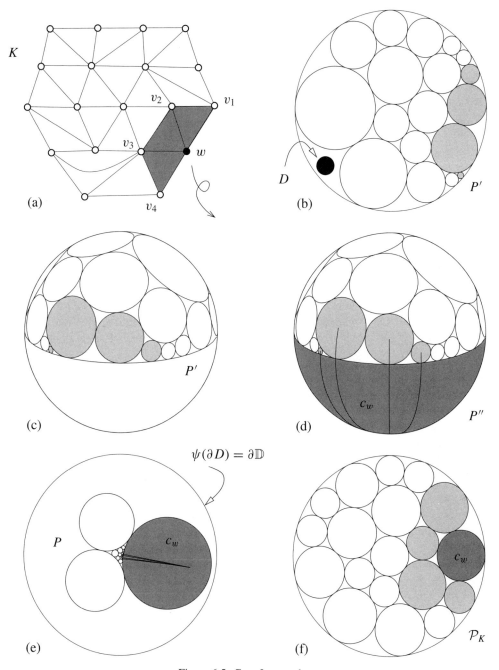

Figure 6.5. Case 2 example.

7

Proof for Combinatorial Spheres

Proposition 7.1 (Koebe–Andreev–Thurston Theorem). *Let K be a combinatorial sphere. Then there exists an essentially unique univalent circle packing $\mathcal{P}_K$ for K on the Riemann sphere $\mathbb{P}$.*

Proof. Designate a vertex v_∞ of K and let $F = \{v_\infty; v_1, \ldots, v_n\}$ denote its combinatorial flower. Remove the corresponding faces from K to obtain a reduced complex K'. One easily verifies that K' is a combinatorial closed disc. By Proposition 6.1, there exists a univalent packing P' for K' whose boundary circles are horocycles. Projecting P' to the sphere, we may add the unit circle $\mathbb{T}$, bounding the southern hemisphere, to form the univalent packing $\mathcal{P}_K = P' \cup \mathbb{T}$. The petals $v_1, \ldots, v_n$ of F are boundary vertices of K', so their circles are tangent to $\mathbb{T}$. The edges and faces induced by these tangencies along with the carrier of P' reconstitute the combinatorics of the original K. In other words, $\mathcal{P}_K$ is a univalent circle packing for K.

Uniqueness is equally easy. Given any other circle packing Q for K, we may assume, by applying a Möbius transformation, that the circle for v_∞ is the southern hemisphere. This implies that the portion of Q associated with the reduced complex K' forms a univalent circle packing in the northern hemisphere with its boundary circles tangent to the equator. Projecting to the hyperbolic plane and appealing to the essential uniqueness in Proposition 6.1 implies that Q' and P' differ by an automorphism of $\mathbb{D}$. But every automorphism of $\mathbb{D}$ is the restriction of a Möbius transformation of $\mathbb{P}$ which maps the southern hemisphere to itself. That is, Q and $\mathcal{P}_K$ differ by an automorphism of $\mathbb{P}$, and we have proved the essential uniqueness of $\mathcal{P}_K$. $\square$

Short chapter! Historically, the sphere was the original setting for circle packing – I alluded in Notes I to the three independent proofs by Koebe, Andreev, and Thurston. These are all quite sophisticated, and reading any one gives an appreciation for the clean simplicity of our back-door approach. Our argument is also a blueprint for practical numerical computation of spherical packings – indeed, bootstrapping the hyperbolic case is currently the *only* method known and is used for all spherical packings in the book. (Thurston introduced the argument we used above, but in the *reverse* direction – he used the spherical case to *prove* the existence of packings in $\mathbb{D}$.)

8

Proof for Combinatorial Open Discs

Proposition 8.1. *Let K be a combinatorial open disc (hence, infinite, simply connected, without boundary). Then there exists an essentially unique univalent circle packing $\mathcal{P}_K$ for K whose carrier fills either the hyperbolic plane $\mathbb{D}$ or the euclidean plane $\mathbb{C}$.*

The proof will be carried out in three sections: existence/univalence, completeness (the "fills" condition), then uniqueness. In the first stage you will quickly see why the split between the geometries occurs; after that, the methods often depend on the setting, hyperbolic *versus* euclidean. Each section begins by introducing key geometric notions needed for that stage – but also important for your general toolkit. The proofs are "elementary" in the mathematical sense, but challenging. All I can say to encourage the reader is that the struggle pays off handsomely in the currency of geometric intuition.

8.1. Existence and Univalence

8.1.1. Geometric Tool

One of the key insights which got the topic of circle packing rolling is the Ring Lemma of Rodin and Sullivan.

Lemma 8.2 (Ring Lemma). *For each integer $k \geq 3$ there exists a constant $\mathfrak{c}(k) > 0$ such that if F is any univalent k-flower of circles in the euclidean or hyperbolic plane having a central circle of radius r_0, then the radius r of each petal satisfies $r \geq \mathfrak{c}(k) \cdot r_0$.*

The proof is surprisingly elementary, but I have put it off until Appendix B because, as with so many geometric observations, there is a lot more there if you squeeze a little. In the case of the Ring Lemma, there are several surprises. First, these sharp values have been established for the constants:

$$\mathfrak{c}(k) = \frac{1}{\left[\left(\dfrac{5-2\sqrt{5}}{5}\right)\left(\dfrac{3+\sqrt{5}}{2}\right)^k + \left(\dfrac{5+2\sqrt{5}}{5}\right)\left(\dfrac{3-\sqrt{5}}{2}\right)^k - 1\right]}, \quad k \geq 3.$$

(14)

Moreover, these constants are all reciprocal integers,

$$\mathfrak{c}(3) = 1, \quad \mathfrak{c}(4) = 1/4, \quad \mathfrak{c}(5) = 1/12, \quad \mathfrak{c}(6) = 1/33, \ldots,$$

and are related to such classics as the golden ratio, Soddy's "bowl of integers," and "Farey" arithmetic. Check out the proof on p. 318, and then visit the rest of the Appendix as your time and inclination allow.

8.1.2. The Fundamental Dichotomy

Choose a nested sequence of finite simply connected complexes $\{K_j : j = 1, 2, \ldots\}$ containing the α-vertex v_1 and exhausting K. More precisely, we require the following properties:

- $v_1 \in K_1$;
- K_j is a combinatorial closed disc for each j;
- $K_j \subset K_{j+1}$, as simplicial complexes, for each j; and
- $K_j \uparrow K$, meaning that if L is any finite subcomplex of K, then $L \subseteq K_j$ for all sufficiently large j.

Any sequence $\{K_j\}$ will serve, but perhaps the simplest defines K_j as the smallest simply connected subcomplex of K containing all vertices of generation j (i.e., combinatorial distance j) from v_1. In this case, K_1 is simply the flower of v_1, and by induction K_{j+1} is obtained from K_j by adding all vertices neighboring vertices of K_j and then, if necessary, filling in any islands this created to get a simply connected result.

Regardless of how the K_j are obtained, for each we may apply Proposition 6.1 to get $R_j = \mathcal{R}_{K_j}$, the maximal hyperbolic label, and $\mathcal{P}_j = \mathcal{P}_{K_j}$, the maximal circle packing in $\mathbb{D}$, normalized in every case so that the α-circle is centered at the origin.

Consider the sequence $\{R_j\}$ of labels. Since $K_j \subset K_{j+1}$, the Homework problem you did at the close of Chap. 6 implies that $R_{j+1}(v) \leq R_j(v)$ for every v; that is, the sequence of labels is monotone decreasing at every vertex. Focusing on v_1, one of two mutually exclusive possibilities must hold:

Type Dichotomy:

(I) $R_j(v_1) \downarrow r_1 > 0$ as j goes to infinity.

(II) $R_j(v_1) \downarrow 0$ as j goes to infinity.

We will find that the outcome is independent of v_1 and the sequence $\{K_j\}$, so it reflects an intrinsic property of K. In Case (I) we prove that there exists a univalent circle packing for K in $\mathbb{D}$ and the complex K will be termed *hyperbolic*. In Case (II), on the other hand, we prove that there exists a univalent circle packing for K in $\mathbb{C}$ and no circle packings for K in $\mathbb{D}$ and K will be termed *parabolic*.

In both cases, existence and univalence follow from diagonalization and "geometric" convergence arguments. The intuition behind the dichotomy is nicely captured in experiments with the heptagonal and hexagonal complexes, $K^{[7]}$ and $K^{[6]}$, whose packings were shown in Figure 1.9. In each case, take $K_j^{[\cdot]}$ to represent the jth generational subcomplex. The 7-degree situation is illustrated in Figure 8.1, with packings $\mathcal{P}_j$, $j = 3, 5, 7, 9$, which are superimposed in the bottom image. The parallel images for the 6-degree case are illustrated in Figure 8.2.

One glance and you would rightly guess that $K^{[7]}$ is hyperbolic while $K^{[6]}$ is parabolic. In Figure 8.1 you can almost see the infinite packing for $K^{[7]}$ emerge as the circles all begin to stabilize in location and size. This does not happen in Fig. 8.2, where all circles are degenerating to the point at the origin. To salvage the hexagonal case, one can convert to euclidean packings Q_j and scale so the radii for the circles at the origin are 1, as in Figure 8.3. When these are superimposed in the bottom picture, one again sees a limit packing emerge. Our task, of course, is to convert these geometric hints into proofs.

The first thing to verify is that the dichotomy between (I) and (II) is independent of v_1 and of the K_j. Suppose $\{L_j\}$ is another sequence of subcomplexes exhausting K, $\{S_j\}$ the associated maximal labels. For each i, $L_i \subset K_k$ for sufficiently large k, and in turn $K_k \subset L_n$ for sufficiently large n. By maximality of the various labels, $S_n(v) \leq R_k(v) \leq S_i(v)$. Thus the sequences $\{R_j(v_1)\}$ and $\{S_j(v_1)\}$ are intertwined, and since both are monotone, they converge to the same limit. For independence from v_1 we need to show that for any vertex v,

$$R_j(v) \downarrow 0 \iff R_j(v_1) \downarrow 0. \tag{15}$$

The key is the Ring Lemma. Fix any vertex $v \in K$ and choose an edge-path γ in K from v_1 to v. If it has, say, N edges, we may describe it as a string of successive neighboring vertices from v_1 to v,

$$\gamma = \langle v_1, v_2, v_3, \ldots, v_N \rangle, \quad \text{with} \quad v_N = v. \tag{16}$$

There is an integer d so that each $v_n \in \gamma$ has at most d petals in its flower and there is an integer J so large that γ is interior to K_j for $j \geq J$. If v_n and v_{n+1} are successive vertices of γ, then in $\mathcal{P}_j$ their flowers are univalent, so two uses of the Ring Lemma give

$$\mathfrak{c}(d)R_j(v_{n+1}) \leq R_j(v_n) \leq \frac{1}{\mathfrak{c}(d)}R_j(v_{n+1}), \quad j \geq J.$$

Applying this reasoning successively from one end of γ to the other gives the inequalities

$$(\mathfrak{c}(d))^N R_j(v_1) \leq R_j(v) \leq \frac{1}{(\mathfrak{c}(d))^N} R_j(v_1), \quad j \geq J. \tag{17}$$

The double implication of (15) follows immediately.

We have now shown that the dichotomy depends only on K. In cases (I) and (II), K is said to have the *type* of the hyperbolic or euclidean plane, respectively. These cases must be treated separately.

Before we go on, let us record an observation here which follows from our construction of $\mathcal{P}_K$, Lemma 6.6, and the Type Dichotomy: *If K is a parabolic combinatorial disc, then there can be no circle packing for K in the hyperbolic plane.*

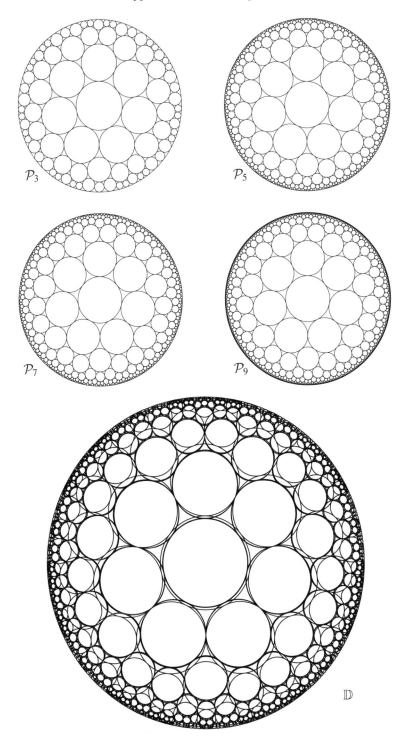

Figure 8.1. Geometric convergence for $K^{[7]}$.

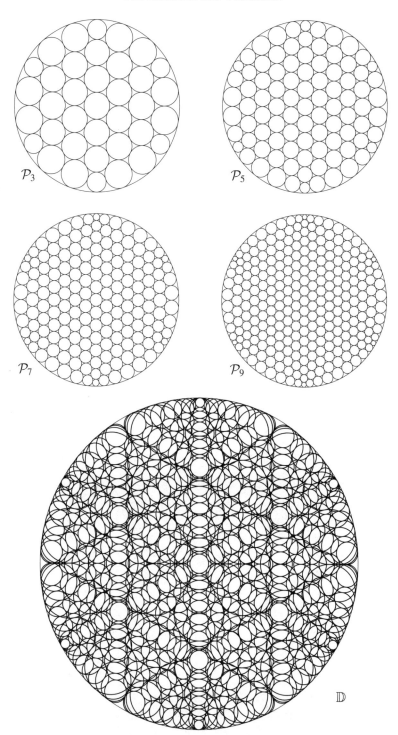

Figure 8.2. Degeneration for $K^{[6]}$.

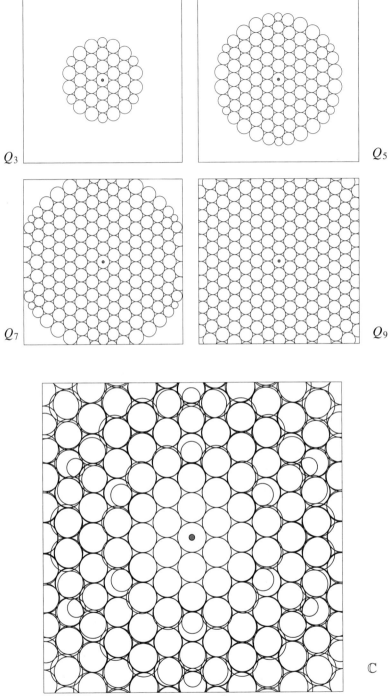

Figure 8.3. Geometric convergence for $K^{[6]}$.

8.1.3. Hyperbolic Existence

Assume K is hyperbolic and an exhaustion $\{K_j\}$ has been chosen. Let $\{Z_j\}$ denote the collection of hyperbolic centers for circles of $\mathcal{P}_j$, so $Z_j(v)$ is the center of the circle for vertex v (for sufficiently large j). Using diagonalization we will identify a subsequence $\{\mathcal{P}_{j_k}\}$ so that both the sequences $\{R_{j_k}\}$ and $\{Z_{j_k}\}$ converge for every vertex v. This involves an infinite family of subsequences inductively defined. The notation can become a mess, so we denote them simply as

$$\left\{\{P_j^{(k)}\} : k = 0, 1, 2, \ldots\right\}, \quad \{P_j^{(k+1)}\} \subset \{P_j^{(k)}\}, \tag{18}$$

where the "$\subset$" indicates "subsequence."

Enumerate the vertices, $\{v_1, v_2, v_3, \ldots\}$. Define the initial subsequence $\{P_j^{(1)}\}$ to be the full sequence $\{\mathcal{P}_j\}$. Note that since v_1 is always centered at the origin, the sequence $\{Z_j^{(1)}(v_1)\}$ converges to 0. Assume by induction that successive subsequences $\{\{P_j^{(k)}\} : k = 1, 2, \ldots, n-1\}$ have been extracted with the property that for each k, $\{Z_j^{(k)}(v_k)\}$ is a convergent sequence of complex numbers.

Give your attention to the last subsequence $\{P_j^{(n-1)}\}$ and to the next vertex v_n. The centers $\{Z_j^{(n-1)}(v_n)\}$ lie in the closed unit disc, so by the Bolzano-Weierstrass Theorem one can extract the next subsequence $\{P_j^{(n)}\} \subset \{P_j^{(n-1)}\}$ so that $\{Z_j^{(n)}(v_n)\}$ converges. This completes the inductive step, giving the infinite nested collection of (18). Note that for each n we have

$$Z_j^{(n)}(v_n) \longrightarrow z_n \quad \text{and} \quad R_j^{(n)}(v_n) \downarrow r_n \quad \text{as } j \to \infty, \tag{19}$$

for some $z_n \in \overline{\mathbb{D}}$ and some $r_n > 0$. The former convergence is by construction and the latter is due to the monotonicity of radii. A detail to note is that $|z_n| < 1$, for if γ is an edge-path from v_1 to $v_N = v_n$ in K (using the notation of (16)), then following (17), the hyperbolic distance from the origin to $Z_j(v_n)$ is no bigger than $2Nr_1/(\mathfrak{c}(d))^N$.

Of course, with infinitely many subsequence extractions, you might well throw out *every* packing – there may be no sequence left. That is the beauty of the Weierstrass diagonalization argument: define the target subsequence $\{\mathcal{P}_{j_k}\}$ by choosing $\mathcal{P}_{j_k}$ to be the kth packing of the kth subsequence $\{P_j^{(k)}\}$. To see that this works, choose any vertex $v \in K$. Then $v = v_n$ for some n, and since the tail of the subsequence $\{\mathcal{P}_{j_k}\}$ is a subsequence of $\{P_j^{(n)}\}$, (19) implies

$$Z_{j_k}(v_n) \longrightarrow z_n \quad \text{and} \quad R_{j_k}(v_n) \downarrow r_n \quad \text{as } k \to \infty, \tag{20}$$

and we may define c_{v_n} as the circle of hyperbolic radius r_n and hyperbolic center z_n. Once we have done this for all n, we define $\mathcal{P}_K$ as the configuration $\mathcal{P}_K = \{c_{v_n} : n = 1, 2, 3, \ldots\}$. We will write $\mathcal{P}_{j_k} \longrightarrow \mathcal{P}_K$ and call this "geometric convergence."

It is elementary to show that $\mathcal{P}_K$ is a univalent circle packing using geometric convergence. Consider distinct vertices v and w. Their circles will have disjoint interiors in $\mathcal{P}_K$ because the circles converging to them have disjoint interiors in each $\mathcal{P}_{j_k}$. Likewise, if

$w \sim v$ in K then c_v will be tangent to c_w because the pairs converging to them are tangent in each $\mathcal{P}_{j_k}$. (It is worth a few moments to verify these statements.)

You might consider the alternate approach of defining $\mathcal{R}_K$ to be the hyperbolic label with $\mathcal{R}_K(v_n) = r_n$ for all vertices v_n (in the notation of (20)). For every vertex v, $\theta_{\mathcal{R}_k}(v) = 2\pi$ because $\theta_{R_j}(v) = 2\pi$ for sufficiently large j and angle sums are continuous in their labels. With this packing label and the Monodromy Theorem, one can lay out the packing $\mathcal{P}_K$ itself. Unfortunately, I do not see a convenient way to confirm univalence in this approach; one must bring in the geometry at some point.

In any case, we have $\mathcal{P}_K \longleftrightarrow K(\mathcal{R}_K)$, and existence and univalence are established in the hyperbolic case. Let us note for later use that $\mathcal{P}_K$ deserves its "maximal" name.

Lemma 8.3. *Suppose K is a hyperbolic combinatorial disc and R is a hyperbolic packing label for K. Then $R \le \mathcal{R}_K$, meaning that for each vertex v, $R(v) \le \mathcal{R}_K(v)$. Equality at even a single interior vertex v implies equality at all, and hence $R \equiv \mathcal{R}_K$.*

Proof. If K is finite, the inequality is Lemma 6.6. Otherwise, choose an exhausting sequence of combinatorial closed discs as we did in the proof above, $\{K_j\} \uparrow K$. Suppose $R(v) > \mathcal{R}_K(v)$ for some vertex v. Since $R_j(v) \downarrow \mathcal{R}_K(v)$, there would exist n so that $R(v) > R_n(v)$. This would contradict the maximality of $R_n = \mathcal{R}_{K_n}$ for the finite complex K_n, so we conclude that $R(v) \le \mathcal{R}_K(v)$ for all v.

Next, let V denote the set of vertices of K for which equality $R(v) = \mathcal{R}_K(v)$ holds. This is an "open" set of vertices in the following sense: if $v \in V$, then in light of the inequality and the strict monotonicity of angle sums from Lemma 6.3, every petal in the flower of v must also be in V. The same argument shows that the set of vertices *not* in V is also open; that is, V is "closed." Since the vertices of K are edge-connected, the empty set and the full set of vertices are the only sets that can be both open and closed. In other words, equality $R(v) = \mathcal{R}_K(v)$ holds at *no* vertices or at *all* vertices. This completes the proof. $\qquad\square$

8.1.4. Parabolic Existence

Assume now that K is parabolic. We certainly have to switch tactics, because the radii in the hyperbolic labels R_j all decrease monotonically to zero. Figure 8.3 suggests the strategy: treat each $\mathcal{P}_j$ as a euclidean packing and apply a euclidean dilation mapping the α-circle at the origin to the unit circle. Write E_j for the new euclidean label and $Q_j \longleftrightarrow K_j(E_j)$ for the euclidean packing.

We now replicate the hyperbolic arguments. Using the euclidean Ring Lemma now, we find as before that for every vertex v there exist constants $0 < a < b < \infty$ depending on v so that $E_j(v) \in [a, b]$ for all large j. Also as before, the centers of the circles for v in Q_j lie within a fixed distance of the origin (depending on v). The same applications of Bolzano–Weierstrass and diagonalization lead to a subsequence of packings $\{Q_{j_k}\}$ which converges geometrically to a univalent circle packing $\mathcal{P}_K$ for K, now in $\mathbb{C}$. The euclidean labels E_{j_k} converge to the euclidean label $\mathcal{R}_K$, and $\mathcal{P}_K \longleftrightarrow K(\mathcal{R}_K)$.

8.2. Completeness

"Completeness" of a maximal packing $\mathcal{P}_K$ refers to metric completeness – the property that carr($\mathcal{P}_K$) fills the underlying space. For K a combinatorial open disc, this will mean $\mathbb{D}$ or $\mathbb{C}$ depending on whether K is hyperbolic or parabolic.

8.2.1. Geometric Tools

Our first need is for a new monotonicity in hyperbolic labeled complexes. Recall that each label R, whether a packing label or not, determines a metric $d_R(\cdot, \cdot)$ on $K(R)$ as described in Section 5.1.

Lemma 8.4. *Let $K(R')$ and $K(R)$ be labeled complexes with hyperbolic labels satisfying $R' \leq R$. Then*

$$d_{R'}(v_1, v_2) \leq d_R(v_1, v_2) \tag{21}$$

for any two vertices v_1, v_2 of K.

Proof. Throughout the following, v_1 and v_2 are fixed vertices of K and Γ denotes the collection of paths between them. Quantities associated with $K(R')$ will be marked with a prime ($'$) to distinguish them from those associated with $K(R)$. For example, $l(\cdot)$ will denote length in $K(R)$ while $l'(\cdot)$ denotes length in $K(R')$.

We may make several assumptions which simplify our work yet do not weaken the outcome. First, we will assume that the hyperbolic labels in R (and hence also in R') are finite, since standard limiting arguments will then give the general result. We will assume that labels R and R' differ only for a single vertex v and write $r = R(v)$ and $r' = R'(v)$, with $0 < r' < r < \infty$. We will be working with a fixed path $\gamma \in \Gamma$, and without affecting the infimum of lengths over Γ, we may assume that γ is simple and that it passes through a finite chain $\mathcal{F} = \{f_0, f_1, \ldots, f_m\}$ of (not necessarily distinct) faces of K. It is not appropriate to compare $l(\gamma)$ and $l'(\gamma)$, since we have not fully specified point maps between $K(R)$ and $K(R')$ – we only know that vertices are identified. Instead, we construct a path $\gamma' \in \Gamma$ satisfying

$$l'(\gamma') \leq l(\gamma).$$

Our approach involves successively laying down triangles for the faces of $\mathcal{F}$ and then measuring lengths of paths via their immersed images in $\mathbb{D}$. To this end, suppose $T_0, T_1, \ldots, T_m$ and $T'_0, T'_1, \ldots, T'_m$ are triangles corresponding to the faces in $\mathcal{F}$ from our two structures on K. Carry out developments along the chain using each of these two sets of triangles. We may loosely refer to these developments of faces as F and F', and with a slight abuse of notation may use the same symbols for quantities in K and in the developments. The path γ is immersed in $\mathbb{D}$ in the development F. We will build our γ' within the development F'.

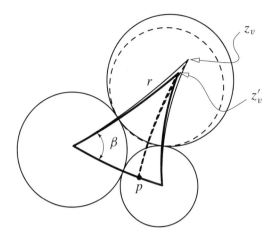

Figure 8.4. Monotonicity of a geodesic segment.

Let $S \subseteq K$ be the union of faces containing v. We start by breaking γ into segments γ_j, $j = 1, 2, \ldots, N$, classified as "good" or "bad." Corresponding segments γ_j' will be built in F' and concatenated to form γ'. The (maximal connected) segments of γ which lie in the complement of the interior of S are termed *good*. Triangles associated with faces not in S are congruent under the structures of $K(R)$ and $K(R')$, so a good segment γ_j may be transplanted via a hyperbolic isometry to a segment γ_j' in F' having the same length. The bad segments γ_j require all the work; for each we must find a corresponding γ_j' so that $l'(\gamma_j') \leq l(\gamma_j)$.

Let us first handle a special case; namely, paths starting (or ending) at v itself. Suppose σ is a path in S from v to $p \in \partial S$. In the development F, the shortest path from z_v to p is clearly the geodesic $[z_v, p]$, which lies in a single face. In other words, $d_{hyp}(z_v, p) \leq l(\sigma)$. The situation is pictured in Figure 8.4. Decreasing the radius at v from r to r' causes the angle β and the length of the edge between β and z_v to decrease; by the hyperbolic Law of Cosines, $d_{hyp}(z_v', p) < d_{hyp}(z_v, p)$. If we define σ' to be the path from v to p in S which is associated with $[z_v', p]$ in the development F', then $l'(\sigma') = d_{hyp}(z_v', p) < d_{hyp}(z_v, p) \leq l(\sigma)$; that is, $l'(\sigma') < l(\sigma)$.

Now fix attention on a bad segment γ_j. If γ_j contains v, then it consists of one (if $v = v_1$ or $v = v_2$) or two segments σ of the type handled above, so we can define γ_j' with the same endpoints as γ_j and satisfying $l'(\gamma_j') < l(\gamma_j)$.

Assume, therefore, that γ_j does not contain v. It runs between points $x, y \in \partial S$, each of which is an endpoint of a contiguous good segment (or is equal to v_1 or v_2). Let $c = \{f_{j_1}, f_{j_2}, \ldots, f_{j_n}\}$ be the subchain of $\mathcal{F}$ through which γ_j passes, with corresponding triangles $t_1, t_2, \ldots, t_n$ and $t_1', t_2', \ldots, t_n'$. In the development F, the t_i will lay out in $\mathbb{D}$ like a fan about a point z_v associated with v; likewise, in F' the t_i' will lay out like a fan about z_v'. (Caution: there is not a well-defined immersion of S if the packing condition does not hold at v, so one always measures lengths face by face.)

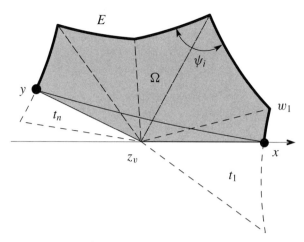

Figure 8.5. The fan about v.

Figure 8.5 depicts the situation in F. By the maximality of γ_j, x lies on the edge of t_1 opposite to v while y lies on the edge of t_n opposite to v. Without loss of generality, we have placed z_v at the origin and x along the positive real axis and we have assumed that the faces proceed in the positive direction about z_v. Since γ_j does not contain v, it does not pass through the origin. As a consequence, we can make a continuous choice of argument $\arg(z)$ for complex numbers z moving from x to y along γ_j. The change in argument $\Delta\theta = \arg(y) - \arg(x)$ is well-defined and positive. If $\Delta\theta \geq \pi$, then the shortest path from x to y (through $t_1, \ldots, t_n$ in succession) would pass through z_v at the origin. We could replace γ_j by a shorter path through v and then we could find a yet shorter path γ'_j in F' by applying our earlier special case.

We are left, then, with the case $0 < \Delta\theta < \pi$, as in Figure 8.5. Let $E = [x, w_1, w_2, \ldots, w_{n-1}, y]$ denote the polygonal path from x to y along the outer edges of the triangles, and let Ω denote the polygonal region consisting of the rays from the origin to points of E. It is clear that the shortest path from x to y within the development F must lie in Ω.

The parallel situation can be arranged within the development F'. Since E is shared with triangles not in S, the corresponding polygonal path E' will have identical edge lengths and will have points x', y' already in place. Our task is to find a path γ'_j in Ω' connecting x' to y' and having length no greater than $l(\gamma_j)$.

It suffices to work with *small* changes in r. From our earlier monotonicity results, here is how this situation will evolve as r decreases:

- the angles the triangles form at v will increase ($\Delta\theta$ grows);
- the distances $d_{hyp}(0, x)$ and $d_{hyp}(0, y)$ will decrease;
- the interior angles ψ_i formed by the edges of E (see Figure 8.5) will decrease.

Suppose first that the hyperbolic geodesic $[x, y]$ lies, except for its endpoints, in the interior of Ω (as in Figure 8.5). We may assume without loss of generality that $\gamma_j = [x, y]$.

By hyperbolic convexity of the triangles, we see that the corners $w_1, w_2, \ldots, w_{n-1}$ of E lie on one side of $[x, y]$. As we decrease r, the edge lengths of E remain unchanged but the interior angles ψ_i decrease. It is not difficult to prove, using for example the Law of Cosines and an induction on the number of edges in E, that the hyperbolic distance between the endpoints of E decreases with r. (This depends on the fact that E lies on *one side* of the geodesic through its endpoints and that the ψ_i are positive by our assumption of finite radii.) Assuming a small decrease in r, the geodesic $[x', y']$ will lie in Ω' and we take it as our choice for γ_j'.

Alternately, if $[x, y]$ does not lie in the interior of Ω we may assume without loss of generality that γ_j consists of a finite number of geodesic subsegments separated by points or subarcs of E. In this case we can apply the previous reasoning to the geodesic subsegments, while any points or subarcs of E don't change in length between F and F'. The result is again a new path γ_j' in F' with $l'(\gamma_j') \leq l(\gamma_j)$.

The good and bad segments of $\gamma \in \Gamma$ have now yielded companion segments in F' of equal or shorter length that link together end to end to form $\gamma' \in \Gamma$. This is the γ' we were looking for, and taking the infimum over Γ gives (21). $\qquad\square$

Our next tool involves a practical way to measure the influence of various sets of boundary vertices of a complex K on its interior vertices. Let me discuss the classical situation that we are going to mimic.

We begin with the unit disc. Suppose a is an arc of $\partial\mathbb{D}$. For $z \in \mathbb{D}$ draw hyperbolic geodesics from z to the endpoints of a, as illustrated in Figure 8.6. The *harmonic measure* of a at z with respect to $\mathbb{D}$, denoted $\omega(z, a, \mathbb{D})$, is the number

$$\omega(z, a, \mathbb{D}) = \theta/(2\pi).$$

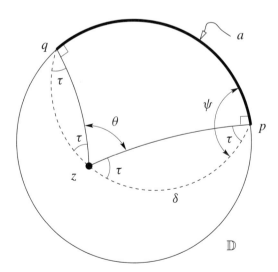

Figure 8.6. The classical situation.

Note that $\omega(z, a, \mathbb{D}) \in (0, 1)$ and because conformal automorphisms of $\mathbb{D}$ preserve angles and geodesics, it is conformally invariant; that is, if $\phi \in \text{Aut}(\mathbb{D})$, then

$$\omega(z, a, \mathbb{D}) = \omega(\phi(z), \phi(a), \mathbb{D}).$$

For later use let us note some additional geometry illustrated in Figure 8.6. Let δ be a circular arc inside $\mathbb{D}$ connecting the endpoints p and q of a and making some angle $\psi \in (0, \pi)$ with $\partial\mathbb{D}$. This breaks $\mathbb{D}$ into three sets: the points on δ, the points "inside" δ, meaning the same side as a, and the points "outside" δ. From Figure 8.6 one can observe that for any $z \in \delta$, $\theta + 2\psi = 2\pi$, implying

$$\omega(z, a, \mathbb{D}) = 1 - \psi/\pi, \quad \text{for } z \in \delta. \tag{22}$$

It is then easy to confirm that

$$\omega(z, a, \mathbb{D}) \begin{cases} > 1 - \psi/\pi, & \text{for } z \text{ inside } \delta \\ < 1 - \psi/\pi, & \text{for } z \text{ outside } \delta. \end{cases} \tag{23}$$

Next, suppose Ω is a Jordan domain in the plane. By the Riemann Mapping Theorem there is a conformal mapping (i.e., a univalent analytic function) $F : \mathbb{D} \longrightarrow \Omega$ of $\mathbb{D}$ onto Ω, and by a result of Carathéodory F extends to homeomorphism of $\partial\mathbb{D}$ onto $\partial\Omega$. Suppose that $z \in \Omega$ and a is an arc of $\partial\Omega$. Then one defines $\omega(z, a, \Omega)$, the harmonic measure of a at z with respect to Ω, by

$$\omega(z, a, \Omega) = \omega(F^{-1}(z), F^{-1}(a), \mathbb{D}). \tag{24}$$

There are many other additive and monotonicity properties of harmonic measure. I will mention two involving b as an arc of $\partial\Omega$:

$$a \subset b \implies \omega(z, a, \Omega) \leq \omega(z, b, \Omega). \tag{25}$$

$$b = \partial\Omega \setminus a \implies \omega(z, b, \Omega) = 1 - \omega(z, a, \Omega). \tag{26}$$

As to the interpretation of harmonic measure, in the disc $\omega(0, a, \mathbb{D})$ is the proportion of the horizon that a occupies. For a general region Ω, harmonic measure is, roughly speaking, a conformally invariant way to measure the proportion of the boundary of Ω which various sets a occupy as "viewed" from z.

We exploit parallel notions in the discrete setting with a direct analogue of the definition just given for Jordan regions. Suppose K is a combinatorial closed disc and α is a boundary edge-path of K. Let $\mathcal{P}_K$ be a maximal packing for K and let a be the arc of the unit circle determined by the centers of the horocycles associated with α (i.e., the arc determined by their tangency points on $\partial\mathbb{D}$).

Definition 8.5. For an interior vertex $v \in K$, define the **discrete harmonic measure** of α at v with respect to K by

$$\omega(v, \alpha, K) = \omega(Z_v, a, \mathbb{D}),$$

where Z_v is the hyperbolic center of the circle for v in $\mathcal{P}_K$.

Note that we use the same $\omega(\cdot, \cdot, \cdot)$ notation for both the classical and discrete cases – one distinguishes by observing whether the last argument is a set or a complex. This shared notation is justified in part by the fact that the classical properties we have described above carry over intact to the discrete setting: discrete harmonic measure is conformally invariant, so its value does not depend on *which* maximal packing $\mathcal{P}_K$ is used; the equality (22) and the inequalities (23) hold whenever z is the center of an interior circle of $\mathcal{P}_K$; and the discrete parallels of (25) and (26) hold for boundary edge paths. A better justification is provided in the forthcoming lemma.

We will use discrete harmonic measure in comparisons involving the intersections of circle packings with discs. Suppose K is a combinatorial closed disc, P is a circle packing for K in the sphere, D is a disc on the sphere, and c_v is an interior circle of P that lies inside D. We need a convenient way to isolate the portion of $P \cap D$ that contains v. Define $\mathcal{C}$ to be the collection of circles of P that can be reached from c_v by connected chains of circle intersecting D. Define K_v (suppressing dependence on P) to be the smallest simply connected subcomplex of K containing the vertices for circles of $\mathcal{C}$. (You may have to doodle a few pictures to see the potential pitfalls in alternate definitions.)

Definition 8.6. The circle packing P_v obtained from P by restricting to the subcomplex K_v is called the **component** of $P \cap D$ associated with v.

In the situations we will face, ∂K_v contains a nonempty positively oriented boundary edge-path $\alpha = \langle w_1, w_2, \ldots, w_n \rangle$ so that P_v and α satisfy the following properties:

(a) α is also an edge-path in ∂K,

(b) the circles for the first and last vertices of α intersect ∂D,

(c) the circles for the other vertices of α lie in $\overline{D}$, and

(d) all circles for vertices of the edge-path $\beta = \partial K_v \backslash \alpha$ intersect the complement of D.

Define the arc $a \subset \partial D$ to be the shortest positively oriented arc which begins at a point of the circle for w_1 and ends at a point of the circle for w_n. The situation is illustrated in Figure 8.7, where the shaded circles comprise P_v and the dashed arc is a. The result we are after relates the discrete harmonic measure of α to the classical harmonic measure of a. (The principal purpose for introducing P_v is to articulate the hypotheses vis-à-vis the relation between α and P. Since we have it, we might as well incorporate it into the inequalities as well.)

Lemma 8.7. *Let P be a univalent circle packing in the plane, $c_v \subset \mathbb{D}$ an interior circle of P with hyperbolic center z_v, and P_v the component of $P \cap \mathbb{D}$ for v. Assuming $\alpha \subset \partial K$*

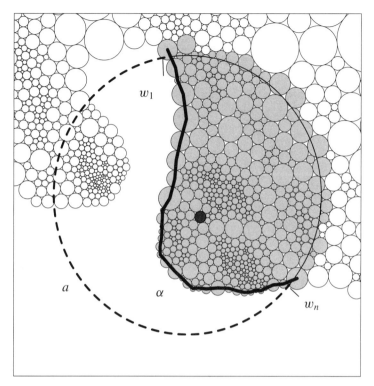

Figure 8.7. Defining the circle packing P_v.

and P_v satisfy conditions (a)–(d) above and that $a \subset \partial\mathbb{D}$ is the arc determined by α as described above, then we have the inequalities

$$\omega(v, \alpha, K) \geq \omega(v, \alpha, K_v) \geq \omega(z_v, a, \mathbb{D}). \tag{27}$$

I will flash my literary license here and delay this proof until the end of the chapter. Instead, I prefer a short intermission. With all the definitions and preliminaries, the reader may feel a growing distance from the core topic. Were we sitting at computers, a few experiments would bring us right back to the real world. Among other things, we might see how counterintuitive this latest lemma actually is. Let me try to make that point with a few static pictures.

Look at the nested packings of Figure 8.8 first. (*Warning*: This is a test: do *not* look ahead to the next figure.) Each packing has the circle c_v centered at the origin, a boundary edge-path α that stretches through the interior of $\mathbb{D}$, and the arc $a = [p, q]$ of the unit circle between the tangency points of the horocycles at the ends of α. The complementary boundary edge-path β varies from one packing to the next: in the first β also lies in $\mathbb{D}$, in the next it lies just outside $\mathbb{D}$, and in the last it extends much further out.

Imagine, now, computing the maximal packing for each of these complexes, with the circle for v placed again at the origin. In the maximal packing let δ denote the arc of the

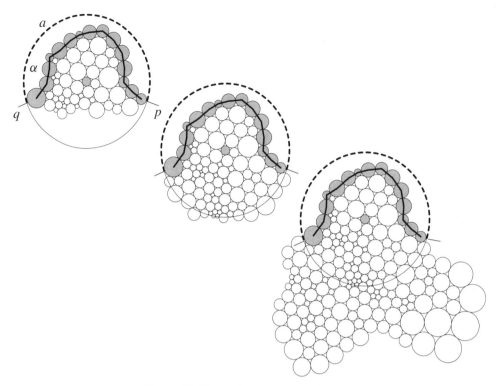

Figure 8.8. Harmonic measure experiments.

unit circle associated with the circles of α. *How does $|\delta|$ (arclength) compare to $|a|$ in each case? How does $|\delta|$ vary from one of these packings to the next?* Treat this as a mental experiment and stake your claims before you check the results presented in Figure 8.9. Note that $|\delta|/2\pi$ is the harmonic measure of α at v in all three cases. However, Lemma 8.7 only applies for the last two packings, since condition (d) fails for the first.

How did you do? If the results are not what you expected, you might think in particular about the normalization that puts c_v at the origin in the maximal packing. We allude to further uses for discrete harmonic measure in Part III of the book, but for now let us get on with the matter of completeness.

8.2.2. Hyperbolic Completeness

Suppose K is a combinatorial open disc and $\mathcal{P}_K \longleftrightarrow K(\mathcal{R}_K)$ is the maximal packing in $\mathbb{D}$ which we constructed earlier. Write $\Omega = \mathrm{carr}(\mathcal{P}_K) \subset \mathbb{D}$. Because K is simply connected and $\mathcal{P}_K$ is univalent, Ω is simply connected, and because K has no boundary, Ω is open. We need to show that $\Omega = \mathbb{D}$.

We begin by showing that Ω is *hyperbolically convex*; that is, if $a, b \in \Omega$, then the hyperbolic geodesic $[a, b]$ lies in Ω. There are two helpful reductions – we leave the standard justifications to the reader. First, since every point of Ω is in the convex hull of a finite number of centers of circles of $\mathcal{P}_K$, it is enough to assume that a and b are themselves centers, say $a = z_v$ and $b = z_u$. Second, given such z_u and z_v, it is enough to show that for any $\epsilon > 0$ there is a curve σ in Ω running from z_u to z_v such that

$$l_{hyp}(\sigma) \leq d_{hyp}(z_u, z_v) + \epsilon, \tag{28}$$

where d_{hyp} denotes hyperbolic distance and l_{hyp} hyperbolic length.

Recall that $\mathcal{P}_K$ was obtained as a geometric limit $\mathcal{P}_j \longrightarrow \mathcal{P}_K$, where the $\mathcal{P}_j \longleftrightarrow K_j(R_j)$ are maximal packings for an exhausting sequence $\{K_j\} \uparrow K$. The carriers of the $\mathcal{P}_j$ are hyperbolically convex – each is the intersection of hyperbolic half planes defined by its boundary geodesics. However, this does not in itself imply convexity of the carrier of their limit $\mathcal{P}_K$. Our proof depends on our ability to move from the carriers to the complexes themselves so that we can apply Lemma 8.4.

Let me remind the reader that a label R for complex K induces a metric d_R on K. If R is a packing label for a univalent packing P, then the metric space (K, d_R) is locally isometrically isomorphic to $\text{carr}(P)$. In particular, the d_R-length of a curve γ in K is identical to the length of its image curve in $\text{carr}(P)$.

Choose j sufficiently large that our designated vertices u, v lie in K_j. We use the metrics on K_j associated with two different labels; namely, $R_j = \mathcal{R}_{K_j}$ and R', defined as the restriction to K_j of the maximal label $\mathcal{R}_K$. With regard to the former, since the geodesic $[z_v^{(j)}, z_u^{(j)}]$ lies in $\text{carr}(\mathcal{P}_j)$, we deduce that

$$d_{\mathcal{R}_j}(u, v) = d_{hyp}\left(z_v^{(j)}, z_u^{(j)}\right).$$

With regard to the label R', note that the maximality of R_j (vis-à-vis K_j) implies that $R' \leq R_j$. By Lemma 8.4 and the previous equality,

$$d_{R'}(u, v) \leq d_{hyp}\left(z_v^{(j)}, z_u^{(j)}\right).$$

The packing $P' \longleftrightarrow K_j(R')$ is simply the restriction of $\mathcal{P}_K$ to K_j. Moving this last inequality to $\text{carr}(P')$ implies existence of a path σ from z_u to z_v which lies in $\text{carr}(P')$, hence in Ω, with the property

$$l_{hyp}(\sigma) \leq d_{hyp}\left(z_v^{(j)}, z_u^{(j)}\right).$$

The points $z_u^{(j)}, z_v^{(j)}$ converge to z_u, z_v, respectively, due to the geometric convergence $\mathcal{P}_j \longrightarrow \mathcal{P}_K$. Given $\epsilon > 0$, we can choose j sufficiently large that

$$d_{hyp}\left(z_v^{(j)}, z_v\right) < \epsilon/2, \quad d_{hyp}\left(z_u^{(j)}, z_u\right) < \epsilon/2.$$

With the triangle inequality, the above inequalities imply

$$\begin{aligned} l_{hyp}(\sigma) &\leq d_{hyp}\left(z_v^{(n)}, z_u^{(n)}\right) \\ &\leq d_{hyp}\left(z_v^{(n)}, z_v\right) + d_{hyp}(z_v, z_u) + d_{hyp}(z_u^{(n)}, z_u) \\ &< d_{hyp}(z_v, z_u) + \epsilon \end{aligned}$$

This establishes (28) and proves the convexity of Ω.

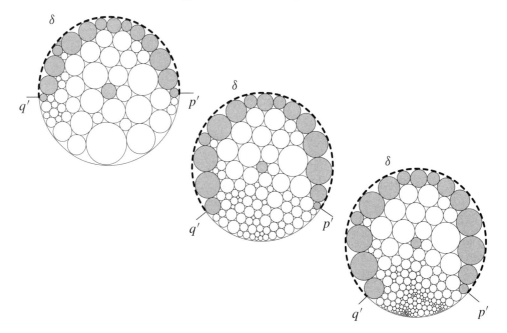

Figure 8.9. The results.

Convexity implies that if $\Omega \neq \mathbb{D}$, then $F = \mathbb{D}\backslash\Omega$ would necessarily have nonempty interior. We show that this would contradict the maximality of $\mathcal{R}_K$. In the following, $D(z, r)$ denotes the disc of euclidean radius r centered at z.

Say the interior of F contains a closed disc D. Applying an automorphism of $\mathbb{D}$ we may arrange that D be centered at the origin, and letting it grow, if necessary, we may further assume that $D \cap \overline{\Omega}$ contains a point p. In other words, $D = D(0, |p|)$. The Möbius transformation

$$\phi(z) = \frac{|p|}{z}$$

maps D onto the exterior of the unit disc; hence it maps $\mathcal{P}_K$ to a new packing $\mathcal{P}' = \phi(\mathcal{P}_K)$ for K in $\mathbb{D}$. As before, we use ($'$) to denote quantities associated with $\mathcal{P}'$.

Since $p \in \partial\Omega$, there are circles of $\mathcal{P}_K$ accumulating at p, and these circles present problems. If c_v lies in a sufficiently small neighborhood of p, the euclidean radii for c_v and for $c'_v = \phi(c_v)$ will be roughly comparable. On the other hand, since c'_v is near the unit circle, its hyperbolic radius will be much larger than that of c_v. That is, we can find a v with $R'(v) > \mathcal{R}_K(v)$, contradicting Lemma 8.3.

We have not done many computations in this book, so let us try our hand at filling in the details here. Fix $\epsilon > 0$ so that $|p| + \epsilon < 1$ and choose a vertex v whose circle $c_v \in \mathcal{P}_K$ lies in $D(0, |p| + \epsilon) \subset \mathbb{D}$. Bounds are available on euclidean and hyperbolic radii for c_v

and c_v'. Let r_v be the euclidean radius of c_v. The form of ϕ tells us that

$$r_v' > \frac{r_v}{|p|^2(|p|+\epsilon)^2}. \tag{29}$$

As for hyperbolic radii, one can bound them by using the euclidean radii in conjunction with the hyperbolic density (see Section 3.2.4). In particular, the hyperbolic radius $\mathcal{R}_K(v)$ for c_v satisfies

$$\mathcal{R}_K(v) < \frac{2r_v}{1-(|p|+\epsilon)^2}. \tag{30}$$

On the other hand, c_v' lies *outside* $D(0, |p|/(|p|+\epsilon))$, so its hyperbolic radius satisfies

$$R'(v) > \frac{2r_v'}{1-\left(\dfrac{|p|}{|p|+\epsilon}\right)^2}. \tag{31}$$

Putting (31) and (29) together gives

$$R'(v) > \frac{2r_v}{|p|^2(|p|+\epsilon)^2 - |p|^4}, \tag{32}$$

so (30) and (32) imply

$$\frac{R'(v)}{\mathcal{R}_K(v)} > \frac{1-(|p|+\epsilon)^2}{|p|^2(|p|+\epsilon)^2 - |p|^4}. \tag{33}$$

The ratio on the right goes to infinity as ϵ goes to zero. In other words, for sufficiently small ϵ we can find a vertex v so that $R'(v) > \mathcal{R}_K(v)$. This contradiction to maximality proves $\Omega = \mathbb{D}$, hence proves completeness for maximal packings of hyperbolic combinatorial discs.

8.2.3. *Parabolic Completeness*

Assuming now that K is a parabolic combinatorial disc, we prove that $\mathrm{carr}(\mathcal{P}_K) = \mathbb{C}$. The key ideas in this proof were distilled from a beautiful paper of He and Schramm on the Koebe Conjecture.

Write $\Omega = \mathrm{carr}(\mathcal{P}_K)$ and note that Ω is a simply connected open subset of $\mathbb{C}$ and that $\partial\Omega$ consists of accumulation points for the circles of $\mathcal{P}_K$. Write $F = \mathbb{P}\backslash\Omega$, so that F is a closed connected subset of the sphere. It suffices to show that F has spherical diameter zero; that is, that $F = \{\infty\}$.

Our approach relies on the following easy restatement of the condition for parabolicity. Here $z_v^{(n)}$ denotes the hyperbolic center of the circle for v in $\mathcal{P}_n$. The complex K is *parabolic if and only if, given any vertices $u, v \in K$, we have*

$$d_{hyp}\left(z_u^{(n)}, z_v^{(n)}\right) \longrightarrow 0, \quad \text{as } n \longrightarrow \infty. \tag{34}$$

We will assume F has positive spherical diameter and find vertices $u = v_1$ and $v = v_2$ that violate (34).

We first dispense with the case where F has nonempty interior. Suppose D is an open disc in F. A Möbius transformation ϕ mapping D to the exterior of the unit disc carries $\mathcal{P}_K$ to a packing Q for K lying in $\mathbb{D}$. But as we know from the Type Dichotomy, when K is parabolic there can exist no packing for K in $\mathbb{D}$.

We assume, then, that F has empty interior and positive spherical diameter. Applying an appropriate Möbius transformation we may assume that F lies in the closed unit disc and that 1 and -1 lie in F. For each $\theta \in \mathbb{R}$ there is a unique circle which goes through 1 and -1 and makes an angle θ with the positive x-axis at 1. Let δ_θ denote the open arc of that circle which leaves 1 at angle θ and ends at -1. For each $z \in \mathbb{P}$, $z \neq \pm 1$, there exists $\theta = \theta(z)$ so that $z \in \delta_\theta$. Because Ω is simply connected, there exists a continuous map $\widehat{\theta} : \Omega \longrightarrow \mathbb{R}$ so that $\widehat{\theta}(z) = \theta(z)(\text{modulo } 2\pi)$ for each $z \in \Omega$. The geometry is perhaps a little clearer if we normalize $\widehat{\theta}$ by adding an appropriate multiple of 2π so that $\widehat{\theta}(\infty) = 0$. Then, because F is connected, lies in the unit disc, and contains ± 1, it is easy to confirm that $\widehat{\theta}(\Omega) \subset (-3\pi/2, 3\pi/2)$. Define

$$\theta_1 = \min\{-5\pi/8, \inf\{\widehat{\theta}(z) : z \in \Omega\}\}$$
$$\theta_2 = \max\{5\pi/8, \sup\{\widehat{\theta}(z) : z \in \Omega\}\}.$$

Because F lies in $\mathbb{D}$,

$$\theta_1 \in [-3\pi/2, -5\pi/8], \quad \theta_2 \in [5\pi/8, 3\pi/2]. \tag{35}$$

Define open discs D_1 and D_2 in $\mathbb{P}$ as follows:

$$D_1 = \bigcup \{\delta_\theta : \theta_1 < \theta < \theta_1 + \pi\} \quad \text{and} \quad D_2 = \bigcup \{\delta_\theta : \theta_2 - \pi < \theta < \theta_2\}.$$

Using the fact that F has empty interior and that its complement Ω is connected, it is easy to see that

$$\theta_1 + 2\pi \leq \theta_2 \leq \theta_1 + 3\pi. \tag{36}$$

If the first inequality were false, F would contain the nonempty open set $\bigcup \{\delta_\theta : \theta_2 < \theta < \theta_1 + 2\pi\}$. The second inequality follows immediately from (35). From (36) we conclude that

$$D_1 \cap D_2 \subset \mathbb{D}. \tag{37}$$

Define $\eta_1 = \delta_{\theta_1 + \pi}$ and $\eta_2 = \delta_{\theta_2 - \pi}$; these are the "outside" arcs of D_1 and D_2, respectively, meaning the arcs in the complement of $\mathbb{D}$.

Finally, the definitions of θ_1, θ_2, and $\widehat{\theta}$, along with the fact that the circles packing Ω accumulate at F, imply that we may choose vertices v_1 and v_2 of K so that the associated circles c_1 and c_2 of $\mathcal{P}_K$ satisfy

$$c_1 \subset \bigcup \{\delta_\theta : \theta_1 < \theta < \theta_1 + \pi/4\}, \tag{38}$$
$$c_2 \subset \bigcup \{\delta_\theta : \theta_2 - \pi/4 < \theta < \theta_2\}. \tag{39}$$

In particular, $c_j \in D_j$, $j = 1, 2$. A typical situation is illustrated on the left in Figure 8.10; the shaded disc is $\mathbb{D}$, the wandering dark curve is F.

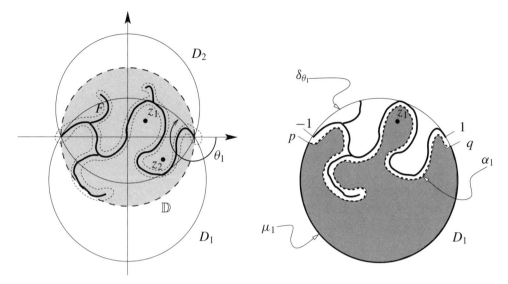

Figure 8.10. A typical situation.

We are not quite at the point where we can apply our harmonic measure ideas. Fix n so large that v_1, v_2 are interior vertices of K_n. Let Q_n denote the restriction of $\mathcal{P}_K$ to K_n. Let us work with D_1 first. The circle for v_1 lies in D_1, and by the definition of θ_1 and D_1, any curve in D_1 starting at c_1 and ending on δ_{θ_1} must leave the carrier of Q_n. That implies the existence of an edge-path α_1 in ∂Q_n which starts at some point $q \in \eta_1$, ends at some point $p \in \eta_1$, and separates c_1 from δ_{θ_1} (within D_1). Let μ_1 be the sub arc of η_1 from p to q. Unfortunately, Q_n may dip back inside D_1 and not cover μ_1. We solve this by augmenting Q_n with a finite number of additional circles so that it remains simply connected, contains μ_1 in its carrier, and so that α_1 remains in its boundary.

The situation for D_1 is further isolated on the right in Figure 8.10. The dot is z_1, the shaded region is $\text{carr}(Q_n) \cap D_1$, and the dashed line is the arc α_1 in the boundary of the carrier. We now have this string of inequalities:

$$\omega(v_1, \alpha_1, K_n) \geq \omega(z_1, a, D_1) \geq \omega(z_1, \delta_{\theta_1}, D_1) > 3/4. \tag{40}$$

The first follows from Lemma 8.7 (after identifying D_1 with the unit disc); the second reflects the fact that the arc $a \subset \partial D_1$ determined by α_1 contains the arc δ_{θ_1}; the third follows from (38) (see Figure 8.6 and (22)).

Identical arguments may be applied to D_2. (We assume that we had the foresight when finitely augmenting Q_n earlier for D_1 to take care of the analogous detail for D_2.) We obtain an edge-path α_2 in ∂K_n that yields the inequality

$$\omega(v_2, \alpha_2, K_n) \geq \omega(z_2, \delta_{\theta_2}, D_2) > 3/4. \tag{41}$$

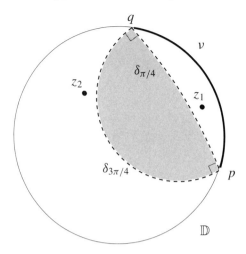

Figure 8.11. A lune angle of $\pi/2$.

Since D_1 and D_2 do not intersect outside the closed unit disc, the edge-paths α_1 from D_1 and α_2 from D_2 will be disjoint in ∂K_n. Inequality (41) therefore implies

$$\omega(v_2, \alpha_1, K_n) < 1/4. \tag{42}$$

Now, let us return to the maximal packing $\mathcal{P}_{K_n}$ for K_n in $\mathbb{D}$. Let v be the arc of $\partial \mathbb{D}$ associated with α_1 and let p, q denote its endpoints. Let $\delta_{3\pi/4}$ (resp. $\delta_{\pi/4}$) be the circular arc in $\mathbb{D}$ through p and q making angle $3\pi/4$ (resp. $\pi/4$) with v. Let z_1, z_2 be the hyperbolic centers for the circles of v_1, v_2 in $\mathcal{P}_{K_n}$. By (22) and (42), z_1 lies on the side of $\delta_{\pi/4}$ toward v, while by (22) and (40), z_2 lies on the side of $\delta_{3\pi/4}$ opposite to v. As shown in Figure 8.11, z_1 and z_2 are separated by a lune of angle $\pi/2$, shaded in the illustration.

Each of the curves $\delta_{\pi/4}$ and $\delta_{3\pi/4}$ is mapped to itself by all Möbius transformations of $\mathbb{D}$ that fix p and q. In particular, they are a fixed hyperbolic distance $h > 0$ apart. Therefore,

$$d_{hyp}(z_1, z_2) > h > 0.$$

This holds for the centers associated with v_1 and v_2 for all sufficiently large n, contradicting (34), and hence contradicting the assumption that K is parabolic. The carrier of $\mathcal{P}_K$ must fill $\mathbb{C}$, proving completeness for parabolic combinatorial discs.

Remarks. Doyle spirals are circle packings for the parabolic complex $K^{[6]}$ whose carriers omit precisely one point of the plane – in particular, they are not complete. An example is shown in Figure 1.2(e) of the Menagerie and more information is provided in Appendix C. These spirals are *locally* but not *globally* univalent, so it is *global* univalence that is essential in the above proof. Also, comparing Doyle spirals to the "penny packing," the maximal packing for $K^{[6]}$, one sees that in the parabolic setting the label $\mathcal{R}_K$ is not maximal in the same sense that it is by Lemma 8.3 in the hyperbolic case. The appellation "maximal" for our parabolic extremals is honorary – a vestige of their construction.

8.3. Uniqueness

Let me be clear on the issue here. The claim is that if P is a univalent circle packing whose carrier fills $\mathbb{G}$ ($\mathbb{D}$ or $\mathbb{C}$, as appropriate), then there exists an automorphism ϕ of $\mathbb{G}$ so that $\phi(P) = \mathcal{P}_K$. The first new tool below adds to our collection of monotonicity results and will let us settle hyperbolic uniqueness using now familiar techniques.

After that, however, there is a major shift as we prepare for parabolic uniqueness. I will dump a batch of new "winding number" and "index" tools into your toolbox. Let me be blunt: *These are among the most powerful, beautiful, and elegant tools you will see in circle packing!* There is a lot of material to be introduced, but once you get the hang of the key ideas, I think you will be impressed by their versatility.

8.3.1. Geometric Tools

Lemma 8.8. *Let P be a closed flower $\{c; c_1, \ldots, c_k\}$ of circles on the sphere P and assume c lies in $\mathbb{D}$. If any petal circles of P which intersect the complement of the closed disc, $\overline{\mathbb{D}}$, are replaced by horocycles, then the angle sum at c can only decrease. In particular, the new angle sum θ satisfies $\theta \leq 2\pi$, with strict inequality if one or more replacements are made.*

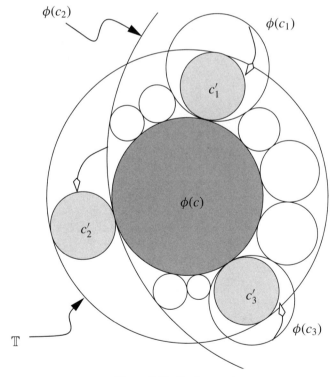

Figure 8.12. Replacements.

Proof. Define label $R = \{r; r_1, \ldots, r_k\}$ for the flower, where r is the hyperbolic label of c and r_j is either the hyperbolic label for petal c_j if c_j is a hyperbolic circle or $r_j = \infty$ if c_j intersects the unit circle $\mathbb{T}$. Apply a Möbius transformation $\phi \in \text{Aut}(\mathbb{D})$ to P so that $\phi(c)$ is centered at the origin. Of course ϕ is a Möbius transformation of the whole sphere, so $\phi(P)$ remains a flower vis-à-vis the sphere; moreover, any petals which intersected $\mathbb{T}$ will continue to do so; those which are hyperbolic circles will keep their hyperbolic radii (as does c). Now we simply compare $\phi(P)$ to the flower one would get from the label R. An illustration will suffice: Figure 8.12 shows $\mathbb{D}$ and the flower $\phi(P)$.

There are three circles that intersect $\overline{\mathbb{D}}$, $\phi(c_j)$, $j = 1, 2, 3$. Each has been replaced by a horocycle c'_j that is tangent to $\phi(c)$ at the same point as $\phi(c_j)$ itself. It is clear that each replacement action can only decrease the angle sum at $\phi(c_v)$, establishing the inequality. This is especially the case for $\phi(c_2)$, which, due to ϕ, encircles ∞. (*Note*: Since $\phi(c)$ is centered at the origin, its angle sum can be measured euclideanly, so euclidean arguments suffice.) $\qquad\square$

Now we shift to the introduction of geometric tools based on winding numbers, which were also brought to circle packing in Zheng-Xu He and Oded Schramm's work on the Koebe conjecture. I could present these as formal mathematical results – cold, clinical, efficient. But, to appreciate these wonderful ideas, the reader needs to dig in and really think them through, so my goal is to bring the reader along as a full participant.

Many readers will already have encountered winding numbers in their purest form. Suppose you are standing at the dot in Figure 8.13 and a long closed loop of rope is lying about you on the ground. *Does the rope wind (or wrap) around you or not?* In other words,

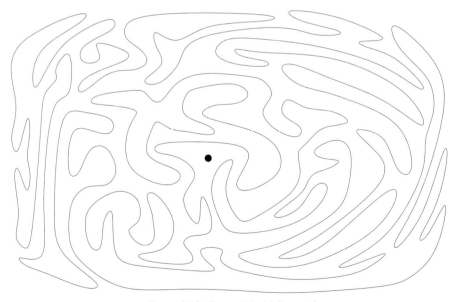

Figure 8.13. Trapped inside? or not?

could a friend hook the rope over her car bumper and drive away without making you move?

The most efficient way to check is to look toward a point on the horizon and count the number of times the rope crosses your line of sight. However, we take a much more dynamic approach. Hold one end of an elastic string, give the other end to your friend, and have her walk around the closed curve defined by the rope. If, when she has returned to her starting point, the elastic string wraps around you, then you are "inside" the rope.

This is the notion we formalize mathematically. Suppose $\gamma : [a, b] \to \mathbb{C}$ is a (continuous) curve in the plane and g is a continuous nonvanishing function $g : \gamma \to \mathbb{C}$. (It is common practice to use γ to refer both to the parameterization and to its trace, the image arc.) For $z \in \gamma$, the complex number $w = g(z)$ may be put in polar form $w = |w|e^{i\theta}$; the argument $\theta = \arg(w)$ is the angle the vector from the origin to w makes with the positive x-axis (measured, by convention, in radians and in the counterclockwise direction). There are infinitely many choices for θ, differing one from another by integral multiples of 2π. However, because w is never zero and varies continuously along γ, one may make a consistent choice of arguments so that the map $t \mapsto \arg(g(\gamma(t)))$ is continuous on $[a, b]$. The following quantity is independent of that choice.

Definition 8.9. If $\gamma : [a, b] \to \mathbb{C}$ is a continuous curve in the plane and $g : \gamma \to \mathbb{C}$ is a continuous nonvanishing function on γ, then the **winding number** of g on γ, denoted $\mathfrak{w}(g; \gamma)$, is the change in the argument of g while transiting γ; that is,

$$\mathfrak{w}(g; \gamma) = \frac{\arg(g(\gamma(b))) - \arg(g(\gamma(a)))}{2\pi}.$$

When γ is a closed curve, then $\mathfrak{w}(g; \gamma)$ will be an integer, since $g(\gamma(a)) = g(\gamma(b))$. This explains the terminology: if you treat $g(\gamma(t))$ as a *vector*, then $\mathfrak{w}(g; \gamma)$ gives the net *winding* of that vector around zero as you transit γ; it is normalized by dividing by 2π and is positive for counterclockwise winding. The details of the parameterization of γ are not important beyond giving γ its direction, so it is generally suppressed in the following. If γ is a simple closed curve, that is, a *Jordan* curve, then unless stated otherwise we assume a positive orientation, which means counterclockwise, so the interior region is on the left as one transits γ.

In our situation, the rope defines the curve γ and g is, for each point z on γ, the vector representing the elastic string from z to your position z_0. This is nonvanishing because z_0 is not on the rope. The winding number $\mathfrak{w}(g; \gamma)$ reflects your relationship to the rope: if nonzero, then you are "inside," otherwise you are "outside."

Let us up the ante a little. Suppose that you are not forced to stand still at z_0 while your friend determines your fate. Unbeknownst to her, there is a second closed curve σ that you get to travel around; as she walks once about γ, you travel once around σ and are back to z_0 when she completes her trip. Does the rope capture you now? The result will depend on γ and σ and on how your positions on them relate to one another.

To make this mathematically precise, we restrict ourselves to positively oriented Jordan curves γ and σ and to orientation-preserving homeomorphisms $f : \gamma \to \sigma$. Thus, for each point z on γ there is a corresponding point $f(z)$ on σ; as z moves once around γ in the positive direction, $f(z)$ moves once around σ in the positive direction. We assume that f is fixed point–free, so $f(z) \neq z$. The complex number $g(z) = f(z) - z$ then represents, for each $z \in \gamma$, the nonzero vector from z to $f(z)$, and so has an integer-valued winding number on γ.

Definition 8.10. Let γ and σ be positively oriented Jordan curves and let $f : \gamma \longrightarrow \sigma$ be an orientation-preserving, fixed point–free homeomorphism. Then the **fixed-point index** of f, denoted $\eta(f; \gamma)$, is the winding number $\mathfrak{w}(g; \gamma)$, where $g(z) = f(z) - z$.

Fixed-point indices are elementary – directly accessible without an army of big theorems – yet they somehow pick out crucial information about the relative positions and interactions of pairs of curves. I want to guide the reader through a few examples and basic computation techniques. Of course, "elementary" is not the same as "easy," but by working through this material carefully, one discovers a kit of true power tools.

Examples: We start with the good news that many indices can be computed by direct observation. Moreover, the index is quite insensitive to much of the detailed behavior of γ, σ, and f, so more complicated or ambiguous situations can often be reduced to these simple ones.

Begin with Figure 8.14. Treat $g(z) = f(z) - z$ as the vector from $z \in \gamma$ to its image $f(z) \in \sigma$ and simply count its windings as z transits γ. In (a) and (b), for example, it is clear that although the argument of g changes, it ultimately backtracks for no net winding, giving $\eta(f; \gamma) = 0$. In (c) it is equally clear that the index is 1. Note that (d) is just (c) with the roles of γ and σ interchanged. You might think at first that these indices should be negatives of one another, but in fact they are equal. $f^{-1} : \sigma \longrightarrow \gamma$ is an orientation-preserving, fixed point–free homeomorphism, and if $w = f(z)$, then $f^{-1}(w) - w = -(f(z) - z) = -g(z)$; a moment's reflection shows that $\mathfrak{w}(-g; \gamma) = \mathfrak{w}(g; \gamma)$.

These first examples are actually a little too simple. We have to prepare for curves that intersect, perhaps several times. Moreover, there are generally additional *constraints* on f (beyond it being an orientation-preserving fixed point–free homeomorphism); typically, some finite set $a, b, \ldots$ of points of γ must be mapped to designated points $a', b', \ldots$ of σ. The index may still be quite easy to compute. I illustrate three key situations in the examples of Figure 8.15.

In Figure 8.15(a), curves γ and σ having two intersection points, p and q. The key to the index lies with the locations of the images $f(p)$ and $f(q)$ on σ. On the right is a cartoon that illustrates how I personally do these computations using pencil and paper. (Can we keep this between you and me?) Place the eraser at some starting point z on γ and point the tip toward $f(z)$ on σ (parallel to the displacement vector $g(z)$). Run the eraser once around γ, keeping the tip moving around σ, making sure that the pencil lines up with any known constraints – in Figure 8.15(a), it must parallel $g(p)$ and $g(q)$ when

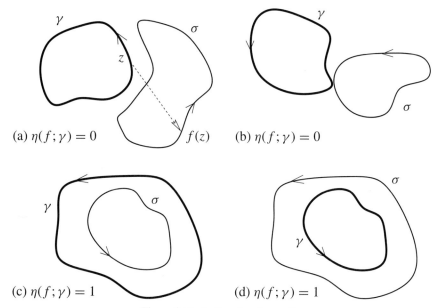

(a) $\eta(f;\gamma) = 0$ (b) $\eta(f;\gamma) = 0$

(c) $\eta(f;\gamma) = 1$ (d) $\eta(f;\gamma) = 1$

Figure 8.14. Indices by observation.

the eraser reaches p and q, respectively. Capture the pencil directions in a cartoon, then simply read off the net winding – in case (a) it is $\eta(f;\gamma) = 0$. Given the locations of $f(p)$, $f(q)$, this is the only possible outcome, irrespective of the rest of the behavior of f.

Figure 8.15(b) juxtaposes two circle packing interstices, with tangeney points (corners) labeled. The constraints are $f(a) = a'$, $f(b) = b'$, $f(c) = c'$. These constraints alone are not enough (in this case) to determine the index of f. However, there are qualitatively only four possibilities, represented in the cartoons; I strongly recommend that the reader work these out.

Finally, try your hand with the topological quadrilaterals of Figure 8.15(c): with corners identified, there can be only one outcome, $\eta(f;\gamma) = -1$.

Additional techniques for computing fixed-point indices exploit their insensitivity to details of behavior.

Lemma 8.11. *Suppose $\phi : \mathbb{C} \longrightarrow \mathbb{C}$ is an orientation-preserving homeomorphism. If $f : \gamma \longrightarrow \sigma$ is an orientation-preserving fixed point–free homeomorphism, then the same is true of the map $f_1 \equiv \phi \circ f \circ \phi^{-1} : \phi(\gamma) \longrightarrow \phi(\sigma)$. Moreover, $\eta(f_1; \phi(\gamma)) = \eta(f;\gamma)$.*

Define $m = \inf\{|f(z) - z| : z \in \gamma\}$, noting that $m > 0$. Suppose ϕ moves every point of γ a distance less than m, then $f_2 \equiv f \circ \phi^{-1} : \phi(\gamma) \longrightarrow \sigma$ is an orientation-preserving fixed point–free homeomorphisms and $\eta(f_2; \phi(\gamma)) = \eta(f;\gamma)$. Likewise, if ϕ moves every point of σ a distance less than m, then $f_3 \equiv \phi \circ f : \gamma \longrightarrow \phi(\sigma)$ is an orientation-preserving fixed point–free homeomorphisms and $\eta(f_3;\gamma) = \eta(f;\gamma)$.

You can confirm the result for f_1 using the fact that there exists an *isotopy* between the identity and ϕ, that is, a continuously parameterized family $\phi_s, 0 \le s \le 1$, of

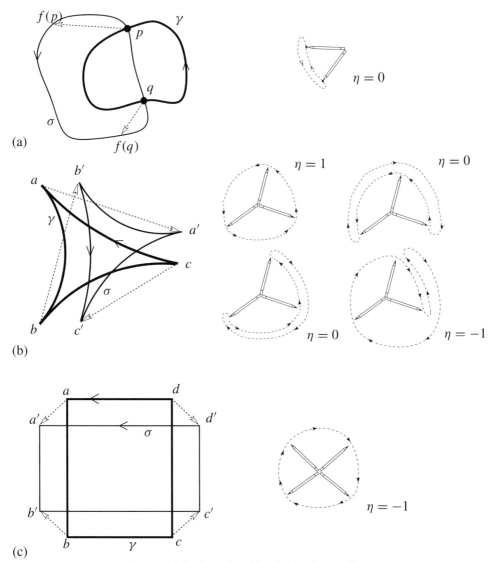

Figure 8.15. Examples of key indices by pencil.

homeomorphisms such that $\phi_0 = \text{id}$ and $\phi_1 \equiv \phi$. Each of the maps $g_s = \phi_s \circ g \circ \phi_s^{-1}$ is a continuous nonvanishing function on $\phi_s(\gamma)$ and one can show its winding number is a continuous function of $s \in [0, 1]$. Since it is integer-valued, it must be constant. Confirming the results for f_2 and f_3 is left for the reader, but here is a hint: this is a "walk-the-dog" situation. Suppose you take a walk with a dog on a leash – you follow a closed path, the dog follows its own closed path. Suppose that on completion of the walk you know that your winding number about a certain fire hydrant came out to be an integer N. Moreover, you know that your distance from that hydrant during the walk was always greater than the length of the dog's leash. Then you can conclude that the dog's winding number about that hydrant was also N. From this you get the result for f_2. (If the dog does the computation, it gets the result for f_3.)

The equality for f_1 in the lemma is useful because homeomorphisms ϕ can deform seemingly complicated situations into transparent ones. For instance, if the regions bounded by γ and σ have mutually disjoint closures, as in Figure 8.14(a), an appropriate homeomorphism ϕ will map γ into the left half plane and σ into the right half plane; the vector $(\phi \circ f \circ \phi^{-1})(z) - z$ always points left to right so it cannot wind around the origin, meaning that $\eta(f; \gamma) = 0$. Similarly, if γ and σ intersect in at most two points, then ϕ may be chosen so that $\phi(\gamma)$ and $\phi(\sigma)$ are both circles, a situation we address shortly. Modifications of the type associated with f_2 and f_3 can be used to jiggle situations without changing the indices.

Now for a bit of bad news: given a pair of curves and mapping constraints, there may be no qualifying fixed point–free maps whatsoever, as, for instance, when $a = a'$. Fortunately we are not at the mercy of fate alone in our curves and constraints and we can avoid problem situations with *general position* arguments.

Definition 8.12. Jordan curves γ and σ with constraint points $\{a, b, \ldots\} \in \gamma$ and $\{a', b', \ldots\} \in \sigma$ are said to be in **general position** if no constraint point of γ lies on σ and no constraint point of σ lies on γ.

The previous lemma allows us to make minor adjustments of the packings we consider to ensure general position – using, for example, small translations and/or rotations of the packings.

Compatibility Results: The particular indices of interest to us are those arising when γ and σ are both circles, both topological triangles, or both topological quadrilaterals. In the latter two cases, the maps f are required to map corners to corners. We say informally that the curves γ and σ are "compatible" if one can find a homeomorphism $f : \gamma \longrightarrow \sigma$ which satisfies the constraints and has a nonnegative fixed-point index. Our first result says that any two circles are compatible.

Lemma 8.13 (Circle Compatibility). *If γ and σ are circles and if $f : \gamma \longrightarrow \sigma$ is any orientation-preserving, fixed point–free homeomorphism, then $\eta(f; \gamma) \geq 0$.*

Proof. The cases of disjoint or tangent circles are covering in Figure 8.14. When there are two intersections, say p and q, then the index is determined entirely by the locations of $f(p)$, $f(q)$. The reader has taken care of one case in Figure 8.15(a) – the remaining five cases require little more and I leave them to the reader as well. $\qquad\square$

The situation with triangles is more delicate. A topological triangle is a Jordan curve γ with three distinguished points a, b, c, called *vertices*, and the resulting three arcs,

$\langle a, b \rangle$, $\langle b, c \rangle$, $\langle c, a \rangle$, called *edges*. We use the notation $\gamma_{\langle a,b,c \rangle}$ for a topological triangle, so our homeomorphism f may be written as $f : \gamma_{\langle a,b,c \rangle} \to \sigma_{\langle a',b',c' \rangle}$, where it is to be understood that $f(a) = a'$, $f(b) = b'$, $f(c) = c'$.

Lemma 8.14 (Triangle Compatibility). *Let $\gamma_{\langle a,b,c \rangle}$ and $\sigma_{\langle a',b',c' \rangle}$ be topological triangles in general position. Then there exists an orientation-preserving, fixed point–free homeomorphism $f : \gamma_{\langle a,b,c \rangle} \to \sigma_{\langle a',b',c' \rangle}$ for which $\eta(f; \gamma) \geq 0$.*

This is quite different from the circle result. Two topological triangles can intersect in any number of complicated ways, with arbitrarily many intersections, even along subarcs. It is easy to build examples with homeomorphisms having *any* prescribed index, positive, negative, or zero. However, note that in the triangle case we have an *existence* statement; we get to choose our f – wisely, we hope.

For completeness's sake, I will outline the formal proof. However, it uses results far deeper than our situations require, so I will then describe the common sense approach.

Proof. The general proof relies on the Riemann Mapping and Carathéodory Theorems of classical analysis. According to these, given any two Jordan regions of the plane, there exists a univalent analytic function F mapping one onto the other which extends continuously to an orientation-preserving homeomorphism between their boundaries. Moreover, F can be chosen to map any three points in positive order on the boundary of the domain to any three points in positive order on the boundary of the range. Applying this result to the Jordan regions defined by $\gamma_{\langle a,b,c \rangle}$ and $\sigma_{\langle a',b',c' \rangle}$, the interior mapping F induces an orientation-preserving homeomorphism $F : \gamma_{\langle a,b,c \rangle} \longrightarrow \sigma_{\langle a',b',c' \rangle}$. If necessary, an arbitrarily small shift of γ or σ will ensure that F has no fixed points on γ.

As for the index, the Cauchy Integral Formula tells us that the winding number of $F(z) - z$ on γ counts the number of zeros of $F(z) - z$ on the interior of γ. In other words, this winding number is nonnegative. Define $f \equiv F|_\gamma$. □

Now I want to put the formalities aside and see how to prove what we need in very practical terms. I will go back to our rope walks with you and a friend connected by an elastic string. As your friend walks once around γ, you get to choose how you walk once around σ. The function f reflects your choice: you are at $f(z)$ when your friend is at z. The only rules are that you must walk forward on σ and that you must reach a', b', c' when your friend reaches a, b, c, respectively.

Challenge. In the triangle situation, show that you can always arrange your movements so that the rope wraps a nonnegative number of times around you.

In the situations we encounter, this is, in fact, an easy challenge to meet, and anyone who tries just a few concrete cases is likely to discover a simple strategy – what I call "rush/hold." Say your friend is about to cover the edge $\langle a, b \rangle$. A "rush" strategy means that the instant your friend leaves the point a, you rush along the edge $\langle a', b' \rangle$ and then hover near b' while your friend transits $\langle a, b \rangle$. A "hold" strategy means that as your friend leaves a, you hold near a' until she has almost arrived at b, and only then do you move along $\langle a', b' \rangle$ to b'. Each of these strategies results in a certain change in the argument, $\Delta_{\text{arg}} = \Delta(\arg(f(z) - z))$, between a and b; you naturally choose to rush or hold depending on which change in argument is largest. Both may, in fact, be negative on a given edge, but if you apply this strategy on each edge, you invariably end up with a nonnegative index.

Claim. *When $\gamma_{\langle a,b,c \rangle}$ and $\sigma_{\langle a',b',c' \rangle}$ are interstices of triples of circles, then applying a rush/hold strategy on each edge will meet the Challenge.*

Perhaps you arrived at this strategy yourself when you worked out the possible indices for the example of Figure 8.15(b). You may have noted, for example, that since the edges $\langle a, b \rangle$ $\langle a', b' \rangle$ do not cross, your strategy on that edge is irrelevant, Δ_{arg} is independent of what you do, rush/hold, or anything else. When corresponding edges intersect once, there are only two possibilities for Δ_{arg} and they differ by 2π. You get one with rush and the other with hold; it is simply a question of whether you choose to cross the intersection point before or after your friend does. When corresponding edges intersect twice, rush and hold again give the same change in argument (at least, when the two edges intersect like two arcs of circles). In this case, a more complicated strategy might make Δ_{arg} even more positive, but that is not needed for the claim.

I will leave the proof of this claim to the reader. Indeed, I am sure there is some clever way to do the counting. I have not found it, however, and my proof devolves into a catalog of special cases and is too inelegant to share here. I hope the reader can do better. I should say that this rush/hold strategy works equally well on the noninterstice topological triangles we introduce shortly.

The next logical step would be to address compatibility of topological quadrilaterals, but we have already seen an incompatible example in Fig. 8.15(c). We may summarize the possibilities for map $f : \gamma \longrightarrow \sigma$:

Compatibility Summary: *Any two circles are compatible, regardless of f; two topological triangles are compatible if one chooses an appropriate f; but two topological quadrilaterals may be incompatible no matter how f is chosen.*

Elements and Cadres: The first targets for our index arguments are called the *elements* of a packing. Suppose that P is a circle packing for complex K. A *circular element* $\mathfrak{e}_v$ for vertex $v \in K$ is just the circle $c_v \in P$ treated as a positively oriented simple closed curve. A *triangular element* $\mathfrak{e}_t$ for a face $t \in K$ is the boundary of the interstice of P associated with that face; this is a curvilinear triangle formed by arcs of the circles forming the interstice and is treated as a positively oriented topological triangle whose three vertices are the three points of tangency of those circles. An *auxiliary element* $\mathfrak{e}$ is a topological triangle that has been adjoined to two boundary circles (or, on occasion, to one) of P as a sort of "boundary interstice." These boundary interstices coat the outer boundary of the

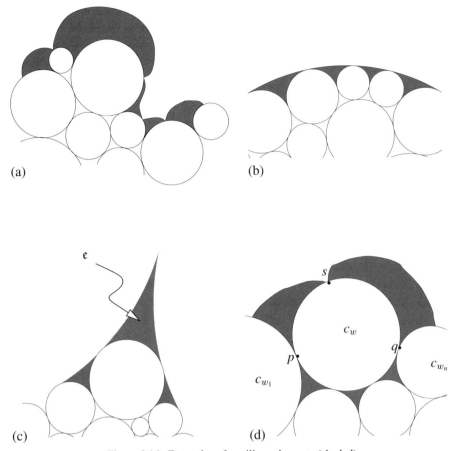

Figure 8.16. Examples of auxiliary elements (shaded).

packing, as will be elaborated momentarily. Assuming for now that they are in place, we have the following definition.

Definition 8.15. Given a circle packing P with complex K, the collection of **elements** of P, denoted $\mathcal{E}_P$, is the collection of Jordan curves $\mathfrak{e}_j$ associated with its circles, interstices, and appropriately constructed auxiliary interstices. The union of the elements and their interiors will be called an **augmented carrier** of P and its oriented boundary, consisting of the outer edges of the auxiliary elements, will be called the **augmented boundary** of P, written Γ_P.

Auxiliary elements are needed only if K has a boundary, and in this case are created by the user, within certain constraints. In many circumstances there will be natural choices, but the properties you build in depend largely on what you want to accomplish. In describing them it does not pay to be overly legalistic – that would only limit future uses – so instead look to Figure 8.16, where the examples of the type we will need pretty much speak for themselves. Let me make three key points, however:

Auxiliary Elements are Topological Triangles: Generically, an auxiliary element bridges two successive boundary circles of P as an exterior interstice. The user chooses an arc connecting the two circles, forming a topological triangle with the endpoints of the arc and the point of tangency of the two circles as its vertices. Figure 8.16(a) shows that you have considerable latitude in forming auxiliary elements; (b) shows the natural auxiliary elements for a maximal packing; (c) shows an auxiliary element $\mathfrak{e}$ which abuts a single boundary circle.

Every Circle of P is a Finite Union of Arcs of Triangular Elements: This is automatic for an interior circle, which is the union of arcs between tangency points with its petals, and each of these arcs is the edge of an interstice element. For a boundary circle this represents a constraint placed on the construction of contiguous auxiliary elements. Look to Figure 8.16(d) for an example. If $w \in \partial K$ has combinatorial (open) flower $F = \{w; w_1, \ldots, w_n\}$, then c_w has c_{w_1} as downstream neighbor with tangency point, say, p, and c_{w_n} as upstream neighbor with tangency point q. The arc $[p, q] \subset c_w$ (positive orientation) is a succession of arcs shared with interstice elements. The remainder of c_w, arc $[q, p]$, will typically be a union of arcs of two auxiliary elements, one bridging to c_{w_n} and one to c_{w_1} with a shared vertex $s \in [q, p]$.

Every Connected Component γ in ∂K Gives Rise to a Connected Component Γ of the Augmented Boundary Γ_P: There are auxiliary elements associated with successive circles in any edge-path $\gamma \subseteq \partial K$, and their outer edges chain together to form an arc in Γ_P. The main point is that when K is simply connected, then Γ_P will be a simple closed curve.

Our methods actually involve working with *pairs* of packings P and P' for the same complex K. We assume P and P' are in *general position*, meaning that every pair of corresponding elements is in general position (see Definition 8.12). In applications this may require some small perturbation of the packings, chosen carefully to avoid invalidating the conclusions. We also assume that we have defined augmented carriers for P and P'. The collections $\mathcal{E} = \mathcal{E}_P$ and $\mathcal{E}' = \mathcal{E}_{P'}$ of elements are in one-to-one, type-for-type correspondence:

- For every $v \in K$, the circular element $\mathfrak{e}_v$ in $\mathcal{E}$ has its corresponding circular element $\mathfrak{e}'_v$ of $\mathcal{E}'$.

- For every face $t \in K$, the interstice element $\mathfrak{e}_t$ has its corresponding interstice element $\mathfrak{e}'_t$. Moreover, the vertices of these topological triangles correspond: each vertex of $\mathfrak{e}_t$, as a tangency point of two circles of P, has a corresponding vertex of $\mathfrak{e}'_t$, namely, the tangency point of the corresponding two circles of P'.

- Each auxiliary element $\mathfrak{e}$ has a corresponding auxiliary element $\mathfrak{e}'$. Generically, $\mathfrak{e}$ bridges two boundary circles and has their tangency point as one of its vertices; $\mathfrak{e}'$ does the same. Should $\mathfrak{e}$ be constructed to abut a single boundary circle, $\mathfrak{e}'$ must do likewise.

We now enumerate the elements, $\mathcal{E} = \{\mathfrak{e}_1, \mathfrak{e}_2, \ldots\}$ and define, for each j, an orientation-preserving fixed point–free homeomorphism $f_{\mathfrak{e}_j} = f_j : \mathfrak{e}_j \longrightarrow \mathfrak{e}'_j$. We always begin with the triangular elements, both interstice and auxiliary. Suppose $\mathfrak{e}_j = \gamma_{\langle a,b,c \rangle}$ and $\mathfrak{e}'_j = \gamma'_{\langle a',b',c' \rangle}$. According to Lemma 8.14, we may choose $f_j : \mathfrak{e}_j \longrightarrow \mathfrak{e}'_j$,

$$f_j : \gamma_{\langle a,b,c \rangle} \longrightarrow \gamma'_{\langle a',b',c' \rangle},$$

so that $\eta(f_j; \mathfrak{e}_j) \geq 0$. When we have done this for all triangular elements, we move on to the circular elements. However, it turns out there is nothing more we need to do. Recall that a circular element $\mathfrak{e}_k$ is a finite chain of edges shared with contiguous triangular elements, and on each of these edges we have already defined a map to the corresponding edge, which is in $\mathfrak{e}'_k$. Simply piecing these together defines $f_k : \mathfrak{e}_k \longrightarrow \mathfrak{e}'_k$. A moment with a picture will establish that f_k is orientation-preserving, fixed point–free, and (for nonbranched packings) a homeomorphism. Because $\mathfrak{e}_k$ is circular, $\eta(f_k; \mathfrak{e}_k) \geq 0$. This sets up our last definition.

Definition 8.16. Let P and P' be circle packings for a complex K that are in general position and let $\mathcal{E}$ and $\mathcal{E}'$ be their sets of elements. A collection

$$\mathcal{F} = \mathcal{F}_{P,P'} = \{f_j : \mathfrak{e}_j \longrightarrow \mathfrak{e}'_j : \mathfrak{e}_j \in \mathcal{E}\}$$

of orientation-preserving, fixed point-free homeomorphisms with the property that

$$\eta(f_j; \mathfrak{e}_j) \geq 0, \ \forall j,$$

is termed a ***cadre*** of maps.

Each auxiliary element $\mathfrak{e} \in \mathcal{E}_P$ has an outer arc τ belonging to the augmented boundary Γ_P and a map from τ to τ', the restriction of $f_{\mathfrak{e}}$. Collectively, these restrictions define a map F on the whole augmented boundary,

$$F : \Gamma_P \longrightarrow \Gamma_{P'}.$$

One can easily tick off the properties of F from those of the auxiliary maps $f_{\mathfrak{e}}$: in particular, *F is an orientation-preserving, fixed point–free homeomorphism between the augmented boundaries.*

Lemma 8.17. *Let P and P' be circle packings for K, let $\mathcal{E}$ and $\mathcal{E}'$ be collections of elements, and let $\mathcal{F}_{P,P'}$ be a cadre of maps. If F is the induced map $F : \Gamma_P \longrightarrow \Gamma_{P'}$, then*

$$\eta(F; \Gamma_P) = \sum_j \eta(f_j; \mathfrak{e}_j), \tag{43}$$

where the sum is over all elements $\mathfrak{e}_j$ of $\mathcal{E}$. In particular, $\eta(F; \Gamma_P) \geq 0$.

Proof. Define the sum of the elements' fixed-point indices,

$$W = \sum_j \eta(f_j; \mathfrak{e}_j). \tag{44}$$

This sum is rife with cancellations. An edge τ shared by a triangular element $\mathfrak{e}_j$ and a contiguous circular element $\mathfrak{e}_k$ of P can affect W only via the two terms $\eta(f_j; \mathfrak{e}_j)$ and $\eta(f_k; \mathfrak{e}_k)$. The former reflects the change in argument of $g_j(z) = f_j(z) - z$ as z transits τ in the positive direction about $\mathfrak{e}_j$; the latter reflects the change in argument of $g_k(z) = f_k(z) - z$ as z transits τ in the opposite direction (the positive direction about $\mathfrak{e}_k$). However, recall that f_k was *defined* on this arc of $\mathfrak{e}_k$ to be equal to f_j: that is, $g_k(z) = g_j(z)$ for $z \in \tau$. Consequently, the contributions to $\eta(f_j; \mathfrak{e}_j)$ and $\eta(f_k; \mathfrak{e}_k)$ from edge τ are equal and opposite in sign – they cancel one another in the sum W.

What is left in W after all such cancellations? Only the contributions by those arcs belonging to a single element – namely, by the outer edges of the auxiliary elements. Refer to these as arcs τ_k. For each k, recall that $g_k(z) = f_k(z) - z$ and define $G(z) = F(z) - z$. Then

$$\eta(F; \Gamma_P) = \mathfrak{w}(G; \Gamma_P) = \sum_k \mathfrak{w}(g_k; \tau_k).$$

The terms on the right are precisely the ones that do not cancel in Eq. (44); in other words, $\eta(F; \Gamma) = W$, as we claimed. The last statement of the lemma follows from the fact that by construction, all maps f_j of a cadre have nonnegative fixed-point indices. $\quad\square$

This completes our collection of new geometric tools. Now we must put them to use in the hyperbolic and parabolic cases.

8.3.2. Hyperbolic Uniqueness

Suppose K is hyperbolic and $P \longleftrightarrow K(R)$ is a univalent circle packing whose carrier fills $\mathbb{D}$. We may assume by way of normalization that the circles for v_1 are at the origin in both P and $\mathcal{P}_K$.

Suppose $\epsilon > 0$. Treat P as euclidean and define Q_ϵ to be the dilated packing $Q_\epsilon = (1 + \epsilon)P$. Let $\{K_n\} \uparrow K$ be the usual exhausting sequence of finite simply connected subcomplexes. Define Q_n as the restriction of Q_ϵ to K_n. By hypothesis, the circles of P can accumulate only at $\partial\mathbb{D}$, so there exists $N = N(\epsilon)$ such that for all $n \geq N$ the boundary circles of Q_n will all intersect the complement of $\overline{\mathbb{D}}$.

Claim. *For any $\epsilon > 0$ and any $n > N(\epsilon)$, the circle for v_1 in Q_n has larger euclidean radius than the circle for v_1 in $\mathcal{P}_n = \mathcal{P}_{K_n}$.*

Assuming we have shown this, consider C_1, the circle for v_1 in $\mathcal{P}_K$. By monotonicity, the circles for v_1 in the maximal packings $\mathcal{P}_n$ converge with *decreasing* radii to C_1. The Claim therefore implies that the circle for v_1 in Q_n is larger than C_1, and therefore, the circle for v_1 in $(1 + \epsilon)P$ is larger than C_1 for every $\epsilon > 0$. This implies $R(v_1) \geq \mathcal{R}_K(v_1)$. However, by Lemma 8.3, this must be equality and $R \equiv \mathcal{R}_K$, meaning that P is an automorphic image of $\mathcal{P}_K$ and we are done.

Proof of Claim Fix ϵ and $n > N(\epsilon)$. The boundary circles of Q_n all intersect the complement of $\mathbb{D}$, so they are, in hyperbolic terms, "larger" than infinite radius. Lemma 8.8 tells
us how to exploit this.

Let S denote the hyperbolic label for K_n defined by

$$S(v) = \begin{cases} \text{radius}(c_v) & \text{if } c_v \subset \mathbb{D} \\ \infty & \text{otherwise.} \end{cases}$$

This may not be a proper label since interior vertices could be assigned infinite labels.
However, we can still make sense of the angle sum function for S, $\theta_S(\cdot)$, and after we do
that we will be able to adjust S to get a legitimate label.

Consider each of the vertices of K_n. If $S(v) = \infty$, we simply define $\theta_S(v) = 0$ (which
is consistent, e.g., with the fact that angles at horocycles are always zero). If $S(v) < \infty$,
then v lies in $\mathbb{D}$; its flower has angle sum 2π in the sphere, and by Lemma 8.8, this angle
sum can only get smaller when we use our new label S, so $\theta_S(v) \le 2\pi$.

We can now adjust S to a new hyperbolic label by using our old friends, continuity
and monotonicity of angle sums. In particular, visit in turn each interior vertex v for
which $S(v) = \infty$ and choose a new value for $S(v)$ which is finite but sufficiently large
to ensure that $\theta_S(v) < 2\pi$. Note that decreasing a label can only decrease the angle sum
at its neighbors, so this inequality will not be invalidated by subsequent changes to other
infinite labels.

Let S denote the adjusted (now legitimate) label and note that it belongs to the following
family Φ of hyperbolic labels for K_n:

$$\Phi = \{R : \theta_R(v) \le 2\pi \text{ for every interior vertex } v\}.$$

We now mimic techniques from the proof of Proposition 6.1 (see Section 6.3). First observe
that the maximal label for K_n, which we denote by $\mathcal{R}_n = \mathcal{R}_{K_n}$, lies in Φ. Also, Φ is closed
under taking minima (note the inequalities here are the reverse of those in Section 6.3).
Let $\widetilde{R} = \min\{S, \mathcal{R}_n\}$. Since $\widetilde{R} \le \mathcal{R}_n$ and $\widetilde{R}(w) = \infty$ for $w \in \partial K_n$, we can duplicate the
area argument of Section 6.3 to conclude that $\widetilde{R} = \mathcal{R}_n$. In other words, $S \ge \mathcal{R}_n$, and in
particular, $S(v_1) \ge \mathcal{R}_n(v_1)$. Indeed, since $n > N(\epsilon)$, this inequality is strict. Now recall
the definition: $S(v_1)$ is the hyperbolic radius of v_1 in Q_n and $S(v_1) > \mathcal{R}_n(v_1)$. We have
proven the claim and thereby settled hyperbolic uniqueness.   □

8.3.3. Parabolic Uniqueness

I will begin with a new proof of essential uniqueness for maximal packings of combinatorial spheres (see Proposition 7.1) because the reasoning sets the stage for the parabolic
case. Thus, suppose K is a combinatorial sphere, $\mathcal{P} = \mathcal{P}_K$ is a maximal packing, and
$Q \subset \mathbb{P}$ is another univalent packing for K. We must show that Q is a Möbius image of $\mathcal{P}$.

A preliminary task is to place Q and $\mathcal{P}$ in appropriate positions. Any triple of mutually
tangent circles on the sphere is the Möbius image of any other triple. Therefore we may
assume for some face $\langle x, y, z \rangle$ of K that each of the three circles c_x, c_y, c_z in Q is identical

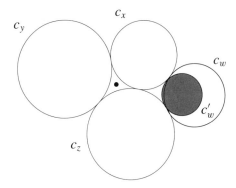

Figure 8.17. Preliminary situation.

to the corresponding circle c_x', c_y', c_z' of $\mathcal{P}$. Starting from this face in the carrier of $\mathcal{P}$, you can step through successively contiguous faces until encountering one which is different in $\mathcal{P}$ and Q. This must happen unless Q and $\mathcal{P}$ are identical. We may clearly assume that this occurs at the first step; in other words, we find ourselves in the situation pictured in Figure 8.17. Namely, the circles c_x, c_y, and c_z are identical in both packings $\mathcal{P}$ and Q, but the circle $c_w \in Q$ defining a contiguous face is different from the corresponding circle $c_w' \in \mathcal{P}$ (the shaded circle).

We are still not quite ready. The Möbius transformation $z \mapsto 1/z$ interchanges the points 0 and ∞, and if we apply this to both Q and $\mathcal{P}$ (observe the dot marking the origin in Figure 8.17), the result is as shown in Figure 8.18(a). In particular, the remaining circles of the two packings now lie *inside* the interstice. A few circles from Q are shown for reference; c_w' is again shaded and overlaps c_w.

The next adjustment involves contracting the packing Q toward the point p to produce the configuration in Figure 8.18(b); this can be done so that $\mathcal{P}$ and Q are in general position and so that zero lies inside one of the circles or interstices of Q. Finally, discard from both packings the circles for vertices x, y, z, and w. It is these reduced packings, denoted $\widetilde{Q}$ and $\widetilde{P}$, to which we apply index arguments leading to a contradiction.

As the auxiliary elements for $\widetilde{Q}$ simply use the necessary triangular elements from the larger packing Q; namely, the interstices trapped between $\widetilde{Q}$ and the four deleted circles. Define the auxiliary elements for $\widetilde{P}$ in parallel fashion. Denote the augmented boundaries of the two packings by $\Gamma = \Gamma_{\widetilde{Q}}$ and $\Sigma = \Gamma_{\widetilde{P}}$

Choose a cadre $\mathcal{F}_{\widetilde{Q}, \widetilde{P}} = \{f_j\}$ of maps and let $F : \Gamma \longrightarrow \Sigma$ be the map it induces between the augmented boundaries. By Lemma 8.17,

$$\eta(F; \Gamma) = \sum_j \eta(f_j; \mathfrak{e}_j) \geq 0. \tag{45}$$

The sum on the right is nonnegative because that is the case for each index involved. A closer look at Γ and Σ will yield the contradiction; see Fig. 8.18(c). The thicker curve is Γ and the lighter is Σ; these are topological quadrilaterals, the four corresponding corners marked as a, b, c, d and a', b', c', d', respectively. Compare this to the crossed

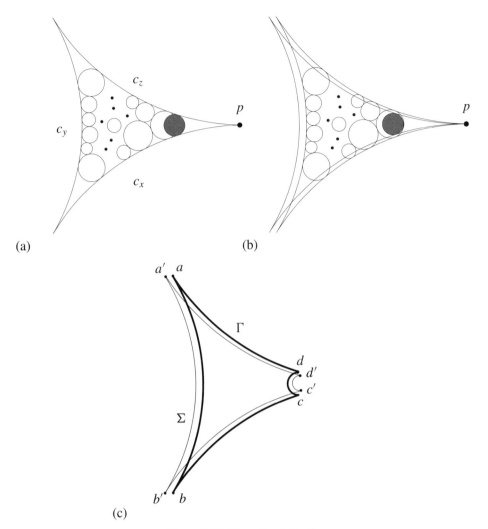

Figure 8.18. Normalized situation.

quadrilaterals of Fig. 8.15 and you will reach the inevitable conclusion, as we did there, that $\eta(F; \Gamma) = -1$. This contradicts (45) and we are done. Backtracking to our original assumptions, we see that Q must be a Möbius image of $\mathcal{P} = \mathcal{P}_K$.

We are now ready to carry this reasoning over to the parabolic case. The potential difficulty lies with the now infinite sum on the right in (45). Each index breaks into pieces, infinitely many of which we expect to cancel against one another. Can one legally do infinitely many cancellations? That involves infinite rearrangement, and hence requires verification of some convergence conditions. We will finesse this difficulty by getting back to the finite case using extra aggregate elements.

Assume now that K is parabolic with maximal packing $\mathcal{P} = \mathcal{P}_K$ and that Q is another univalent circle packing for K in the plane. Treating these both as packings in the sphere, we can apply the same manipulations as before to reach the packings $\widetilde{Q}$ and $\widetilde{P}$ situated as in Figure 8.18(b). The origin lies, by our positioning arguments, inside some element $\mathfrak{e}$ of $\widetilde{Q}$. (This actually is a little more work in this case because $\widetilde{Q}$ has infinitely many circles; still, one can show that an arbitrarily small translation will put the origin in the open interior of an element and will put $\widetilde{Q}$ and $\widetilde{P}$ in general position.)

Since $\mathrm{carr}(\mathcal{P}_K) = \mathbb{C}$ by our earlier completeness results, the circles of $\mathcal{P}_K$ accumulate only at $\{\infty\}$. After the inversion $z \mapsto 1/z$, therefore, the circles of $\widetilde{P}$ accumulate only at the origin. This is an important fact since it allows us to choose a finite chain $\{\tau_n' : n = 1, 2, \ldots, N\}$ of edges of elements of $\widetilde{P}$ which forms a Jordan curve σ inside $\mathfrak{e}$ that separates $\mathfrak{e}$ from the origin. The corresponding chain of edges for $\widetilde{Q}$ necessarily forms a Jordan curve γ. All but finitely many elements of $\widetilde{P}$ are inside σ, and likewise all but finitely many elements of $\widetilde{Q}$ are inside γ.

Next, we modify our cadre by treating σ and γ as new elements and discarding the former elements of $\widetilde{P}$ and $\widetilde{Q}$, respectively, which lie inside them. When maps already defined between the edges forming γ and the corresponding edges forming σ are used, σ and γ inherit, edge by edge, the maps defined between those former elements to give a homeomorphism $f : \gamma \longrightarrow \sigma$. This is orientation-preserving, and fixed point–free and inherits the constraints from the edges involved. We are left with a finite number of elements in the usual one-to-one correspondence between $\widetilde{Q}$ and $\widetilde{P}$ – the only difference from normal being that we include the new multiedge elements γ and σ in the collection.

Now, simply observe that σ, since it lies inside a single element $\mathfrak{e}$ of $\widetilde{Q}$, is disjoint from γ, and therefore, $\eta(f; \gamma) \geq 0$. Having reduced to only finitely many elements, all with nonnegative indices, we can proceed to a contradiction exactly as we did with the sphere above.

This completes the proof of Proposition 8.1 as stated (modulo proof of Lemma 8.7 in the next section). Before we leave this chapter, however, there is one final simply connected situation not covered by our accumulated existence and uniqueness results: a simply connected complex K can be infinite yet have one or more boundary components – the analogue of the unit disc $\mathbb{D}$ with one or more open arcs of $\partial\mathbb{D}$ attached. Such complexes K are always hyperbolic, and the maximal packing $\mathcal{P}_K$ will have horocycles associated with the vertices on its boundary. If one adjoins the external interstices between neighboring boundary horocycles and the unit circle, then the augmented carrier will be all of $\mathbb{D}$; we will say that $\mathcal{P}_K$ *fills* $\mathbb{D}$. Here is the statement. I leave the proof for the reader with this one hint: by the Ring Lemma, a horocycle n generations from the α-vertex cannot be too small in euclidean radius.

Proposition 8.18. *Let K be a combinatorial disc with boundary. Then there exists an essentially unique univalent circle packing $\mathcal{P}_K$ for K which fills the hyperbolic plane $\mathbb{D}$; in particular, all circles associated with boundary vertices of K are horocycles.*

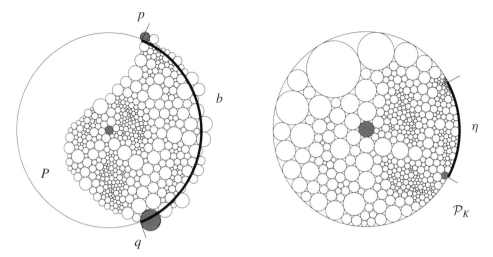

Figure 8.19. Comparing harmonic measures.

8.4. Proof of Lemma 8.7

Let us finish the lemma we experimented with in Figures 8.8 and 8.9; I delayed the proof because we will use the index arguments of Section 8.3.1. You should review the situation as described in Definition 8.6, we have the finite univalent packing P, $c_v \subset \mathbb{D}$, the component P_v of $P \cap \mathbb{D}$, the edge-path α belonging to the boundaries of both K and K_v, and the arc $a \in \partial \mathbb{D}$ defined by α. We proceed in two steps.

Step I: As a notational convenience, let us assume that $K = K_v$, $P = P_v$. Let C_v and c_v be the circles for v in $\mathcal{P}_K$ and P, respectively and by applying automorphisms of $\mathbb{D}$ arrange that both are centered at the origin. Let p and q be the endpoints of a, so $a = [p, q]$ and $b = [q, p]$. Let $\eta \subset \partial \mathbb{D}$ denote the arc associated with the boundary edge-path $\beta = \partial K \setminus \alpha$ in the maximal packing, so by definition, $\omega(v, \beta, K) = \omega(0, \eta, \mathbb{D})$. We arrange by rotations that the arcs b and η be centered at 1. Figure 8.19 illustrates a typical situation.

We will argue by contradiction, assuming $\omega(v, \alpha, K) < \omega(0, a, \mathbb{D})$.

$$\omega(v, \alpha, K) < \omega(0, a, \mathbb{D})$$
$$\implies 1 - \omega(v, \alpha, K) > 1 - \omega(0, a, \mathbb{D})$$
$$\implies \omega(v, \beta, K) > \omega(0, b, \mathbb{D})$$
$$\implies \omega(0, \eta, \mathbb{D}) > \omega(0, b, \mathbb{D}).$$

In other words, with our normalization the assumption is now that $b \subset \eta$, and we will use index arguments to obtain a contradiction. (This assumption is, of course, contrary to what one observes in Fig. 8.19 – I could not convince the picture to lie.)

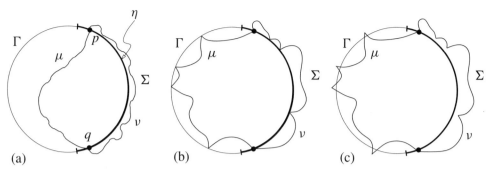

Figure 8.20. The model case and perturbations.

We now develop a cadre of maps. As auxiliary elements for $\mathcal{P}_K$, choose its natural boundary interstices, so that the augmented boundary $\Gamma = \Gamma_{\mathcal{P}_K}$ is just the unit circle. As for the augmented boundary $\Sigma = \Gamma_P$ of P, recall that the circles for α lie in $\overline{\mathbb{D}}$ while those for β intersect the complement of $\mathbb{D}$. With the latitude we have in choosing auxiliary interstices, it is not difficult to arrange that the subarc of Σ, call it μ, running along the auxiliary elements associated with α goes from p to q in $\overline{\mathbb{D}}$, while the subarc, call it ν, associated with β lies in the complement of $\mathbb{D}$. Our hypothesis is that η properly contains b. An arbitrarily small rotation of P will preserve this and ensure that $\mathcal{P}_K$ and P are in general position. A cadre $\mathcal{F}_{\mathcal{P}_K, P}$ then induces the fixed point-free homeomorphism $F : \Gamma \longrightarrow \Sigma$.

We have more details to assemble regarding Γ and Σ, but it might help to see the punchline right now. The model situation is shown in Figure 8.20(a). Precisely because of our assumption that $b \subset \eta$, this juxtaposition of curves is isotopic to that in Fig. 8.15(a), where we observed that the index is zero. By Lemma 8.17,

$$\eta(F; \Gamma) = 0 \implies \eta(f_j; \mathfrak{e}_j) = 0, \ \forall j. \tag{46}$$

However, C_ν and c_ν are both centered at the origin; assuming they have different radii, the associated element contributes index 1. This contradiction finishes the argument for Step I in this model case. (Note: we work with the "component" rather than the full packing in order to get this uncluttered situation; this Σ is based on the packing on the left of Fig. 8.19, which in turn comes from the parent packing of Figure 8.7.)

And what are the missing details? Our hypotheses do not preclude circles of α or of β being horocycles. We can choose the auxiliary elements along α so that they lie *inside* $\mathbb{D}$ except for isolated intersections $\mu \cap \Gamma$ at any points of tangency of horocycles. Likewise, we can choose those along β to lie *outside* $\overline{\mathbb{D}}$ except for isolated intersections $\eta \cap \Gamma$. The situation is now that suggested by Figure 8.20(b). One can, by means of a homeomorphism $\phi : \mathbb{C} \longrightarrow \mathbb{C}$, make arbitrarily small adjustments to Σ drawing any tangency points of μ into the interior of $\mathbb{D}$ while pushing any tangency points of ν outside $\overline{\mathbb{D}}$. This returns us to the model situation of Figure 8.20(a) so $\eta(\phi \circ F; \Gamma) = 0$; the last statement of Lemma 8.11 implies $\eta(F; \Gamma) = 0$ once again.

The other detail concerns the circles C_v and c_v. If they have the same radius, then $\mathcal{P}_K$ and P are not in general position and the element map for v in our cadre may be defective (i.e., may have a fixed point). An arbitrarily small dilation of P could fix this while leaving the qualitative situation unchanged, *except* that it might push any tangency points along μ outside Γ, giving a juxtapostion like that of Figure 8.20(c). So we have to proceed more carefully. Rotate P slightly, if necessary, so that the elements of P and $\mathcal{P}_K$ other than c_v and C_v are in general position. We get a fixed-point free map $F : \Gamma \longrightarrow \Sigma$ as before, with $\eta(F; \Gamma) = 0$. Since $m = \inf\{|F(z) - z| : z \in \Gamma\}$ is positive, a dilation of P by λ, for $1 < \lambda < 1 + m/2$, will not change the index of F according to Lemma 8.11. The circles λc_v and C_v are now nested, giving an index of 1 and we arrive yet again at our contradiction to (46). This completes Step I; we have shown that

$$\omega(v, \alpha, K_v) \geq \omega(0, a, \mathbb{D}).$$

Step II: Extending this result to the parent packing P follows from a more general fact which the reader may recognize as the discrete version of the *extension of domain* principle for harmonic measure.

Lemma 8.19. *Suppose v is an interior vertex of L and $L \subset K$ as a subcomplex, where L and K are combinatorial closed discs. Suppose further that α is an edge-path that lies in the boundaries of both L and K. Then*

$$\omega(v, \alpha, K) \geq \omega(v, \alpha, L). \tag{47}$$

Proof. Let Q be the restriction of $\mathcal{P}_K$ to L. We can arrange that the circles for v in $\mathcal{P}_L$ and $\mathcal{P}_K$ are at the origin, and of course the circles for α are horocycles in each.

Let $[p', q']$ be the arc of the unit circle defined by α in $\mathcal{P}_L$, and let $[p, q]$ be the arc defined by α in $\mathcal{P}_K$. Then by definition

$$\omega(v, \alpha, L) = \omega(0, [p', q'], \mathbb{D}) \quad \text{and} \quad \omega(v, \alpha, K) = \omega(0, [p, q], \mathbb{D}).$$

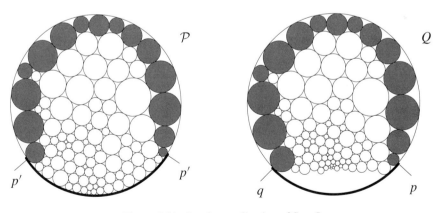

Figure 8.21. Another application of Step I.

It suffices, therefore, to show that

$$\omega(0, [p, q], \mathbb{D}) \geq \omega(0, [p', q'], \mathbb{D}). \tag{48}$$

But wait; this is just another use of Step I! Let me illustrate using packings from Figure 8.9. The second and third packings from that experiment are shown on the left and right, respectively, in Figure 8.21. On the right are the circles for L taken from the larger maximal packing $\mathcal{P}_K$, while on the left is the maximal packing for L itself; the shaded circles are those of α. You can apply Step I to get (48), but now you apply it to the complementary edge-path $\beta = \partial L \backslash \alpha$ of Q, whose circles lie *inside* $\mathbb{D}$. I leave the details for the reader to work out. □

Combining this extension of domain for $L = K_v$ with the conclusion of Step I completes the proof of Lemma 8.7. There are natural conjectures on when equality holds in these results, but they are "slightly" false. In the result just above, for instance, equality does not imply $L = K$; it instead implies that all the boundary vertices of L are boundary vertices of K. For instance, K might be obtained from L by simply adding a neighbor to two of its boundary circles. When we get to function theory in Part III, we put a natural added condition on our complexes to avoid such petty annoyances.

9

Proof for Combinatorial Surfaces

Proposition 9.1. *Let K be a multiply connected combinatorial surface. Then there exist a Riemann surface $\mathcal{S}_K$ and a circle packing $\mathcal{P}_K$ for K in the associated intrinsic metric on $\mathcal{S}_K$ such that $\mathcal{P}_K$ is univalent and fills $\mathcal{S}_K$. The Riemann surface $\mathcal{S}_K$ is unique up to conformal equivalence and $\mathcal{P}_K$ is unique up to conformal automorphisms of $\mathcal{S}_K$.*

The variety encompassed here is impressive: K can be finite or (countably) infinite, with or without boundary, have holes, handles, and so forth. In every case, the combinatorics of K determine, through circle packing, a rigid geometry in the form of a conformal structure, converting K from a topological to a Riemann surface. The complex K itself is termed *parabolic* or *hyperbolic* depending on whether the intrinsic metric of $\mathcal{S}_K$ is euclidean or hyperbolic, respectively. (Refer to Section 3.4 for background and terminology, and to the literature for an appreciation of the geometric richness of Riemann surfaces.)

The proof of this proposition is routine, at least for those steeped in the methods of surface theory. However, I would do even those readers a disservice if I did not take advantage of the very concrete and accessible nature of circle packing. Were you to have access to software (such as `CirclePack`), you could enter some combinatorics, compute the packing label, and *bingo*, there would be a packing on your screen! It just works! I want to begin by trying to see that geometry in action.

9.1. A Discrete Torus

The most familiar multiply connected surface is the torus. Why don't you get us started by drawing a triangulation K on a torus? Label its vertices from 1 to N and determine the flower of each vertex, that is, the closed list of neighbors oriented counterclockwise as you view them on the torus.

Humm...? I guess I will have to serve as your computer avatar. Suppose for argument's sake that the torus you have marked is that of Figure 9.1. The first thing I will do is put your vertices and their flowers in the format required by `CirclePack`. Since the back of the torus is out of sight, you will just have to trust that I have interpreted the combinatorics faithfully. You can check that K has no boundary, 20 faces, 30 edges, and 10 vertices, hence euler characteristic $10 - 30 + 20 = 0$, so at least this is, in fact, a torus.

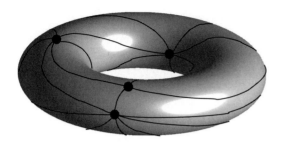

```
NODECOUNT: 10
GEOMETRY: euclidean
ALPHA/BETA/GAMMA: 1 2 2
FLOWERS:
 1 8    2 6 4 3 9 10 7 8 2
 2 5    1 8 3 5 6 1
 3 6    1 4 5 2 8 9 1
 4 7    1 6 7 10 9 5 3 1
 5 7    2 3 4 9 8 7 6 2
 6 5    1 2 5 7 4 1
 7 6    1 10 4 6 5 8 1
 8 6    1 7 5 9 3 2 1
 9 6    1 3 8 5 4 10 1
10 4    1 9 4 7 1
END
```

Figure 9.1. A triangulated torus and its data.

Now we are ready to run `CirclePack`. It takes three commands – `read`, `repack`, `post` – and about 2 seconds, including typing, to produce the postscript image underlying Figure 9.2(a). Oh, you had a question about the geometry? It cannot be spherical (a univalent circle packing filling $\mathbb{P}$ would have a simply connected carrier, the carrier is simplicially equivalent to K, so K would have to be a sphere), so it is either euclidean or hyperbolic. If you try the latter, `CirclePack` will display the unit disc with a tiny dot at the origin – repacking will have driven the radii to zero. We will see later that K is necessarily euclidean; that is why I specified "euclidean" in the file and why `CirclePack` successfully computed a euclidean packing label R.

You will note how poorly the 3-dimensional image of the torus served our purposes. On the other hand, the flat embedding of K in Figure 9.2(a) relies on a conventional drawing device; namely, the embedding has "ghost" vertices (the open dots) and edges (dashed) showing the identifications where K has been cut open. Take a few moments to familiarize

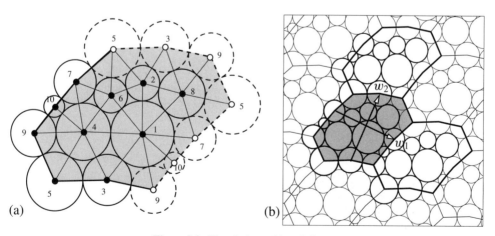

(a)   (b)

Figure 9.2. The circle packing P for K.

yourself with this type of layout. You might check, for instance, that the combinatorics are indeed those of K. This may take a moment for vertices on the edges; e.g., the faces in the flower for circle #5 lie in three "fans." (There are many other ways that P could be laid out; we discuss what `CirclePack` does in the Practicum.)

Checking that P satisfies the packing condition (angle sum 2π) at every circle takes a little more work. It is clear for a circle surrounded in place by its neighbors, circle #1, for instance. For a circle along the cut, the angle sum is distributed among the two or more fans of the flower. Elementary geometry tells us that these pieces can be brought together *via* rigid motions to reconstitute a full flower. In Figure 9.2(b), for example, it appears that a translation by vector (complex number) w_1 will reconstitute the flower for #7, while translation by w_2 does likewise for #3. Using translations of the form $\{z \mapsto z + nw_1 + mw_2 : n, m \in \mathbb{Z}\}$ appears to fill the plane and reconstitute every flower.

This wraps up our hands-on experiment, but I hope it leaves you with many questions. Will *translations* actually bring P and its copies into precise alignment, as Figure 9.2(b) suggests? Where is the torus on which P supposedly lives? Why euclidean geometry? Is P unique? And let me leave you with this: does Figure 9.2(b) not look an awfully lot like the parabolic maximal packing of Figure 1.3(e)?

If you are not familiar with the "theory of covering surfaces," you may not fully appreciate this experiment and the ideas it suggests. Yet I need to quote standard results from that theory. What to do? Let me provide a bridge by reviewing the torus from a classical perspective; the key ideas are all there and you can draw parallels with the experiment. After that, we should be ready for the general proof.

9.2. A Classical Torus

There are three standard geometric representations for a torus T. Most concrete is the embedded image of Figure 9.3(a). Of course, we've noted the shortcomings of the 3-dimensional image. It is common to open the torus along two fundamental curves to get a simply connected sheet which lays out as a flat euclidean parallelogram, as with Ω in Figure 9.3(b). Keep in mind that opposite sides are identified; if you exit on one edge of Ω, you instantaneously reenter on the opposite.

Representing T as a parallelogram makes the points on the edges appear special some-how when in fact they are not. One can fix this with a further abstraction. The translations identifying the sides of Ω, say $\phi_1(z) = z + w_1$ and $\phi_2(z) = z + w_2$, generate a *lattice group*

$$\Lambda = \{\phi : z \mapsto z + mw_1 + nw_2 : m, n \in \mathbb{Z}\} \subset \text{Aut}(\mathbb{C})$$

and one easily verifies that translated copies of Ω fit together to tile the plane,

$$\mathbb{C} = \bigcup_{\phi \in \Lambda} \phi(\Omega).$$

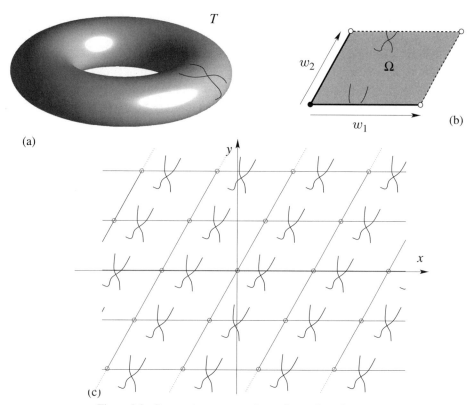

Figure 9.3. Geometric representations of a conformal torus.

Λ is said to be *properly discontinuous* in $\mathbb{C}$, meaning that the nonidentity elements have no fixed points and that for any point $z \in \mathbb{C}$, the *orbit* of z, $\mathcal{O}_z = \{\phi(z) : \phi \in \Lambda\}$, accumulates only at ∞. The orbit of zero is the *lattice* of dots shown in Figure 9.3(c), and for any z, $\mathcal{O}_z$ is just the translation by z of $\mathcal{O}_0$. The collection of orbits is denoted $\mathbb{C}/\Lambda = \{\mathcal{O}_z : z \in \mathbb{C}\}$ and called "$\mathbb{C}$ modulo Λ". The map $\pi : z \mapsto \mathcal{O}_z$ is the *projection* from $\mathbb{C}$ to $\mathbb{C}/\Lambda$. Because the group is properly discontinuous, every point of $\mathbb{C}$ has an open neighborhood U such that the map $z \longrightarrow \mathcal{O}_z$ is one-to-one on U. Let $\mathcal{O}_U = \{\mathcal{O}_z : z \in U\}$. The collection of sets $\mathcal{O}_U$ for open U defines a Hausdorff topology on $\mathbb{C}/\Lambda$, making $\mathbb{C}/\Lambda$ into a topological surface.

What does this have to do with T? Every point $p \in T$ is identified with a unique point $z_p \in \Omega$, and hence with a unique orbit $\mathcal{O}_{z_p} \in \mathbb{C}/\Lambda$. One can show that the map $p \mapsto \mathcal{O}_{z_p}$ is a homeomorphism; that is, T and $\mathbb{C}/\Lambda$ are topologically equivalent. The projection π can now be interpreted as a continuous surjection $\pi : \mathbb{C} \longrightarrow T$.

With this model of T, you enjoy the advantages of working in a familiar geometric environment rather than directly on the surface. I have drawn a pair of curves on T and "lifted" them (under π^{-1}) to $\mathbb{C}$ to suggest how this representation works.

$\mathbb{C}/\Lambda$ inherits not only a topology from $\mathbb{C}$, but also conformal and metric structures. Given $p, q \in T$, the distance $d(p, q)$ is defined as the distance between orbits $\mathcal{O}_{z_p}$ and $\mathcal{O}_{z_q}$, that is, the minimum of the distances between pairs of points, one from each orbit. This defines a locally euclidean metric on T with element of arclength $ds = |dz|$; you measure the length of a curve in T by lifting it to $\mathbb{C}$ and measuring its euclidean length there. It is immediate then that the angle between curves γ_1 and γ_2 meeting at a point $p \in T$ is just the angle at which lifted curves $\tilde{\gamma}_1$, $\tilde{\gamma}_2$ meet at a point of $\pi^{-1}(p)$. In particular, $T = \mathbb{C}/\Lambda$ is now seen to be a *conformal torus* and I will switch my notation to $\mathcal{T}$. I might also point out (summarizing 150 years of some very deep mathematics in a single sentence) that the link between the conformal torus $\mathcal{T}$ and the group Λ is very tight: two conformal tori, $\mathcal{T} = \mathbb{C}/\Lambda$ and $\mathcal{T}' = \mathbb{C}/\Lambda'$, are conformally equivalent if and only if Λ and Λ' are conjugate subgroups of $\mathrm{Aut}(\mathbb{C})$.

Thus, at the cost of some abstraction, representing $\mathcal{T}$ as $\mathbb{C}/\Lambda$ pulls the topological, metric, and conformal structures of $\mathcal{T}$ into a common setting, with conformal classification thrown into the bargain! This comprehensive approach goes by the name *covering theory*: $\mathbb{C}$ is the *universal covering surface* of $\mathcal{T}$, π is the *universal covering projection* (or *map*), Ω is a *fundamental domain* for π (one of many possible), Λ is the *universal covering group*, and the individual automorphisms $\phi \in \Lambda$ are *covering projections*.

Let us see what covering theory does for us in wrapping up the circle packing experiment of the previous section. All the elements are put together in Figure 9.4. There is the triangulation on T in (a). The layout that `CirclePack` produces and its two side-pairing translations are in (b). These generate a discrete group Λ; the translated copies of (b) fit together in (c) to form a circle packing $\widetilde{\mathcal{P}}$ of the plane – this will be known as the universal covering packing. The classical theory tells us that $\mathcal{T} = \mathbb{C}/\Lambda$ becomes a conformal torus. Moreover, the intrinsic metric on $\mathcal{T}$ is just the euclidean metric projected under $\pi : \mathbb{C} \longrightarrow \mathcal{T}$, so the circles of $\widetilde{\mathcal{P}}$ project to actual metric circles on $\mathcal{T}$. We already confirmed that once we mod out Λ, these circles are tangent in the pattern of K and have mutually disjoint interiors. In other words, P is a circle packing for K in the intrinsic metric of the Riemann surface $\mathcal{T}$. This is precisely the first conclusion of Proposition 9.1.

9.3. The Proof

Now for the proof itself. It may save you from notation overload (and brain shutdown) to note that we have essentially parallel structures in three settings: combinatorial first, then circle packing, and finally, classical. I keep the notation consistent from one setting to the next – you will be seeing the same ideas develop in three incarnations!

Combinatorics: Let K be a triangulation of a multiply connected surface S. There is a canonical way to construct $\widetilde{K}$, a universal covering complex of K, the combinatorial analogue of building the plane out of translated copies of a fundamental parallelogram for the classical torus. We use the chains of faces and notions of homotopy introduced in Chap. 3.

Fix a base face f_0 of K and consider the collection $\mathcal{C}$ of chains $c = \{f_0, f_1, \cdots, f_n\}$. (Recall that in a chain, each face shares an edge with its predecessor.) For each chain c,

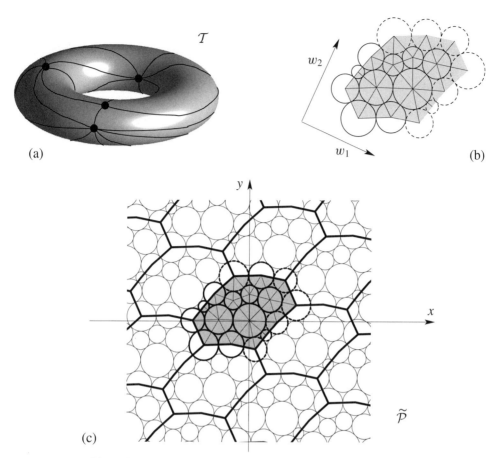

Figure 9.4. Geometric representation of a discrete conformal torus.

create an abstract copy f_c of the last face f_n of c and let L denote the disjoint union of these faces, $L = \bigcup_{c \in \mathcal{C}} f_c$. Denote by $\pi : L \longrightarrow K$ the projection that maps the face f_c for the chain $c = \{f_0, \cdots, f_n\}$ to the face f_n of K.

Define a new simplicial complex $\widetilde{L}$ by making two types of identifications, whenever appropriate, among the faces in L:

(1) Whenever a chain c' is obtained from a chain c by adding one additional face sharing an edge e with its last face, then $f_{c'}$ and f_c are attached to one another by identifying this common edge e.

(2) Whenever chains c' and c end at the same face of K and are homotopic to one another in K, then the faces $f_{c'}$ and f_c are identified. (Recall, our homotopies are fixed-endpoint; see Section 3.1.3)

Note that the projection π is well defined after these identifications, so π induces a projection $\pi : \tilde{L} \longrightarrow K$.

There are several details about $\tilde{L}$ that can be confirmed based on (1). First, that $\tilde{L}$ is a connected and locally planar simplicial complex and that π is locally one to one. Second, that because K is orientable, an orientation can be given to the faces of L which is preserved under the identifications. Together these facts imply that $\tilde{L}$ is (simplicially equivalent to) a triangulation of an orientable surface $\tilde{S}$.

The identifications of (2) guarantee that $\tilde{L}$ (and hence $\tilde{S}$) is simply connected. This is an exercise that I would not want to deny to the interested reader, but let me set up some of the machinery. Recall that homotopies between chains of faces were defined in Section 3.1.3 as finite sequences of local modifications that convert one chain into the other. You can learn to use the projection π to move between chains of faces in K and in $\tilde{L}$; every chain in $\tilde{L}$ projects to one in K, and conversely, every chain in K can be lifted to a chain in $\tilde{L}$ that starts at some designated lift of its starting face. You can now study how to carry homotopies from one setting to the other and you will see that the punchline is almost a tautology: if a "closed" chain in $\tilde{L}$ were not null homotopic there, then the first and last faces would not satisfy the very definition of being "the same."

The facts that $\tilde{L}$ is simply connected, π is locally one to one, and every chain of faces of K lifts to $\tilde{L}$ together imply that $\tilde{L}$ qualifies as the *universal covering complex* for K, with π as the covering projection. These objects are uniquely defined up to simplicial equivalence. We will refer to $\tilde{L}$ henceforth as $\tilde{K}$, with projection $\pi : \tilde{K} \longrightarrow K$.

The key to our work lies with the internal symmetries in $\tilde{K}$ that arise from the topological structure of K. Recall that the *fundamental group* $\pi_1(S)$ of a surface S is the group of equivalence classes, modulo homotopy, of closed paths starting and ending at some base point, the group operation being "concatenation." In our triangulated surface these equivalence classes can be represented by closed chains of faces starting and ending at f_0, with the null chain representing the identity element id of the group. $\pi_1(S)$ is isomorphic to a group of covering transformations of K. A *covering (or deck) transformation* $\phi : \tilde{K} \longrightarrow \tilde{K}$ is a (orientation-preserving) simplicial automorphism satisfying $\pi \circ \phi \equiv \pi$.

Each element $\gamma \in \pi_1(S)$ gives rise to a covering transformation ϕ_γ as follows. Let c_γ be a chain of faces of K representing γ. Because c_γ is closed, we can define a map from $\mathcal{C}$ to itself by replacing each chain c by the concatenation $c_\gamma \cdot c$. In other words, instead of walking from f_0 along c, you first walk around c_γ back to f_0, then walk along c. Observe that these chains end up at the same face of K, namely the last face of c. Since every chain is associated with a face of $\tilde{K}$, this induces a simplicial self-map ϕ_γ of $\tilde{K}$; this ϕ_γ is one-to-one and onto, and since concatenating with c_γ does not change the last face, ϕ_γ does not change the projection to K, that is $\pi \circ \phi_\gamma \equiv \phi_\gamma$.

The simplicial covering transformations of $\tilde{K}$ form a group Λ_K, and it is straightforward to verify from our description that the map $\gamma \mapsto \phi_\gamma$ is a group isomorphism between $\pi_1(S)$ and Λ_K. Since π is locally one-to-one, one can mod out the group and show that $\tilde{K}/\Lambda_K$ is a simplicial complex. One can then verify that the set of faces $\pi^{-1}(f_0)$ in $\tilde{K}$ is in one-to-one correspondence with Λ_K. To see this, note that each path $\gamma \in \pi_1(S)$, as a member of $\mathcal{C}$, determines a face $\tilde{f}_\gamma$ of $\tilde{K}$. If γ_1 and γ_2 determine the same face, then the concatenation of γ_1 with γ_2^{-1} must be null homotopic, meaning $\gamma_1 \circ \gamma_2^{-1} = id$ and hence $\gamma_1 = \gamma_2$. Similar reasoning applies to the preimages of any other face of K. The upshot is that $\widetilde{K}/\Lambda_K$ is simplicially equivalent to K.

This, then, is the combinatorial situation: Given a complex K, there exists a simply connected universal covering complex $\widetilde{K}$ with simplicial projection $\pi : \widetilde{K} \longrightarrow K$ and a group Λ_K of simplicial automorphisms ϕ of $\widetilde{K}$ satisfying $\pi \circ \phi \equiv \phi$ so that $\widetilde{K}/\Lambda_K \sim K$. If K triangulates a topological surface S, then $\widetilde{K}$ triangulates a simply connected surface $\widetilde{S}$ and, interpreting these objects topologically, $\widetilde{S}$ is the topological universal covering surface of S with covering group Λ_K and covering projection $\pi : \widetilde{S} \longrightarrow S$.

Note in wrapping up the combinatorial situation that because K is (by hypothesis) multiply connected, the elements of Λ_K have infinite order, so whether K itself is finite or infinite, the universal covering surface $\widetilde{K}$ is infinite.

Circle Packing: At this point in the book, when confronted with $\widetilde{K}$, I hope your first impulse is to "circle pack it!" Being simply connected, $\widetilde{K}$ is either parabolic or hyperbolic; write $\widetilde{\mathcal{P}}$ for a fixed maximal packing $\mathcal{P}_{\widetilde{K}}$ that is a univalent circle packing filling the geometric space $\mathbb{G}$, one of $\mathbb{C}$ or $\mathbb{D}$. Denoting the maximal label by $\widetilde{R}$, we have $\widetilde{\mathcal{P}} \longleftrightarrow \widetilde{K}(\widetilde{R})$.

Our first job is to exploit the essential uniqueness of $\widetilde{\mathcal{P}}$ to convert the purely combinatorial symmetries of $\widetilde{K}$ encoded in Λ_K to conformal symmetries of $\widetilde{\mathcal{P}}$. Write $\widetilde{\mathcal{P}} = \{c(v) : v \in \widetilde{K}\}$, so $c(v)$ is the circle of $\widetilde{\mathcal{P}}$ associated with vertex v. Fix $\phi \in \Lambda_K$ and define $Q = \{b(v) : v \in \widetilde{K}\}$, where $b(v) = c(\phi(v))$, $v \in \widetilde{K}$. Note that $v \sim u$ in $\widetilde{K}$ if and only if $\phi(v) \sim \phi(u)$ and $\phi : \widetilde{K} \longrightarrow \widetilde{K}$ is onto, so Q is a univalent circle packing for $\widetilde{K}$ filling $\mathbb{G}$. In particular, by the essential uniqueness of $\widetilde{\mathcal{P}}$ there exists a map $\lambda_\phi \in \text{Aut}(\mathbb{G})$ so that $Q = \lambda_\phi(\widetilde{\mathcal{P}})$.

There are now many observations to make. Given a vertex v, the definition of λ_ϕ implies that

$$\lambda_\phi(c(v)) = b(v) = c(\phi(v)), \tag{49}$$

which is the key relationship between the combinatorial and the conformal automorphisms. Suppose ψ is another element of Λ_K. Then for each v,

$$\lambda_\psi(\lambda_\phi(c(v))) = \lambda_\psi(c(\phi(v))) = c(\psi(\phi(v))) \implies \lambda_\psi \circ \lambda_\phi \equiv \lambda_{\psi \circ \phi}.$$

It is immediate that the collection of automorphisms

$$\Lambda = \{\lambda_\phi : \phi \in \Lambda_K\}$$

is a subgroup of $\text{Aut}(\mathbb{G})$. Moreover, one easily confirms that the map $\phi \mapsto \lambda_\phi$ is a group isomorphism $\Lambda_K \longrightarrow \Lambda$. By (49), each $\lambda_\phi \in \Lambda$ is a one-to-one map of $\widetilde{\mathcal{P}}$ onto itself – that is, $\widetilde{\mathcal{P}}$ is invariant under the automorphisms in Λ. Since $\widetilde{\mathcal{P}}$ fills $\mathbb{G}$, its centers cannot accumulate at an interior point of $\mathbb{G}$, hence the orbit under Λ cannot accumulate in the interior of $\mathbb{G}$. That implies that Λ is a properly discontinuous subgroup of $\text{Aut}(\mathbb{G})$.

With invariance we can now mod out the group Λ. For example, define $P = \widetilde{\mathcal{P}}/\Lambda$. This consists of "circles" in a (temporarily) abstract sense; each element represents an abstract equivalence class of circles of $\widetilde{\mathcal{P}}$. Of course, the carrier of $\widetilde{\mathcal{P}}$ is also invariant under Λ, and with (49) one easily confirms that carr($\widetilde{\mathcal{P}}$)/Λ is simplicially equivalent to K, so P is a circle packing for K. Likewise, the label $\widetilde{R}$ is invariant under Λ and we can define the label $R = \widetilde{R}/\Lambda$ for K. This is easily confirmed to be a packing label for K; given the combinatorial flower of some vertex v, laying out circles in $\mathbb{G}$ with the radii from R will result in a geometric flower. Using projection π in its several roles, we have $\pi(\widetilde{\mathcal{P}}) = P$, $\pi(\widetilde{K}) = K$, and $\pi(\widetilde{R}) = R$. Everything seems to be in place, and $P \longleftrightarrow K(R)$. The final hurdle is to convert the abstract "equivalence classes" composing P into legitimate circles in some Riemann surface.

Classical: The classical geometric surface supporting P is now easily at hand: Λ is a properly discontinuous subgroup of Aut($\mathbb{G}$), so $\mathcal{S}_K = \mathbb{G}/\Lambda$ is our Riemann surface.

Recall from classical covering theory that the projection $\pi : \mathbb{G} \longrightarrow \mathbb{G}/\Lambda$ carries the topological, metric, and conformal structure from $\mathbb{G}$ to $\mathcal{S}_K$. We now have, in addition, the invariant objects $\widetilde{\mathcal{P}}$, carr($\widetilde{\mathcal{P}}$), and $\widetilde{R}$. The metric of $\mathbb{G}$ projects to the intrinsic metric of $\mathcal{S}_K$ (this is, in fact, how we defined the intrinsic metric), so the circles of $\widetilde{\mathcal{P}}$ project to legitimate circles of $\mathcal{S}_K$ in this metric; carr($\widetilde{\mathcal{P}}$) projects to a triangulation of $\mathcal{S}_K$ that is simplicially equivalent to K. The collection P therefore represents a circle packing for K in $\mathcal{S}_K$ with carr(P) = π(carr($\widetilde{\mathcal{P}}$)). The packing label for P is just $R = \widetilde{R}/\Lambda$. In other words, $P = \pi(\widetilde{\mathcal{P}})$ is the circle packing for K which the Proposition claimed. We denote it by $\mathcal{P}_K$ and call it the *maximal packing* for K.

There are a few things to check in wrapping up the proof, but I leave the verifications to the reader. The univalence of $\widetilde{\mathcal{P}}$ implies the univalence of $\mathcal{P}_K$, and the fact that $\widetilde{\mathcal{P}}$ fills $\mathbb{G}$ implies that $\mathcal{P}_K$ fills $\mathcal{S}_K$. There are two parts to uniqueness. As to the uniqueness (up to conformal equivalence) of $\mathcal{S}_K$, if P' were a univalent circle packing for K filling some Riemann surface $\mathcal{S}'$, then P' would lift, in the reverse of the process we carried out above, to a univalent circle packing filling the universal covering space $\mathbb{G}'$ of $\mathcal{S}'$ and invariant under the covering group Λ'. This would be a maximal packing for the universal covering complex $\widetilde{K}$, so the essential uniqueness of $\widetilde{\mathcal{P}}$ would imply that $\mathbb{G}' = \mathbb{G}$ and that Λ' is conjugate to Λ in Aut($\mathbb{G}$), and hence that $\mathcal{S}'$ is conformally equivalent to $\mathcal{S}_K$. As to uniqueness of $\mathcal{P}_K$, suppose P' were another packing in $\mathcal{S}_K$; lifting as before to $\mathbb{G}$ gives a univalent packing Q' filling $\mathbb{G}$. The essential uniqueness of $\widetilde{\mathcal{P}}$ implies that there exist $\psi \in$ Aut($\mathbb{G}$) with $\psi(Q') = \widetilde{\mathcal{P}}$. The fact that the same holds for any lift of P' implies that $\psi \Lambda \psi^{-1} \equiv \Lambda$. This means ψ is in the *centralizer* of Λ, and ψ is associated with a conformal self-map, or automorphism, of $\mathcal{S}_K$. In other words, $\mathcal{P}_K$ is unique in $\mathcal{S}_K$ up to automorphisms of $\mathcal{S}_K$. This completes the proof of Proposition 9.1.   □

A comment about classification. The geometry of a maximal packing is *almost* fore-ordained by the topology of K. Namely, spherical geometry applies only when K is a combinatorial sphere; euclidean geometry, when K is compact and genus 1 and hence a combinatorial torus; hyperbolic geometry, when K has boundary or genus greater than 1. For infinite complexes, there is ambiguity only for combinatorial open discs or open

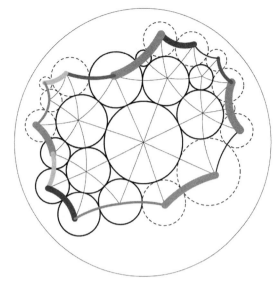

```
NODECOUNT: 12
GEOMETRY: hyperbolic
ALPHA/BETA/GAMMA: 2 12 12
FLOWERS:
1 10     10 9 11 12 8 7 5 6 3 4 10
2 8      4 5 3 6 7 11 8 10 4
3 6      5 7 4 1 6 2 5
4 7      1 3 7 6 5 2 10 1
5 6      3 2 4 6 1 7 3
6 6      4 7 2 3 1 5 4
7 10     1 8 9 10 11 2 6 4 3 5 1
8 7      9 7 1 12 10 2 11 9
9 5      10 7 8 11 1 10
10 8     11 7 9 1 4 2 8 12 11
11 7     9 8 2 7 10 12 1 9
12 4     1 11 10 8 1
END
```

Figure 9.5. A genus 2 example.

annuli, where the geometry depends on the parabolic/hyperbolic dichotomy for K or for $\widetilde{K}$, respectively; that is, in these cases the outcome depends on the combinatorics of K.

9.4. One Final Example

Now, back to reality with a concrete example needed in the sequel. I have generated from scratch a simple genus 2 complex having 12 vertices; I entered it in a file as shown on the left in Figure 9.5, read that file into `CirclePack`, and then issued three commands to max-pack, layout, and print the result. The packing defines a fundamental domain for $\mathcal{S}_K$ with the ghost circles and side-pairings displayed, and, in fact, `CirclePack` supplies (approximations to) the side-pairing Möbius transformations that generate the covering group Λ.

We will be putting Riemann surfaces aside until we get to their function theory in Chap. 16. However, I wanted to leave you with this example because the process is so slick, so refreshing. `CirclePack` works directly on K, no universal cover, no covering group.

Just as rocks do not analyze gravity as they fall, so circles do not think about theory as they pack.

All the involved machinations of the formal proof melt away – the circle packings just "spontaneously" pop up on your computer screen. It is great fun.

Practicum II

Let us peek behind the curtain again for more of the practical side of circle packing. My initial interest rested with the packing algorithm: How does a circle packing morph into its maximal packing? Early fumbling with computer code led to the incremental approach that has become the theoretical base for the topic – the various monotonicities, subpackings, boundary adjustments, and so forth. Fascinating as the packing algorithm is, the most technically demanding part of the coding turns out to be tracking the combinatorics.

A case in point is the task of packing layout. The hint at the end of Practicum I concerning "face-based" layouts foreshadowed our proof of the Monodromy Theorem (the reader may want to revisit Section 5.2) and evolved into an algorithm that is now implemented in software. CirclePack creates a "drawing order" $\mathcal{D}$, which is simply an ordered list of faces $\mathcal{D} = \{f_0, \cdots, f_k\}$. The base face is f_0, every face f_j shares an edge with f_i for some $i < j$, and every vertex belongs to at least one of the faces. A layout is generated by placing f_0 and then iterating through $\mathcal{D}$ in succession to place the remaining circles. Here is meta-code for construction of $\mathcal{D}$ using a companion closed list $\mathcal{L}$.

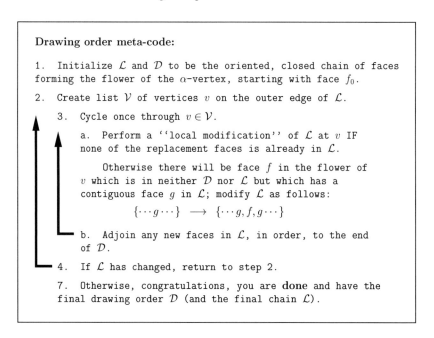

Drawing order meta-code:

1. Initialize $\mathcal{L}$ and $\mathcal{D}$ to be the oriented, closed chain of faces forming the flower of the α-vertex, starting with face f_0.

2. Create list $\mathcal{V}$ of vertices v on the outer edge of $\mathcal{L}$.

3. Cycle once through $v \in \mathcal{V}$.

  a. Perform a ''local modification'' of $\mathcal{L}$ at v IF none of the replacement faces is already in $\mathcal{L}$.

    Otherwise there will be face f in the flower of v which is in neither $\mathcal{D}$ nor $\mathcal{L}$ but which has a contiguous face g in $\mathcal{L}$; modify $\mathcal{L}$ as follows:

    $$\{\cdots g \cdots\} \longrightarrow \{\cdots g, f, g \cdots\}$$

  b. Adjoin any new faces in $\mathcal{L}$, in order, to the end of $\mathcal{D}$.

4. If $\mathcal{L}$ has changed, return to step 2.

7. Otherwise, congratulations, you are **done** and have the final drawing order $\mathcal{D}$ (and the final chain $\mathcal{L}$).

The closed list $\mathcal{L}$ continually morphs during construction, with local modifications pushing it outward until it has eaten its way through all the faces of K as it helps builds $\mathcal{D}$. Why make this business so complicated? The naive approach of Practicum I is, in fact, fine for simply connected complexes – by monodromy, there is only one possible layout. Try your hand at a non–simply connected complex, however, and you very quickly learn that a poorly conceived drawing order will paint itself into a corner, leaving an incomplete and/or inconsistent layout. The algorithm in the meta-code has evolved over time so that it can now handle any finite complex. Let us see the payoff with a small non–simply connected example. (This can be time well spent for the reader new to Riemann surfaces – the juxtaposition of the mathematics of covering surfaces and the dictates of coding can bring subtleties into sharp focus.)

Consider how the circle packing for the genus 2 surface K is laid out in the disc in Figure 9.5. The data file there (and the earlier instance in Figure 9.1) shows how I encode complexes for CirclePack. (No proprietary formats or invisible control codes here!) Each line following the keyword "FLOWERS:" lists a circle index, the number of faces to which it belongs, and its flower of petal vertices listed in

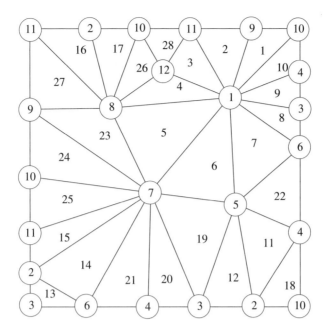

Figure II.1. Schematic of vertex and face indices.

counterclockwise order (all these flowers are closed). Face indices are generated only as a packing is read in or modified; however, the schematic of K in Figure II.1 sets the face indices for our run.

You can see the meta-code in action in the sequence of snapshots in the figure. The meta-code itself, of course, is purely combinatorial, but I have computed the hyperbolic packing label $\mathcal{R}_K$ and have used its radii to endow each stage with its correct geometry. Stage (a) is the initial closed path $\mathcal{L}$ of faces around α-vertex #2; the base face $f_0 = 18$ is shaded. Snapshots (b)–(g) show the situation after each completion of Step 3 in the meta-code. The final image displays the fundamental domain in $\mathbb{D}$ with the face indices; you might refer to the schematic to verify some of the indicated side-pairings.

The dashed closed loop in each snapshot runs through successive faces of the current chain $\mathcal{L}$. The meta-code drives $\mathcal{L}$ to expand until it runs into another part of itself. (For example, after reaching (c), you cannot do a local modification at #3 because face 20, which is next to 19, is already in $\mathcal{L}$; the code moves instead to the local modification at #11, yielding (d).) As a consequence, $\mathcal{L}$ always surrounds a simply connected collection of faces, so there is never any ambiguity about its circle locations. $\mathcal{L}$ finally squeezes into every nook of the complex until it is pushing against itself along its full length – at this point, we are done, as the layout has taken account of all faces.

I emphasize the fact that face-based layouts are not a convenience – they are crucial. Looking at the stages, note the "ghosts" – the dashed circles. These were used for laying out a face, but were later required in a new spot for laying out another face. (I have marked circle #1 with a heavy dot; it was moved three times!) The ghost circles (and edges) are precisely the mechanism for estimating the side-pairing Möbius transformations of the fundamental domain, which provide concrete information about the geometry of the Riemann surface $\mathcal{S}_K$ supporting the circle packing.

Some final practical details: The faces that result from the "otherwise" option of Step 3a in the layout meta-code can be especially aggravating in the actual coding, but the resulting redundancies in $\mathcal{L}$ (such as the $\{\cdots, 9, 8, 9, \cdots\}$ that you can see at stage (g)) are necessary in general so that the faces are laid out unambiguously. CirclePack keeps both the lists $\mathcal{D}$ and $\mathcal{L}$. Starting with f_0, it lays out faces from $\mathcal{D}$ in order, as needed, until it first encounters a face in $\mathcal{L}$. It next lays out all faces of $\mathcal{L}$, encircling the fundamental domain. Then it returns to $\mathcal{D}$ and uses faces in order, as needed, to locate all remaining

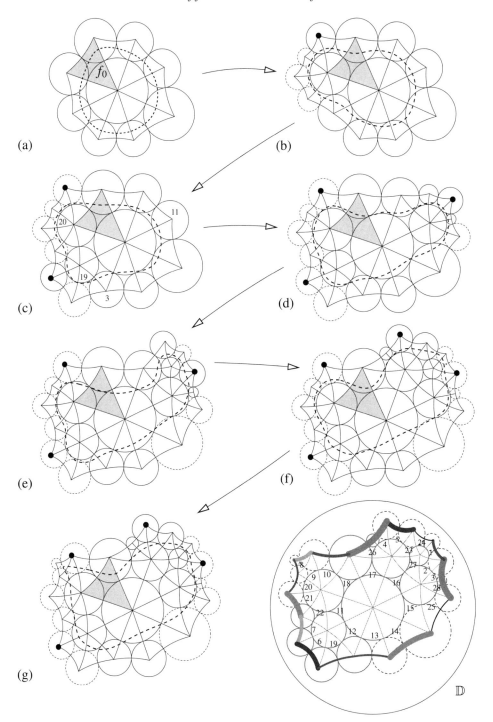

Figure II.2. Layout meta-code in action.

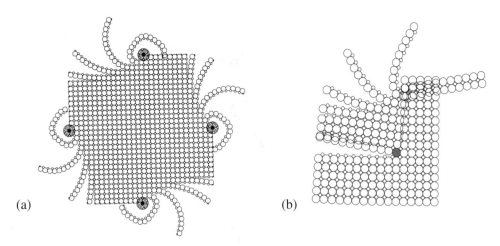

(a) (b)

Figure II.3. Layout breakups.

circles. There are many additional twists and turns in the actual coding, but with due diligence, it all works. (Much of this diligence is due to my Master's student Woodrow Johnson.)

Let me close Practicum II with two "broken" packings. Image (a) of Figure II.2 results from a (short-lived) programming mistake – it is indeed a nice topic when even the errors look this pretty. In layout (b), on the other hand, the code is doing things just the way we intend. It is easy to forget that we are using *computer* arithmetic; the breakup here is very common for larger packings due to inevitable roundoff errors. In (b) I introduced a 10% error in the shaded circle. This results in a gap between two circles that are supposed to be tangent; then a new circle tries to locate itself; using those two; another uses those corrupted results; etc. Depending on the vagaries of indexing and drawing order, crevasses appear and grow, the new circles like ferry passengers having one foot on the pier and one on the departing ferry.

The larger the packing, the more likely such horrible looking breakups become. One's first impulse is to decrease the error tolerance ϵ in the packing algorithm. However, the extra computation can be expensive. There are some alternatives, however; we will be revisiting the practices of repacking and layout in Practicum IV for industrial size packings.

Notes II

Chap. 3. Preliminary material can be found in numerous books. For example, see Massey (1997) for topology, triangulations, fundamental groups; Singer and Thorpe (1967) and Stillwell (1992) for simplicial complexes, surfaces, and covering spaces; Jones and Singerman (1981) and Beardon (1983) for hyperbolic geometry; and Ahlfors and Sario (1960) and Beardon (1984) for Riemann surfaces.

Chap. 4. Maximal packings were addressed first on the sphere (Andreev, 1970, 1970b; Koebe, 1936; Thurston) and then in the disc (Thurston, 1985); these were known as "Andreev" packings in the early literature (see Notes I). The term "maximal" packing came with proof of uniformization (Theorem 4.3) by Beardon and Stephenson (1990) (under a bounded degree assumption).

Chap. 6. Thurston (1985) observed Proposition 6.1 with an argument reverse to that of Chap. 7; the approach via hyperbolic geometry, with associated monotonicities, Perron arguments, and monodromy, is from Beardon and Stephenson (1991). Compare the subpackings to *subharmonic* functions in classical Perron arguments; see, e.g., Helms (1975).

Chap. 8. For the Ring Lemma, see Appendix B. The parabolic/hyperbolic dichotomy was established in Beardon and Stephenson (1991) and hyperbolic uniqueness in Rodin (1991), but bounded degree and quasiconformality were required. The keys to eliminating degree conditions and the heavy machinery of quasiconformality lie principally in the beautiful ideas of Schramm and Zheng-Xu He; see especially (He and Schramm, 1993) for winding number arguments and the proof of completeness. Lemma 8.8 and the elementary general argument behind Section 8.3 are found in Dubejko and Stephenson (1995).

Chap. 9. Packings in hyperbolic surfaces are mentioned in some versions of Thurston's *Notes* (Chap. 13); the covering theory approach and Proposition 9.1 are in Beardon and Stephenson (1990).

Part III

Flexibility: Analytic Functions

The maximal packings of Part II display superrigidity – combinatorics alone determine these packings up to Möbius transformations. This would seem to be at odds with the limitless variability of packings displayed in the Menagerie. The explanation, of course, lies with the univalence and completeness conditions placed on maximal packings; remove these and there is a wealth of new possibilities to study. Vestiges of rigidity remain, however, in a local form, and that variability-within-rigidity, the local-to-global transition, is what we study here in Part III.

The approach we take in studying a general packing is to compare it to the maximal packing having the same combinatorics. In particular, there is a natural identification of the circles in one with the corresponding circles in the other, which on first impulse might be called a circle packing map. However, the local rigidity of circle packings manifests itself in numerous and surprising ways, forcing behavior that strongly mirrors that of classical analytic maps. In fact, it is my contention that their behavior is tantamount to discrete analyticity, so I have chosen to describe the mappings as "discrete analytic functions."

The classical theory thus gives the context for developments in this part of the book. We first establish the basic properties of packings in a function-theoretic form, and then proceed, following the classical model, to study discrete analytic functions on the disc, discrete entire functions, discrete rational and meromorphic functions, and finally discrete analytic functions on Riemann surfaces. Do not be concerned if you are not familiar with the classical theory of analytic functions, for although that is a source of motivation and a general context, the discrete theory will stand perfectly well on its own – one feels, in fact, that the fundamental geometry emerges with new clarity and in a natural and intuitive way. In this part we also establish the foundations for discrete conformal structure and discuss connections with random walks on circle packings.

131

10

The Intuitive Landscape

Analytic functions are known to almost all mathematicians in one form or another. Many of us start with the familiar complex polynomials and rational functions, then perhaps complex versions of the trigonometric functions, and certainly the complex exponential. It might be in trying to understand the logarithm that we first realize there is something quite fascinating going on here. Then as we travel our separate paths we may encounter analytic functions again in the elliptic curves of number theory, or in conformal mapping and fluid flow, in the geometry of surfaces, operator theory, or potential theory and partial differential equations – we certainly encounter the footprints of analytic theory, whether in Fourier analysis, differential geometry, integral formulas, or the notion of manifolds. For engineers or physicists it might be in electrostatics, fluid flow, or wave functions.

For a topic to gain such ubiquity, it must have some fundamental and simple principle at its heart. This is the case with analytic functions. From garden-variety complex polynomials to painfully obscure special classes, all analytic functions are defined by one simple and entirely local condition. If Ω is an open subset of the complex plane and F is a complex (i.e., complex-valued) function on Ω, then F is an *analytic function* if for each $z \in \Omega$, F has a (complex) derivative at z; that is, the following limit exists as a complex number:

$$F'(z) = \lim_{h \to 0} \frac{F(z+h) - F(z)}{h}. \tag{50}$$

The study of analytic functions largely concerns what I call the *local-to-global* transition – finding the global implications of the locally defined derivative.

Circle packings are defined by an equally simple local condition, namely the *packing condition* on flowers of tangent circles. In our established notation, given a label R, the labels for the flower of an interior vertex v must satisfy, for some positive integer n, the angle sum condition

$$\theta_R(v) = \sum_{\langle v,u,w \rangle} \alpha(R(v); R(u), R(w)) = 2\pi n.$$

Taking this local packing condition as our sole imperative and painting with the broadest brush first, we have this intuitive definition.

Definition 10.1. A **discrete analytic function** is a map $f : Q \longrightarrow P$ between circle packings Q and P that preserves tangency and orientation.

It should be clear that we are to take a very geometric view of functions: functions are *mappings* carrying points from one setting to another. Our study of discrete analytic functions revolves around the global geometric mapping consequences associated with the local packing condition, and our aim in the remainder of the book is to uncover the connections between the classical models and these discrete manifestations – connections going far beyond mere analogy. Of course, I first alluded to these in Section 2.2.2, but with more circle packing under our belts, many readers will already have begun the recognition process. I briefly recap the classical theory in Appendix A; among the numerous books on analytic functions, I would recommend especially Ahlfors (1978), Beardon (1979), and Needham (1997) for their geometric views.

There is little loss in generality in narrowing our context slightly. In the sequel, for functions $f : Q \longrightarrow P$ it will be assumed that the domain packing Q is univalent and that Q and P share the *same* complex K: that is, for each vertex $v \in K$, f identifies the circle for v in P with the circle for v in Q.

10.1. Think Geometry!

Most of us think about analytic functions (if at all) in terms of the *arithmetic* properties of the complex numbers; hence we have power series, integrals, and limit processes, and the functions we work with can be added, multiplied, and composed. You can forget much of that here since arithmetic is not available. Fortunately, analytic functions reflect equally profound *geometric* properties of the complex numbers, and it is those that we will see in the discrete setting.

Local geometry, of course, begins with the derivative. Existence of the complex derivative in (50) is a much stronger statement than existence of an ordinary derivative on the real line because h can approach 0 from any direction. Treat h as a small change in z, $h = \Delta z$. By (50), the change in F satisfies

$$\Delta F = F(z + \Delta z) - F(z) \approx F'(z)\Delta z,$$

so the change in F is roughly the change in z rotated by $\arg(F'(z))$ and scaled by factor $|F'(z)|$. As this action is independent of direction, a circle c of small radius r centered at z is carried to a curve $F(c)$ which approximates a circle of radius $|F'(z)|r$ centered at $F(z)$. Moreover, a point moving counterclockwise on c has image traveling counterclockwise around $F(c)$. This geometry is captured in one of my favorite classical sayings: *an analytic function is one which maps infinitesimal circles to infinitesimal circles.*

Over on the discrete side, the domain for a discrete analytic function $f : Q \longrightarrow P$ is discretized as a pattern Q of circles rather than points and each is carried to an *actual* circle in the image P. The requirement that f preserve tangencies is simply the discrete version of *continuity*, and the requirement that it preserve orientation means that petal

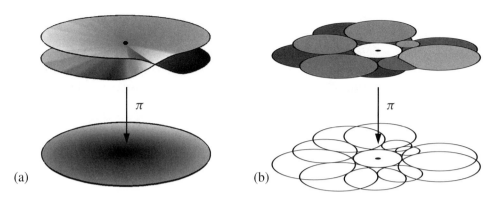

Figure 10.1. Classical and discrete branching of order 1.

circles wrapping counterclockwise about a circle c in Q have image circles wrapping counterclockwise around $f(c)$ in P. We do not have a discrete derivative per se, but the *ratio* function $f^{\#}$ is defined for $c \in Q$ by $f^{\#}(c) = $ radius $(f(c))/$radius (c) and plays the role of $|F'(z)|$ – it is the scale factor for the radius in going from c to $f(c)$.

Classical mapping behavior is a little more subtle at branch points, that is, points z with $F'(z) = 0$. In this case, a small circle c about z has image $F(c)$ wrapping $n \geq 2$ times about $F(z)$, where $n - 1$ is the *order* of branching. Moreover, the image shrinks drastically as c gets small because $|F'(z)| = 0$. Locally, a small neighborhood of z gets mapped n-to-one onto a much smaller neighborhood of $F(z)$. Figure 10.1(a) illustrates the image surface for a branch point of order 1 ($n = 2$) in a style one might see in the classical literature: this image cannot actually be embedded in 3-space since the two discs are "cross-connected" at their centers, so it is displayed as though they pass through one another.

On the discrete side $f^{\#}$ can, of course, never vanish since radii in packings are never zero. However, if the flower of some circle $c \in Q$ has petals whose image circles wrap $n \geq 2$ times around $f(c)$, then the local geometric behavior is very much the same and we call c a *branch circle* of order $n - 1$. Figure 10.1(b) suggests the wrapping of the circles as though on a two-sheeted surface. Note that although $f^{\#}(c)$ does not vanish, the center circle must decrease in size from the univalent flower for its petals to reach twice around it.

To sum up the local situation: *classical analytic functions map infinitesimal circles to infinitesimal circles; discrete analytic functions map real circles to real circles.*

10.2. Fundamentals

With this initial analogy in hand, we can move to global issues. We spend Chap. 11 making various definitions mathematically precise, extending existence results to get a larger variety of packings, and confirming fundamental mapping properties of our discrete functions.

It is important to start teasing apart the combinatorics, topology, and geometry you will be seeing. On the combinatorics front there are discretization issues to settle, such as necessary and sufficient conditions on branch sets, and various and sundry bits of terminology that need interpretation, things like *boundary conditions* and notions of *convergence*. Here, for example, is a minor additional condition that we will henceforth assume for all our complexes K.

Standing Assumptions. *(1) The set of interior vertices of K is nonempty and edge-connected. (2) Every boundary vertex of K has at least one interior vertex among its neighbors.*

These are needed for certain uniqueness statements and are completely natural; with them, K is the discrete analogue of a "domain" in $\mathbb{C}$, namely, it has nonempty and connected interior and every boundary point is in the closure of the interior.

After combinatorics comes topological considerations. Complexes are topological spaces, and we will shortly be refining the definition of discrete analyticity so that our functions are topological mappings. In fact, they become oriented *light interior* mappings, and therefore by a famous result of Stoïlow they are *topologically* equivalent to analytic functions. Many notions commonly associated with analyticity – the argument principle, monodromy, the Riemann–Hurwitz formula, and large parts of covering theory – are available to us because they are in fact topological notions.

Mixed in this combinatoric/topological stew are the truly *geometric* properties that come with the circle packing condition. This is where we look for behaviors associated with analyticity. In the next chapter, for example, we will see maximum principles, monotonicity results, and normal families arguments that depend on the geometry. A paradigm for circle packing effects is provided by complexes $K^{[7]}$ and $K^{[6]}$: each triangulates an open disc, but imposing *geometries* via circle packing, we have proven that one becomes hyperbolic, the other euclidean – one supports bounded discrete analytic functions, the other does not. Keep your eyes open for these circle packing effects as distinct from the combinatorics and topology.

10.3. Discrete Analytic Functions

One of the things I love about complex analysis is its great nomenclature – classes of functions, named lemmas and theorems, named techniques. We will take full advantage of that here. We start as the classical theory does with functions on the standard domains, $\mathbb{D}, \mathbb{C}, \mathbb{P}$, and then on Riemann surfaces. For us this means we study discrete analytic functions of the form $f : \mathcal{P}_K \longrightarrow P$, and we take the domains in turn in Chaps. 13 to 16. So you will see *rational* functions mapping the sphere to itself, *meromorphic* functions mapping to the sphere, *entire* functions with the plane as domain and range, "self-maps" of the disc and so forth.

We begin each chapter with the pieces of the discrete theory as we know them, and we then describe explicit examples of key classes of functions – again, to the extent that examples are known. Nearly every construction involves one or more of the specialized construction tools that I have gathered ahead of time in Chap. 12 – cutting/pasting, doubling, welding, refinement, and so on. In the disc we will see Thurston's Riemann mappings and other univalent functions, finite Blaschke products – my personal favorites – disc algebra functions, and others. In the euclidean plane, we have polynomials, exponentials, and sines/cosines. The sphere and Riemann surfaces are tough environments, depending as they do largely on branching; we have a few examples and experimental approaches to others.

There are many open questions – the classical theory tells us what *should* be true, perhaps we can even see the behavior in tailored examples, but the general theory is often still open. For example, we will see the Liouville Theorem for general entire functions, but the deeper Picard theorem is known only in the hexagonal case. I will try to note the gaps, questions, and opportunities as we go along.

10.4. Discrete Conformal Structures

Chapter 17 changes character. It is based on a second local property of classical analytic functions associated with (50), a dual to the *infinitesimal circle* intuition. Namely, when $F'(z)$ is not zero, then F is a *conformal* mapping at z, meaning that it preserves angles (magnitude and orientation) between curves intersecting at z. Many of the more directly geometric features of classical function theory occur in the context of these so-called *conformal* mappings. This geometry is again reflected discretely by maps between circle packings. Here, however, the typical mapping $f : Q \longrightarrow P$ is one between univalent packings Q and P in different settings, as suggested in Figure 10.2.

In Chap. 17 we will formulate the notion of *discrete conformal structure* and describe *discrete conformal mappings*. It will be important to keep in mind that these are *not*

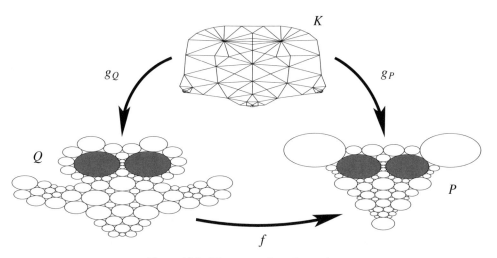

Figure 10.2. Discrete conformal mapping.

conformal in the strict sense – local angles are *not* preserved. Rather it is the *ensemble* and *global* properties associated with this local behavior that carry over. We will see discrete versions of several key concepts, such as conformal structure, extremal length, moduli of rings, and type conditions. We end Part III by introducing random walks, probabilistic methods, and their links to harmonic functions; these are emerging and target-rich areas.

11

Discrete Analytic Functions

In the next few chapters we will be working with discrete analytic functions of the form $f : \mathcal{P}_K \longrightarrow P$, where the maximal packing $\mathcal{P}_K$ packs $\mathbb{P}$, $\mathbb{C}$, $\mathbb{D}$, or the Riemann surface $\mathcal{S}_K$. The image packing P, while having the same combinatorics as $\mathcal{P}_K$, may live in another of the standard spaces and may be branched and nonunivalent.

11.1. Formal Definitions

Although a discrete analytic function f is morally a map between circle packings, the operative definition for f will be as a topological mapping between their carriers.

Definition 11.1. A **discrete analytic function** f from $\mathcal{P}_K$ to P is a continuous orientation-preserving simplicial mapping $f : \text{carr}(\mathcal{P}_K) \longrightarrow \text{carr}(P)$. In particular, it maps the center of each circle of $\mathcal{P}_K$ to the center of the corresponding circle of P, each edge of $\text{carr}(\mathcal{P}_K)$ one-to-one onto the corresponding edge of $\text{carr}(P)$, and each face of $\text{carr}(\mathcal{P}_K)$ one-to-one in an orientation-preserving way onto the corresponding face of $\text{carr}(P)$. The associated **ratio function** $f^{\#}$ is defined on the vertices of $\text{carr}(\mathcal{P}_K)$: if C_v is a circle of $\mathcal{P}_K$ and c_v is the corresponding circle of P, then at the center Z_v of C_v,

$$f^{\#}(Z_v) = \frac{\text{radius}(c_v)}{\text{radius}(C_v)}.$$

We will be liberal in our abuse of notation. We continue to write $f : \mathcal{P}_K \longrightarrow P$, and if v is a vertex of K, we may write any of $f(v)$, $f(C_v)$, $f(Z_v)$ depending on the situation. We may refer to $f(C_v)$ as the circle c_v of P, even though as a *point set* $f(C_v)$ is not necessarily an actual circle. If the circles for the same vertex are centered at the origin in both $\mathcal{P}_K$ and P, we will write $f(0) = 0$; if they have the same radius we write $f^{\#}(0) = 1$. Here are several other observations.

(1) The definition requires that f map the center Z_v of each circle of $\mathcal{P}_K$ to the center z_v of the corresponding circle of P, but the extension of f to the rest of $\text{carr}(\mathcal{P}_K)$ is largely a matter of convenience. Often one can use what are known as *barycentric coordinates*; we will define this when the need arises.

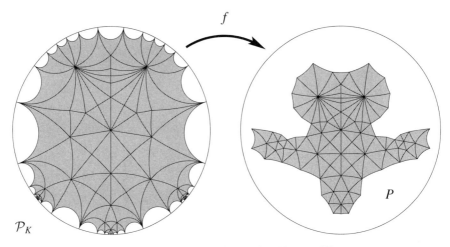

Figure 11.1. A univalent analytic self-map of $\mathbb{D}$.

(2) If K is a topological disc with boundary, then carr($\mathcal{P}_K$) has boundary vertices lying in the ideal boundary of $\mathbb{D}$ and it omits hyperbolic half planes defined by boundary edges (see Figure 11.1). Despite this detail, we refer to f as having domain $\mathbb{D}$.

(3) Unless stated otherwise, the metric used with each packing (in particular, for its label) is the intrinsic metric (i.e., curvature 0, 1, or -1) for the space in which it resides. In some cases we will use an *ambient* metric inherited from a larger space, as when we use the euclidean metric for packings in the unit disc.

(4) Adhering to our usual notation, $\mathcal{P}_K \longleftrightarrow K(\mathcal{R}_K)$ and $P \longleftrightarrow K(R)$, the ratio function is $f^{\#}(Z_v) = R(v)/\mathcal{R}_K(v)$. In particular, its value depends on the geometries of the labels, not directly on the packings. Although $f^{\#}$ is defined at circle centers, it is convenient for certain uses to extend it to the carrier.

(5) $f : \mathcal{P}_K \longrightarrow P$ is said to be an *automorphism* (or to be *Möbius*) if P is an automorphic image of $\mathcal{P}_K$. There is a bit of subtlety here if P and Q have spherical labels: $\phi \in \mathrm{Aut}(\mathbb{P})$ will typically not map centers to centers, so $\phi \circ f$ and $g : \mathcal{P}_K \longrightarrow \phi(P)$ will generally differ as point maps.

(6) There is no natural discrete parallel for the notion of a *constant* analytic function.

11.2. Standard Mapping Properties

Defining our functions as orientation-preserving simplicial maps means that certain mapping properties hold automatically. Thus, discrete analytic functions are continuous, open, and discrete (i.e., *light interior*) mappings, meaning that the preimage of any point is a *discrete* set, a set with no limit point in the interior of the domain carrier. Stoïlow's Theorem tells us that such maps are topologically equivalent to analytic functions.

Theorem 11.2 (Stoïlow's Theorem). *Let $f : D_1 \longrightarrow D_2$ be an orientation-preserving light interior mapping between two domains on the sphere $\mathbb{P}$. Then there exist domains $D_1', D_2' \subset \mathbb{P}$, homeomorphisms $h_1 : D_1 \longrightarrow D_1'$ and $h_2 : D_2 \longrightarrow D_2'$, and analytic function $F : D_1' \longrightarrow D_2'$ so that $f \equiv h_2^{-1} \circ F \circ h_1$.*

From this, for example, we can see that the modulus $|f|$ of a discrete analytic function can have no interior local maxima nor nonzero interior local minima. Also, each point z in the carrier of the domain has a neighborhood in which f will be either (locally) one-to-one or n-to-one for some integer $n > 1$.

A particularly important property to note is the *argument principle*. This originated in complex analysis and we could appropriate it via Stoïlow's Theorem. However, it is elementary in our simplicial setting. Recall from Section 8.3.1 that you can determine if a point ξ lies inside a triangle T by standing at ξ holding an elastic string attached to a friend who walks counterclockwise around the boundary curve ∂T. You are inside T if and only if the string winds once around you, that is, if and only if $\mathfrak{w}(g; \partial T) = 1$, the winding number for $g : z \mapsto z - \xi$ (see Definition 8.9).

Now, suppose K is a combinatorial closed disc and $f : \mathcal{P}_K \longrightarrow P$ is a discrete analytic function with image packing in $\mathbb{C}$. In how many faces of carr(P) does ξ lie? We simple sum the winding numbers of all the faces: $\sum_{T \in \mathrm{carr}(P)} \mathfrak{w}(g; \partial T)$. Of course, this sum collapses with cancellations due to triangles T sharing edges, and the final result is just $\mathfrak{w}(g; \Gamma)$, where Γ is the boundary curve of carr(P). Minor adjustments are left for the reader regarding points ξ on interior edges, but the upshot is this:

Lemma 11.3 (Discrete Argument Principle). *Suppose K is a combinatorial closed disc, $f : \mathcal{P}_K \longrightarrow P$ is a discrete analytic function mapping to the plane, and Γ is the boundary curve of* carr(P). *If $\xi \in \mathbb{C} \backslash \Gamma$, then the number of times, counting multiplicities, that ξ is assumed as a value by f on* carr$(\mathcal{P}_K)$ *is given by*

$$\mathrm{Card}\,(f^{-1}(\xi)) = \mathfrak{w}(g; \Gamma). \tag{51}$$

11.3. Branching

Discrete analytic functions defined using locally univalent circle packings can mimic only locally univalent analytic functions. To go after the broader theory we need branching. Here is the terminology and notation.

Recall from Definition 5.3 that in a *packing label R* for a complex K, every interior vertex v has angle sum $\theta_R(v) = 2\pi(\beta_v + 1)$ for some integer $\beta_v \geq 0$, called the *order* of branching. An interior vertex v with $\beta_v > 0$ is called a *branch vertex* of order β_v. The *branch set $br(R)$* is the set of branch vertices, $\{b_1, b_2, \ldots\}$, each repeated according to its order. If $br(R)$ is nonempty then R is *branched*; otherwise it is *unbranched* or *locally univalent*. (We attach the same terms/notations to the circle packings P themselves.)

The notion of branch point is generally associated with functions, and that is the case here, for if $f : \mathcal{P}_K \longrightarrow P$, then we may associate the branching of P with f. Thus $\mathrm{br}(P) = \mathrm{br}(f)$ and a branch vertex v is associated with the *branch point* Z_v of f, the center of $C_v \in \mathcal{P}_K$. The center z_v of the image circle $c_v \in P$ is called the branch *value*, and f maps a neighborhood of Z_v m-to-one onto a neighborhood of z_v, where $m = \beta_v + 1$ is called the *multiplicity* of the branch point. We will abuse notation, using branch vertex, branch point, and branch circle interchangeably. In any case, note that branch *points* are in the domain, while branch *values* are in the range.

I should remind the reader that the Monodromy Theorem, 5.4, applies to packing labels, whether branched or unbranched: *If K is simply connected and R is packing label for K, then there exists an essentially unique circle packing $P \longleftrightarrow K(R)$ in the geometry associated with the label.*

In branching we find a major discretization effect. Angles in the faces of carrier triangles must be less than π, so at least $2m + 1$ faces are required in a flower if its petals are to wrap m times about its center. In other words, $2\beta_v + 3 \leq \deg(v)$. When one moves beyond single flowers, conditions become more involved; here is a definition.

Definition 11.4. Given a combinatorial closed or open disc K, a set $\beta = \{b_1, b_2, \ldots\}$ of interior vertices of K, including possible repetitions, is said to be a **branch structure** (for K) if the following condition holds: for each simple, closed, positively oriented edge-path $\gamma = \{e_1, e_2, \ldots, e_k\}$ in K the inequality $k > (2N + 2)$ holds, where N is the number of points of β inside γ, counting repetitions.

The crucial first step in the study of branched circle packings is this necessary condition on branch sets, proved independently by Tomasz Dubejko and Phil Bowers; this is Dubejko's formulation.

Theorem 11.5. *Let K be a combinatorial closed or open disc, $\beta = \{b_1, b_2, \ldots\}$ a list of interior vertices of K, possibly with finite repetitions. If there exists a circle packing P for K with br $(P) = \beta$ in the euclidean or hyperbolic plane, then β is a branch structure for K.*

Proof. Assume that P is a euclidean or hyperbolic circle packing for K with $\mathrm{br}(P) = \beta$. Given a simple closed positively oriented edge-path γ in K, let L be the simply connected simplicial subcomplex of K formed by γ and its interior, so γ is the boundary of L. Let k be the count of edges in γ and N the total of branch points of P interior to γ. The claim is that $k > 2N + 2$.

Suppose F, E, V denote the number of faces, edges, and vertices of L; furthermore, write $V = V_\partial + V_o$, where V_∂ and V_o are the numbers of boundary and interior vertices, respectively. Of course, $\partial L = \gamma$, so $V_\partial = k$. Observe that

$$E = (3F - k)/2 + k = (3F + k)/2. \tag{52}$$

This results from counting the interior edges face by face: three edges per face minus the number of boundary edges. This double counts; divide by 2 and add back the number of boundary edges to get E. The Euler characteristic of L is 1, so we have

$$\chi(L) = V - E + F = V_{\text{int}} + k - (3F/2 + k/2) + F = 1 \tag{53}$$
$$\Longrightarrow V_{\text{int}} + k/2 - F/2 = 1$$
$$\Longrightarrow F\pi - (k-2)\pi - 2V_o\pi = 0.$$

In case $k \leq (2N+2)$, we conclude that

$$F\pi - 2N\pi - 2V_o\pi \leq 0. \tag{54}$$

Denote by Q the restriction of P to L. We may assume Q is hyperbolic, shrinking into $\mathbb{D}$ with a scaling in the euclidean case, if necessary (this does not affect the interior angle sums). We arrive at a contradiction by computing area, working face by face, as we have done before (see page 68). Summing over all F faces and reorganizing their angles into angle sums of the vertices we get

$$\text{Area}(Q) = F\pi - \sum_{v \in L} \theta(v) = F\pi - \sum_{v \in \text{int}(L)} \theta(v) - \sum_{w \in \partial L} \theta(w).$$

The total of interior angle sums is $(V_o + N)2\pi$, giving

$$\text{Area}(Q) = F\pi - 2N\pi - 2V_o\pi - \sum_{w \in \partial L} \theta(w).$$

Since area is positive and $\theta(w) \geq 0$ for $w \in \partial L$, we obtain

$$F\pi - 2N\pi - 2V_o\pi > \sum_{w \in \partial L} \theta(w) \geq 0,$$

contradicting (54). We conclude that $k > 2N + 2$ and hence that $\text{br}(P)$ is a branch structure. $\qquad\square$

11.4. Boundary Value Problems

Fundamental existence results were established in Part II, but those depended on the extremal nature of the packings involved. Now we need methods which will ensure a wider selection of packings. The flexibility we have may be something of a surprise considering the rigidity we have encountered up until now. This is a type of boundary value problem – with branching.

Theorem 11.6. *Let K be a combinatorial closed disc, $\beta = \{b_1, \ldots, b_m\}$ a branch structure for K, $g : \partial K \longrightarrow (0, \infty)$ an assignment of values to boundary vertices. Then there exists a unique euclidean packing label R for K so that $R \equiv g$ on ∂K and $\text{br}(R) = \beta$. The same result holds in hyperbolic geometry, with the added feature that g may take infinite boundary values.*

Proof. As we have seen before, there are advantages to starting in the hyperbolic setting. We construct a branched hyperbolic packing label $\widehat{R}$ using a Perron method analogous to that of Section 6.3, except that here we will approach $\widehat{R}$ by *decreasing* labels from above. The Monodromy Theorem (which applies equally well to branched as to unbranched labels) gives us the circle packing $P \longleftrightarrow K(\widehat{R})$.

For each interior vertex v let β_v be the number of times v occurs in the set β. Define the family Φ of β-*superpacking* labels for K as follows:

$$\Phi = \{R : \theta_R(v) \leq 2\pi(\beta_v + 1) \text{ for interior } v, \ R(w) \geq g(w) \text{ for boundary } w\}.$$

We claim this is a Perron family. Φ is nonempty because it contains the maximal packing label $\mathcal{R}_K$ for K, and using the standard monotonicity arguments as in Sect. 6.3, Φ is closed under taking minima. Define

$$\widehat{R} = \inf_{R \in \Phi} \{R\}.$$

If we assume for the moment that $\widehat{R}(v) > 0$ for each interior v, then $\widehat{R}$ is a packing label, for if the inequality $\theta_{\widehat{R}}(v) < 2\pi(\beta_v + 1)$ were to hold at an interior vertex, then a reduction in $\widehat{R}(v)$ could preserve that inequality while only decreasing the neighboring angle sums, giving a new label in Φ and contradicting the minimality of $\widehat{R}$. A similar argument applies if $\widehat{R}(w) > g(w)$ for $w \in \partial K$. We conclude that $\widehat{R}$ is a packing label.

It only remains, therefore, to prove that the infimum $\widehat{R}$ does not degenerate at interior vertices. To that end, consider the collections

$$\mathcal{D}_s = \{v \in K : \widehat{R}(v) > 0\}, \qquad \mathcal{D}_o = \{v \in K : \widehat{R}(v) = 0\},$$

which we will refer to, for visualization purposes, as "solid" dots and "open" dots, respectively. Note that $\partial K \subseteq \mathcal{D}_s$ because of the prescribed boundary labels given by g. Moreover, using monotonicity, continuity of angle sums, and the definition of Φ, it is easy to establish that

$$\theta_{R_n}(v) \longrightarrow 2\pi(\beta_v + 1), \text{ as } n \longrightarrow \infty, \quad \text{for } v \text{ interior, } v \in \mathcal{D}_s. \tag{55}$$

We must prove that $\mathcal{D}_o$ is empty. First a bit of terminology. An edge-connected set $\mathcal{C}$ of vertices of K will be called *simply connected* if no edge-path in $\mathcal{C}$ encloses any vertices not in $\mathcal{C}$. Also, $\partial \mathcal{C}$ will denote vertices not in $\mathcal{C}$ but which have one or more neighbors in $\mathcal{C}$. We establish the following results about such collections of solid or open dots.

Lemma 11.7. *Let $\mathcal{C}$ be a simply connected set of vertices of K and let R_n and $\widehat{R}$ be as defined above. Then the following situations cannot occur:*

(a) $\mathcal{C} \subseteq \mathcal{D}_o$ and $\partial \mathcal{C} \subseteq \mathcal{D}_s$.

(b) $\mathcal{C} \subseteq \mathcal{D}_s$ and $\partial \mathcal{C} \subseteq \mathcal{D}_o$.

(c) $\mathcal{C} \subseteq \mathcal{D}_s$ and $\partial \mathcal{C} \subseteq \mathcal{D}_o$, save for one solid dot.

Proof of Lemma. Assume first that the union of closed stars of vertices in $\mathcal{C}$ forms a subcomplex $L \subset K$ with $\partial L = \partial \mathcal{C}$. In particular, L is simply connected, $\chi(L) = 1$, and (restricting the notations to L) (53) gives

$$F - (V_\partial - 2) - 2V_o = 0. \tag{56}$$

Let N_L denote the sum of branch orders at vertices interior to L, write $\theta_n(\cdot)$ for $\theta_{R_n}(\cdot)$ and Area_n for Area_{R_n}, and use notations v and w, respectively, for interior and boundary vertices of L.

Our proof is based on area computations:

$$\text{Area}_n(L) = F\pi - \sum_v \theta_n(v) - \sum_w \theta_n(w)$$

$$\implies F\pi - \sum_v \theta_n(v) = \text{Area}_n(L) + \sum_w \theta_n(w). \tag{57}$$

Situation (a): Because the interior of L consists of open dots and its boundary of solid dots, it is easy to check that $\text{Area}_n(L) \longrightarrow 0$ and angle sums $\theta_n(w)$ (computed using only the faces of L) go to zero. We conclude that

$$F\pi - \sum_v \theta_n(v) \longrightarrow 0. \tag{58}$$

On the other hand, since $\theta_n(v) \leq 2\pi(\beta_v + 1)$,

$$F\pi - \sum_v \theta_n(v) \geq F\pi - 2N_L\pi - 2V_o\pi, \tag{59}$$

and because V_∂ is strictly greater than $2N_L + 2$ (because β is a branch structure),

$$F\pi - 2N_L\pi - 2V_o\pi > F\pi - (V_\partial - 2)\pi - 2V_o\pi.$$

The quantity on the right is zero by (56), so

$$F\pi - \sum_v \theta_n(v) \geq \pi.$$

This contradicts (58), so situation (a) is impossible.

Situation (b): Consider boundary angle sums first. An edge in ∂L is pictured in Figure 11.2. Since the angle θ goes to zero, the sum $\alpha_1 + \alpha_2$ goes to π as n grows, and summing over all boundary vertices w implies

$$\sum_w \theta_n(w) > (V_\partial - 1)\pi, \quad \text{for large } n; \tag{60}$$

hence

$$\text{Area}_n(L) + \sum_w \theta_n(w) \geq (V_\partial - 1)\pi, \quad \text{for large } n. \tag{61}$$

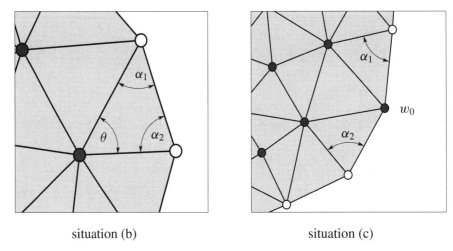

situation (b)   situation (c)

Figure 11.2. Details for situations (b) and (c).

In the interior, on the other hand, (55) implies

$$\sum_v \theta_n(v) \longrightarrow 2N_L\pi + 2V_o\pi. \tag{62}$$

In other words,

$$F\pi - \sum_v \theta_n(v) \longrightarrow (V_\partial - 2)\pi + 2V_o\pi - (2N_L\pi + 2V_o\pi) = (V_\partial - 2 - 2N_L)\pi. \tag{63}$$

Since (61) and (63) together contradict (57), we conclude that situation (b) is impossible.

Situation (c): If you look at ∂L near its single solid dot, w_0, you will see, as suggested in Figure 11.2, that the angles α_1 and α_2 each go to π. Thus the counting in (60) remains valid, and the rest of the argument there yields a contradiction in situation (c).

Finally, what of our assumption that L is a nice, simply connected complex? Figure 11.3 suggests how the combinatorics can go wrong; here $\mathcal{C} \subseteq \mathcal{D}_o$ and $\partial\mathcal{C} \subseteq \mathcal{D}_s$, but $\partial\mathcal{C}$ runs into itself or encircles some peninsula of solid dots or engulfs some solid dots as interior vertices. One could address these problems one at a time, but it is perhaps easier to rethink L.

L does not have to be a subcomplex of K in the arguments above, it simply has to inherit labels and branch structure from K. Therefore, we may build an appropriate L so that it has a simplicial *projection* to K. Think of the connected graph G made from edges with both endpoints in $\mathcal{C}$ as a skeleton (like the darkened edges in Figure 11.3). Let $\{t_j : j = 0, 1, \ldots, k\}$ be a set of copies of the closed triangles in the stars of the vertices of $\mathcal{C}$. Attach each t_j to G along any closed edge it shares with G or at any vertex it shares with G; now travel through G, and at each $v \in G$ reconstitute star(v) by going around the flower attaching its triangles t_j edge-to-edge. The faces form L; vertices not in G are *not* identified, so, for example, each dashed edge in Figure 11.3 represents two edges of L. Because $\mathcal{C}$ is simply connected (as this was defined above), the resulting L will be simply connected and its boundary will be a simple closed curve.

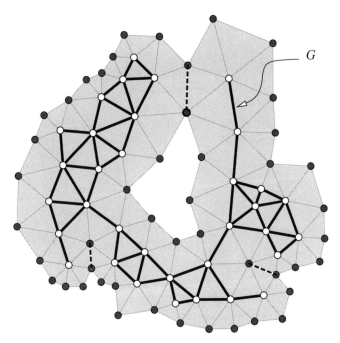

Figure 11.3. Combinatoric complications (------, non identified edges).

Every vertex of L is associated with a vertex of K, so the labels R_n and the limit $\widehat{R}$ lift to L; we partition its vertices into open and solid dots. The total branching N_L (at points of G) is unchanged and V_∂ is at least as great as the number of edges in the outer boundary of the projection of L to K – in other words, the inequality $V_\partial > (2N_L + 2)$ continues to hold. We can now apply our earlier reasoning to L to conclude that situations (a), (b), and (c) lead to contradictions. This completes the proof of the lemma. $\qquad\square$

Returning to the proof of the theorem, we are assuming that $\mathcal{D}_o$ is nonempty. Let $\mathcal{C}$ be an edge-component of $\mathcal{D}_o$. If it were simply connected, it would fall into situation (a), so it must instead surround an edge-component of $\mathcal{D}_s$. However, if this were simply connected, it would run afoul of situation (b), implying it must surround an edge-component of $\mathcal{D}_o \ldots$ and so forth. K is finite, so as we go deeper we eventually encounter a simply connected edge-component of one or the other, giving a contradiction in either case. That proves that K has no open dots, i.e., $\widehat{R}$ is does not degenerate, and the proof of our theorem for the hyperbolic case is complete.

One can extend to the euclidean setting by using the fact that at an infinitesimal level, hyperbolic objects look euclidean. Here is the idea. Given $g : \partial K \longrightarrow (0, \infty)$ and $n > 0$, let $P_n \longleftrightarrow K(R_n)$ be a (branched) hyperbolic packing with boundary labels $R_n(w) = g(w)/n$, $w \in \partial K$. Normalize P_n to put the α-vertex at the origin, convert P_n to a euclidean packing and define Q_n as the scaled packing $2nP_n$. As n grows, the packings P_n shrink toward the origin in $\mathbb{D}$, where the density for the hyperbolic metric is 2, so a boundary circle c_w has euclidean radius roughly $g(w)/2n$. Thus the euclidean label of Q_n at w is approximately $g(w)$ for each $w \in \partial K$. By standard diagonalization arguments, the Q_n converge as $n \longrightarrow \infty$ to a euclidean branched packing P with boundary values precisely g. (It is worthwhile noting that one has to have the *packings* P_n as intermediaries – *sans* location, there is no way to convert the labels.) $\square$

Combining this last result with Theorem 11.5, we have a characterization of branch sets for combinatorial closed discs. Also, the Perron argument gives a broad global monotonicity result.

Corollary 11.8. *Let K be a combinatorial closed disc and $\beta = \{b_1, b_2, \ldots, b_m\}$ a list of interior vertices of K, possibly with repetitions. Then there exists a circle packing P for K with $\mathrm{br}(P) = \beta$ if and only if β is a branch structure for K.*

Corollary 11.9 (Global Monotonicity). *Let K be a combinatorial closed disc and let $f : \mathcal{P}_K \longrightarrow P$ and $g : \mathcal{P}_K \longrightarrow Q$ be discrete analytic functions whose image packings have labels R_f and R_g, respectively, in the geometry of $\mathbb{G}$ (either $\mathbb{C}$ or $\mathbb{D}$). Suppose $\mathrm{br}(f) \subseteq \mathrm{br}(g)$, counting multiplicities, and suppose $R_f|_{\partial K} \geq R_g|_{\partial K}$. Then $R_f \geq R_g$ on K. Equality at a single interior vertex implies $\mathrm{br}(f) = \mathrm{br}(g)$ and $g \equiv \phi \circ f$ for $\phi \in \mathrm{Aut}(\mathbb{G})$; that is, Q is an automorphic image of P.*

In creating packings we often need to specify boundary angle sums rather than boundary radii. Here is the hyperbolic result.

Theorem 11.10. *Let K be a combinatorial closed disc and assume that a hyperbolic packing label R_0 results in boundary angle sums $\phi(w) \in [0, \infty)$, $w \in \partial K$. Then given any nonnegative function $\psi \leq \phi$ on ∂K, there exists a unique packing label R sharing the branch structure of R_0 such that $\theta_R(w) = \psi(w)$ for all $w \in \partial K$. Moreover, $R \leq R_0$, and equality at a single interior vertex v implies equality at all.*

Proof. This is yet another Perron argument, using the family

$$\Phi = \{R : \theta_R(v) = \theta_{R_0}(v) \text{ for interior } v, \ \theta_R(w) \geq \psi \text{ for boundary } w\}.$$

Φ is nonempty since $R_0 \in \Phi$ – indeed, this is precisely why an initial comparison label is needed. I will let the reader do the rest: monotonicity implies that Φ is closed under maxima; the supremum label $\widehat{R}$ is nondegenerate due to the bound of Lemma 6.5 (even more so with branching); $\theta_{\widehat{R}}(v) = \theta_{R_0}(v)$ for interior v by continuity of angle sums; and $\theta_{\widehat{R}}(w) = \psi(w)$ on the boundary, else we could increase $\widehat{R}(w)$. $\square$

Euclidean situations are entirely different. Suppose F, V_o, V_∂ are the numbers of faces, interior vertices, and boundary vertices of K, and N is the number of branch points of R. Each face has angles summing to π, so the total of all angles in the carrier of any euclidean packing $P \longleftrightarrow K(R)$ will be $F\pi$, while the total of angles at the interior vertices will be

$(V_o + N)2\pi$. We arrive at this *necessary condition on euclidean boundary angle sums:*

$$\sum_{w \in \partial K} \theta_R(w) = (F - 2(V_o + N))\pi. \tag{64}$$

Though this is only a necessary condition, things are far from hopeless in the euclidean case. Univalent packings with rectangular or triangular carriers satisfy the necessary condition, for example, and we will see methods for constructing them in Sect. 13.2. In fact, I do not recall a case in which the computer has failed to provide a euclidean packing label for boundary angle sums satisfying this necessary condition! Yet, I doubt that (64) is sufficient. As for branched euclidean packings, there are some nice bits of geometry to note in (64). Each branch point added in the interior requires a 2π decrease in boundary angle sums – that is, a full 2π increase in "left turns" as you transit the outer boundary. Also, the maximal value of N consistent with (64) and with $\chi(K) = 1$ yields the count V_∂ that appears in the branch structure definition.

We have had to put spherical packings aside in this section. Spherical branching and boundary value problems remain largely open, and absent a packing mechanism, are not subject to our usual experimentation. Packings in $\mathbb{P}$ are in practice projections from $\mathbb{C}$ (or $\mathbb{D}$). However, when down in $\mathbb{C}$, spherical data – radii, centers, boundary angle sums – depend on both radii *and* centers, so one generally cannot transfer problems to $\mathbb{C}$, solve them there, and project back to the sphere.

11.5. The Maximum Principle

Our concern here is not with the traditional maximal principle on $|f|$ (this is automatic because f is an open mapping). Our concern is with *comparisons* of different packing labels. *Caution*: there is no general maximum or minimum principle on a *single* label!

Lemma 11.11. *Let R_1 and R_2 be euclidean labels for K. Assume that $\theta_{R_1}(v) \leq \theta_{R_2}(v)$ for every interior vertex v and assume that R_1 is not a scalar multiple of R_2. Then R_2/R_1 has no interior maximum. If R_2/R_1 has an interior minimum at v, then $\theta_{R_1}(v) < \theta_{R_2}(v)$.*

Proof. No "interior local maximum" (resp. minimum) is interpreted to mean that if L is any finite subcomplex of K, then the maximum (resp. minimum) of R_2/R_1 on L must occur on ∂L.

Consider the closed combinatorial flower $F = \{v; v_1, \ldots, v_k\}$. Define t to be the ratio $t = R_2(v)/R_1(v)$ and suppose $t \geq R_2(v_j)/R_1(v_j)$ for $j = 1, \ldots, k$. Then by monotonicity of angle,

$$\begin{aligned}
\theta_{R_1}(v) &= \theta(R_1(v); R_1(v_1), \ldots, R_1(v_k)) \\
&= \theta(t R_1(v); t R_1(v_1), \ldots, t R_1(v_k)) \\
&\geq \theta(t R_1(v); R_2(v_1), \ldots, R_2(v_k)) \\
&= \theta_{R_2}(v).
\end{aligned}$$

In particular, these must all be equalities, so the maximum of R_2/R_1 for any interior flower must occur at one of its petals. In a larger subcomplex L, it is easy to see that one can chase the maximum to a boundary vertex.

The same monotonicity argument works for the minimum of R_2/R_1 in the flower when $\theta_{R_2}(v) = \theta_{R_1}(v)$; therefore you can chase the minimum to the boundary of L or to a point where inequality of the angle sums holds. $\square$

We record the consequence of this result for euclidean ratio functions; there is no spherical version, and the hyperbolic version, which is a little more involved, will not be needed in the sequel and is left for the reader. Note the discretization effect: the best a ratio function can do in mimicking a vanishing derivative is to have a local minimum. You have undoubtedly noticed the shrinking effect branching has on circles in various earlier images.

Theorem 11.12 (Euclidean Maximum Principle). *Let $f : Q \longrightarrow P$ be a discrete analytic function between euclidean packings. Assume that Q is locally univalent and that P is not the image of Q under an automorphism of the plane. Then $f^{\#}$ can have no interior local maximum. It can have an interior local minimum only at a branch vertex.*

11.6. Convergence

One may encounter occasional infinite circle packings that are known through explicit construction, but those are the exceptions. In general, infinite packings result from sequences of finite packings and an appeal to some notion of convergence. An excellent model is provided by classical "normal families," but in our setting convergence comes in three stages: first in combinatorics, then in label, and finally in the geometric convergence of packings themselves.

We have used combinatoric convergence with the finite exhaustions $K_n \uparrow K$ earlier in the book. We need more flexibility, however, and that requires some care. We use *rooted* convergence, meaning that each complex has a designated α-vertex and when (portions of) complexes are identified it is assumed that their α-vertices are identified.

Definition 11.13. Let K be a complex with α-vertex v_1, $\{K_n\}$ a sequence of finite complexes, each with its α-vertex v_1^n. We say that the sequence $\{K_n\}$ **converges** to K if for each finite subcomplex $L \subset K$ containing v_1 there exists an N such that for every $n > N$ there exists a one-to-one simplicial map $s : L \longrightarrow K_n$ with the following properties:

(1) $s(v_1) = v_1^n$,

(2) if $w \in L$ is a boundary vertex of K, then $s(w)$ is a boundary vertex of K_n.

Reference to the α-vertices is generally suppressed and we write $K_n \longrightarrow K$.

We have carefully avoided any assumption in general that the K_n are equivalent to subcomplexes of K; rather, one pictures increasingly large pieces of the complexes ("centered" on v_1) being identical. The extra latitude is justified by this combinatorial normal families result.

Theorem 11.14 (Combinatorial Normal Families). *Let $\{K_n\}$ be any sequence of finite complexes with a common bound d on their degrees and with designated α-vertices v_1^n. Then there exists a subsequence that converges. That is, there exists a complex K, $\deg(K)$ no larger than d, and a subsequence $\{K_{n_j}\}$ so that $K_{n_j} \longrightarrow K$.*

The proof is left to the reader. The key ingredients are the "pigeonhole" principle – there are only finitely many complexes with a given number k of vertices – and diagonalization. The bound on degree is unfortunate but necessary, as shown by easy examples (such as arranging that $\deg(v_1^n)$ grow without bound). Note that there are no topological restrictions on the complexes here.

We next throw in labels. If the complexes K_n have labels R_n in a fixed geometry, then in extracting a subsequence $K_{n_j} \longrightarrow K$, a further diagonalization argument gives a subsequence whose labels $\{R_{n_j}\}$ converge to some function $\widehat{R}$ on the vertices of K. Of course, one would generally start with packing labels R_n, and if the limit $\widehat{R}$ is a legal label, then the continuity of the angle sums will ensure that $\widehat{R}$ is a packing label. The challenge, in general is avoiding degeneracy, such as labels going to zero or interior labels going to infinity (or to π in the spherical case). In the construction of infinite maximal packings in Part II we used the Ring Lemma and when necessary we switched geometry and rescaled to extract a limiting label. In more general packings, with nonunivalence and branching, one does not always have these options.

Finally, we get to the convergence of packings themselves. One normally avoids certain problems, such as having the α-circle drift off to infinity, with normalizations. However, I have phrased this to leave some additional latitude. We call this convergence "geometric" (not to be confused with the geometric series) and write $P_n \longrightarrow P$.

Theorem 11.15 (Normal Families of Packings). *Let $\{P_n\}$ be a sequence of circle packings, $P_n \longleftrightarrow K_n(R_n)$, $n = 1, 2, \ldots$, that are all hyperbolic or all euclidean, and assume there exist constants $c_1 > 1, c_2 > 0, c_3 \geq 3$, and $c_4 > 1$ so that the following conditions hold for all n:*

(1) $1/c_1 < R_n(v_1^n) < c_1$.

(2) The centers z_1^n of α-circles satisfy $|z_1^n| < c_2$.

(3) $\deg(K_n) \leq c_3$.

(4) $R_n(v_j) \leq c_4 R_n(v_k)$ whenever v_j, v_k are neighboring interior vertices of K_n.

Then there exists a subsequence $\{P_{n_j}\}$ which converges to a packing $P \longleftrightarrow K(R)$. In particular, $K_{n_j} \longrightarrow K$, $R_{n_j} \longrightarrow R$, and $P_{n_j} \longrightarrow P$.

In practice, condition (4) is the rub. We are not assuming univalence, so (4) does not follow from (3) and the Ring Lemma. Also, condition (3) does not follow from (4) since the packings may be branched.

As you might expect, various properties of the packings and complexes will filter through the convergence process to yield information about their limits. If $K_n \longrightarrow K$ and the K_n are simply connected, then K is simply connected; if $P_n \longrightarrow P$ and the P_n are univalent, then P is univalent. One must be cautious, however. Suppose, for example, that K_n, $n = 1, 2, \ldots$, is the combinatorial open disc that has n generations of 7-degree vertices centered on v_1^n followed by infinitely many generations of 6-degree vertices. Each K_n is parabolic, yet $K_n \longrightarrow K^{[7]}$, the heptagonal complex, which is hyperbolic. With this type of example and the role of the root in mind, we make the following definition for later use.

Definition 11.16. A sequence $\{K_n\}$ of simply connected complexes is termed **asymptotically parabolic** if it is the case that whenever a subsequence $\{K_{n_j}\}$, regardless of specified α-vertices, converges to a combinatorial open disc K, then K is parabolic.

12

Construction Tools

We will shortly face the construction of various and sundry circle packings P and/or complexes K, always with some desired geometric or combinatorial properties in mind. Here is a collection of hands-on techniques that we will be able to call on. You need not digest these all at once; return here when a method is called for or when you need definitions.

These are all rather straightforward techniques, but that does not mean they are trivial – to the contrary, it is all too easy in constructions to accidentally disconnect the interior of a complex, to leave a vertex orphaned (with no interior neighbor), or to stumble into a nonorientable result. I will describe the methods briefly; they are all sound and have all been implemented in software (for finite cases, of course), but they require various conventions, compatibility checks, special cases, and so forth so that the results are guaranteed to be legal complexes. Keep this in mind when you want to use them in practice.

12.1. Cutting Out

Suppose $\widetilde{K}$ is a complex which is a simplicial subcomplex of a second complex K. A circle packing P for K *restricts* to a circle packing $\widetilde{P} \subseteq P$ for $\widetilde{K}$. We would say that $\widetilde{P}$ is a *cutout* from P. In practice, cutouts are generally described geometrically in situ on P, as when one applies a "cookie cutter" to a packing (and do not forget pruning that may be needed).

A special case of cutting out is *puncturing* K at a vertex v – that is, removing the open star of v (as we did, for example, in Chap. 7). The operation is restricted to interior vertices v that are combinatorial distance at least 3 from ∂K, so that $\widetilde{K}$ is a complex.

12.2. Slitting

This is slightly more subtle than cutting out. Suppose γ is a simple interior edge-path of K. We can slit K open along γ to obtain a new complex $\widetilde{K}$. As when you mathematically slit a surface along a path, there is a complication here in that one side of the slit is closed – that is, it inherits the points from the path – while the other side is open. In the discrete setting this is handled by *cloning* the vertices along the edge-path to provide copies on

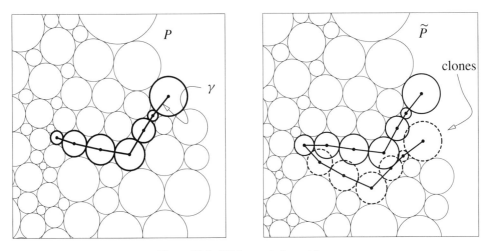

Figure 12.1. Slitting a circle packing.

each side of the opened slit. If P is a circle packing for K, then by also cloning the circles, we get a circle packing $\widetilde{P}$ for $\widetilde{K}$. An end vertex of γ that is interior is called a "tip" of the slit, and as viewed from that tip the slit has a "right" and "left" sides. Slitting is a combinatorial operation, but we generally apply it to packings, as shown in Figure 12.1; I have modified the radii to open the slit and show the cloned circles (dashed).

As for problems, note that a slit can disconnect the surface or its interior and can orphan boundary vertices. Assuming that you have taken care of these problems, if K is simply connected and the slit has at least one end in ∂K then $\widetilde{K}$ will be simply connected, as well.

12.3. Combinatorial Pasting

One can create new complexes by pasting known complexes together along specified boundary edge-paths. Suppose K_1, K_2 are complexes with edge-paths $\gamma_1 = \langle w_0^{(1)}, \ldots, w_n^{(1)} \rangle$ and $\gamma_2 = \langle w_0^{(2)}, \ldots, w_n^{(2)} \rangle$, respectively. (Here n is at least 2, but in theory the edge-paths could be singly or doubly infinite.) We assume γ_1 is negatively oriented in ∂K_1 while γ_2 is positively oriented in ∂K_2. Form $\widetilde{K}$ from the disjoint union of K_1 and K_2 by identifying corresponding vertices and edges of γ_1 to γ_2; orientations ensure that $\widetilde{K}$ will be orientable, hence a complex.

The schematics of Figures 12.2(a) and (b) illustrate two uncomplicated examples. To ensure that the result is a complex, γ_1 and γ_2 must both be open or both be closed; (a) illustrates the former, (b) the latter. If the complex K_2 in (b) were a simple 11-flower for vertex v, then this pasting would add an *ideal boundary vertex* $v = v_\infty$ to K_1 and would be the reverse of the puncturing that we described earlier.

We can paste a complex to *itself* as long as the boundary edge-paths γ_1 and γ_2 are oppositely oriented and do not share any edges. If they share one or both end vertices, then this pasting is simply the reverse of a slit operation with the common end(s) as tip(s) of the

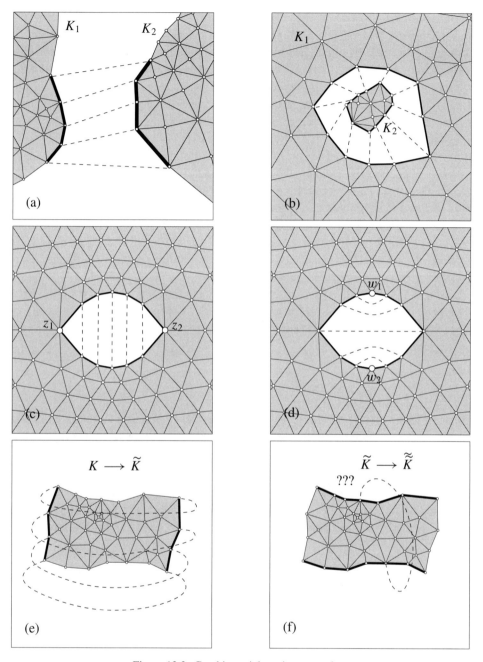

Figure 12.2. Combinatorial pasting examples.

slit. For example, a closed boundary component γ of K can be "sutured" together if it has an even number of edges, the resulting combinatorics depending on where you start the suturing. Figures 12.2(c) and (d) illustrate. I have slit a regular hexagonal complex K along six edges from v_1 to v_2, opening the slit in (c) and showing dashed lines identifying vertices with their clones; suturing these would reconstitute the hexagonal combinatorics. In (d), dashed lines indicate an alternate suturing with vertices w_1, w_2 as ends. A combinatoric change like this could make a big change in the geometry imposed when the complex is circle packed.

Of course, one can also paste complexes together along a collection of boundary edge-paths with a succession of individual pastings. As to the topological effects, it is important to recall van Kampen's Theorem from general topology: *If K_1 and K_2 are distinct simply connected complexes pasted along connected boundary edge-paths, then the result $\widetilde{K}$ will again be simply connected. Likewise, if K is simply connected and is pasted to itself along boundary edge-paths which share one or two end vertices, then $\widetilde{K}$ will be simply connected.* Multiply connected complexes or multiple pastings are another matter. Figure 12.2(e) shows identification of two segments of boundary of a disc K; the result $\widetilde{K}$ is an annulus. Both the (now closed) boundary components of $\widetilde{K}$ happen to contain six vertices and edges, and if we identify them as in (f), the result $\widehat{K}$ is a torus. Note that there are six places to start the pasting in (f), leading to a sort of "Dehn twist" family of resulting combinatorial tori.

12.4. Doubling

Given complex K, let K' denote a copy of K having the reverse orientation. If one or more disjoint, oriented boundary edge-paths γ of K are given, let γ' denote the corresponding edge-path(s) of K'.

Definition 12.1. The complex $\widetilde{K}$ obtained by pasting K to K' along γ and γ' is called the **double** of K across γ. There is an orientation-reversing (simplicial) automorphism $\sigma_{\widetilde{K},\gamma} : \widetilde{K} \longrightarrow \widetilde{K}$ that interchanges K and K' and fixes γ, called the **doubling reflection**.

If K is simply connected and γ is a single edge-path, then $\widetilde{K}$ is simply connected. The double of a combinatorial closed disc K across its full boundary gives a combinatorial sphere. In general, if K, is a finite, planar complex with $n + 1$ boundary components and K is doubled across its full boundary, then $\widetilde{K}$ is a compact complex of genus n. (These are combinatorial analogues of classical *hyperelliptic* Riemann surfaces.)

12.5. Combinatorial Welding

This is very close in spirit to combinatorial pasting except that it accommodates modification of the combinatorics in the neighborhoods of the edge-paths γ_1 and γ_2. Welding is generally done in response to geometric dictates, but Figure 12.3 is a purely combinatorial illustration. The dark dashed lines indicate a set of desired vertex identifications, but since

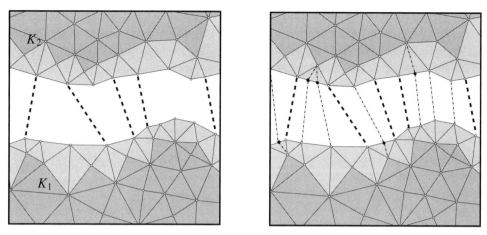

Figure 12.3. Hypothetical combinatorial welding.

the counts of boundary vertices between these do not agree, a direct combinatorial pasting is impossible. We add extra boundary vertices (and necessary interior edges) as needed in K_1 and K_2. As the totality of dashed lines indicates on the right, the modified complexes can now be pasted to get the welded complex $\widetilde{K}$. Note that the changes in combinatorics are limited to the first generation of boundary faces in each complex.

12.6. Geometric Pasting

A *geometric* pasting of circle packings P_1 and P_2 involves the *combinatorial* pasting of their complexes K_1 and K_2, except that we also require geometric compatibility of the circles for the vertices being identified. Look to Figures 12.4(a) and (b): six circles form a chain associated *simultaneously* with edge-paths $\gamma_1 \subset \partial K_1$ and $\gamma_2 \subset \partial K_2$ (negatively oriented and positively oriented, respectively). These six have *the same centers and radii in each of the two packings*. A geometric pasting yields the circle packing $\widetilde{P}$ of (c) by identifying these six circles. The pasting is both geometric and combinatoric: we get both $\widetilde{P}$ and its complex $\widetilde{K}$.

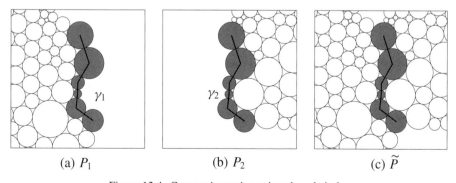

(a) P_1 (b) P_2 (c) $\widetilde{P}$

Figure 12.4. Geometric pasting using shared circles.

Geometric pasting is a very rigid process. It is not enough, for example, that the boundary circles to be identified have the same radii – they must also be in essentially the same *locations*. We say "essentially" here because in the typical situation we allow an isometry to be applied to P_1 and/or P_2 to bring the circles we want to identify into coincidence for the pasting. We will nevertheless use the term *geometric pasting*.

12.7. Schwarz Doubling

As with pasting, there is a *geometric* version of doubling. This is the basis for discrete *Schwarz reflection* later in our work. It is best described in the sphere.

Theorem 12.2. *Let Q be a circle packing for K in the sphere and assume that all the circles associated with some edge-path $\gamma \subset \partial K$ are orthogonal to a common circle $E \in \mathbb{P}$. Then there is a circle packing $\widetilde{Q}$ for $\widetilde{K}$, the double of K across γ, which extends the packing Q; that is, $\widetilde{Q}|_K \equiv Q$.*

The unique anticonformal Möbius transformation of $\mathbb{P}$ that fixes points of E is called a *geometric doubling reflection*; it maps $\widetilde{Q}$ to itself, fixes the circles for γ, and realizes the combinatoric doubling reflection of $\widetilde{K}$.

This doubling process works in the other geometries as well; one simply has to ensure that the reflected circles remain in the geometry. This is guaranteed to be the case, for example, if the circles of γ are centered on a straight line in the euclidean case or on a geodesic in the hyperbolic case; in both these instances, radii are preserved under the doubling.

12.8. Refinement, etc.

Definition 12.3. Given a complex K, the **hexagonal refinement** of K, denoted δK, is the complex obtained from K by adding a new vertex to the interior of each edge and then, for each face, connecting with new edges the three new vertices in its boundary.

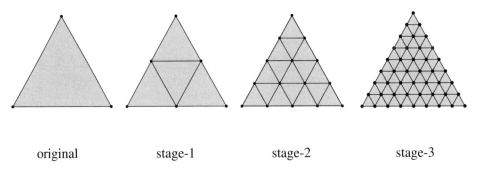

original   stage-1   stage-2   stage-3

Figure 12.5. Three stages of hexagonal refinement.

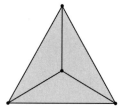 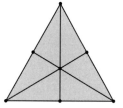

Figure 12.6. A barycenter and barycentric subdivision.

Figure 12.5 shows several stages of hexagonal refinement of a single initial face. There can clearly be a geometric side to the process, and that will be important in applications, but by definition it is a purely combinatorial operation. Note that δK is again a complex: it triangulates the same topological surface as K, all the new interior vertices have degree 6, and the degrees of the original vertices remain unchanged. Hexagonal refinement is especially practical and efficient. It is easy to implement; vertices of K retain their original indices; and under repeated refinement, vertices are encircled by growing rings having hexagonal combinatorics. (Also, the vertices are stratified by their indices; since real-time displays generally draw in vertex order, one often sees a history of the nested structures as the circles are painted on the screen.)

Two other combinatorial changes we will see are shown in Fig. 12.6: on the left a face gets a *barycenter* and on the right it is *barycentrically* subdivided.

13

Discrete Analytic Functions on the Disc

A discrete analytic function $f : \mathcal{P}_K \longrightarrow P$ on the disc is one whose complex K is simply connected and hyperbolic, meaning $\mathcal{P}_K$ packs $\mathbb{D}$. Every circle packing P conjured up for K yields a discrete analytic function, and we can conjure these with wild abandon: branching, boundary labels, and boundary angle sums can be specified willy-nilly and CirclePack will spit out the results faster than one can type. But understanding is our goal, so we look to the construction and analysis of examples with *prescribed* behaviors.

I introduce two basic approaches which apply both here and in later chapters: In a *domain* construction one starts with K, imposes various boundary and branch conditions, uses existence and uniqueness results for a corresponding packing label, and then lays out the image packing P. In a *range* construction, one directly builds an image P having the desired properties, reads off its complex K, then obtains the domain $\mathcal{P}_K$ via the Discrete Uniformization Theorem.

While we cannot mimic the full richness of the classical theory, we do garner some key examples which will be important later – especially discrete Blaschke products – and we formulate several open questions *vis-a-vis* the classical theory. In the main, we restrict to finite complexes K, but results and directions for further work with infinite K are in Section 13.6. The chapter closes with some experimental challenges.

13.1. Schwarz and Distortion

Let us begin by stating some results proven in Part II in their classical formulations. The first is rather anticlimactic at this point in the book, but represents the "Finite Riemann Mapping Theorem" with which Thurston launched the topic.

Theorem 13.1 (Discrete Riemann Mapping Theorem). *If P is a simply connected and univalent circle packing in the plane whose carrier is a proper subset of the plane, then there exists a discrete analytic function f on the unit disc with image packing P; namely, $f : \mathcal{P}_K \longrightarrow P$. Moreover, f is unique up to precomposition with automorphisms of $\mathbb{D}$.*

This is, of course, a special case of the Discrete Uniformization Theorem (Theorem 4.3). In both classical and discrete settings, the real force behind it is the Schwarz Lemma, which for us serves as a sort of ensemble monotonicity statement. Here K can be finite or infinite.

Lemma 13.2 (Discrete Schwarz–Pick Lemma). *Let K be a hyperbolic combinatorial disc and $f : \mathcal{P}_K \longrightarrow P$ a self-map of $\mathbb{D}$. If v, u are interior vertices of K with hyperbolic centers Z_v, Z_u in $\mathcal{P}_K$, and d_{hyp} denotes distance in the hyperbolic metric, then*

$$d_{hyp}(f(Z_v), f(Z_u)) \leq d_{hyp}(Z_v, Z_u).$$

In particular, if $f(0) = 0$ then

$$f^{\#}(0) \leq 1.$$

Finite equality for any pair of distinct vertices in the first instance or equality in the second implies that f is a conformal automorphism of $\mathbb{D}$.

This result follows from Lemmas 8.3 and 8.4 and can equally well be stated as the *discrete hyperbolic contraction principle*: *Discrete analytic functions from the disc into itself are contractions in the hyperbolic metric (on the set of centers).* This is the invariant formulation due to Pick; the inequality for $f^{\#}$ is the Schwarz Lemma, and for us it simply means that if corresponding circles are at the origin in $\mathcal{P}_K$ and P, then the former is the larger.

Here is a companion to the Schwarz Lemma, the function theory form of a result we proved in Section 8.3.2; we will prove a branched version shortly.

Lemma 13.3 (Discrete Distortion Lemma). *Let K be a combinatorial closed disc and $f : \mathcal{P}_K \longrightarrow P$ a discrete analytic function with $f(0) = 0$. Suppose that P is euclidean and that for some $t > 0$ each of its boundary circles intersects the complement of the disc $\{|z| < t\}$. Then the euclidean ratio function $f^{\#}$ satisfies $f^{\#}(0) \geq t$, with equality if and only if $P = \lambda \mathcal{P}_K$ for some λ with $|\lambda| = t$.*

13.2. Discrete Univalent Functions

Univalent functions on the disc are among the best studied in classical function theory. For us $f : \mathcal{P}_K \longrightarrow P$ is *univalent* precisely when P is a univalent packing. We begin with the two principal construction techniques; further ad hoc examples will come up later.

Cookie-cutting. The simplest method uses *cutting* (see Section 12.1) – that is, the packing P is cookie-cut from a larger univalent packing. This is certainly important, as it underlies Thurston's Conjecture (see Figure 2.8), but we do not need to say more about it here.

Prescribed angle sums. Univalent packings can be created using prescribed boundary angle sums. We will have special uses for *rectangular* packings motivated by this classical result: *Given points $\zeta_1, \ldots, \zeta_4$ on the unit circle, there exists a conformal mapping F of $\mathbb{D}$ onto an essentially unique euclidean rectangle $\mathfrak{R}$ such that the points ζ_j are mapped*

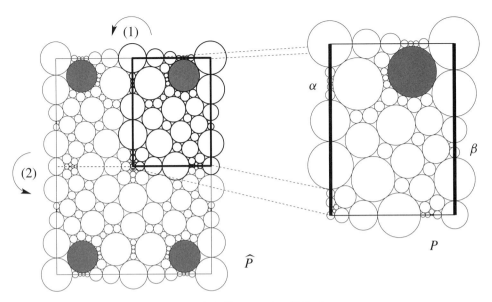

Figure 13.1. "Rectangulated" Owl.

to the rectangle's corners. For discrete versions we work with a *combinatorial rectangle*, which is a combinatorial closed disc K with two designated disjoint edge-paths α, β in its boundary.

Theorem 13.4. *Let $K = \langle K; \alpha, \beta \rangle$ be a combinatorial rectangle. There exists an essentially unique discrete univalent function $f : \mathcal{P}_K \longrightarrow P$ mapping $\mathbb{D}$ onto a euclidean rectangle $\mathfrak{r} = \mathrm{carr}(P)$ so that the edge-paths α and β are are mapped to the two ends of $\mathfrak{r}$. The rectangle $\mathfrak{r}$ is unique up to conformal automorphisms of $\mathbb{C}$.*

I leave the details for the reader, but the ideas are illustrated with Owl in Figure 13.1. Using the combinatorial *doubling* operation (see Section 12.4), one can (1) double K across one side, say α, to get $\widetilde{K}$, and then (2) double $\widetilde{K}$ across a doubled edge between α and $\widetilde{\alpha}$ to get a complex $\widehat{K}$. This has opposite edges that match, and identifying these creates a combinatorial torus L. Its universal covering packing, call it $\mathcal{Q}$, tiles the plane. Within $\mathcal{Q}$ is a univalent packing $\widehat{P}$ for $\widehat{K}$, and within that, a univalent packing P for K itself. The combinatorial doublings imply the existence of reflective symmetries in L, which in turn lift to anticonformal reflections in $\mathbb{C}$ mapping $\mathcal{Q}$ to itself. These reflections commute combinatorially, so their lifts commute, and hence have fixed lines which meet at right angles; these fixed lines define the carrier of P. *Voilà*, it is a rectangle! I have shaded Owl's left eye to help you keep track of orientations.

I should point out that this proof has nothing to do with the practical computation of this Owl example. `CirclePack` computes labels using assigned "target" angle sums. Typically 2π is the assignment for interior angle sums, but one can also specify boundary-angle-sum targets – in this case, $\pi/2$ at the four corners, π at the remaining boundary

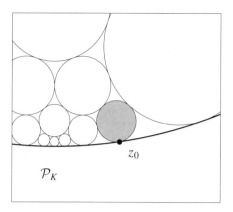
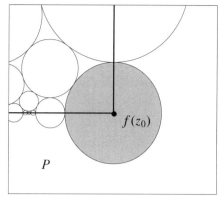

Figure 13.2. Square root behavior at a corner.

vertices. The repacking algorithm does its magic in computing R, and then `CirclePack` lays out P. (Formula (64) works out to $\sum_{w \in \partial K} \pi - \theta_R(w) = 2\pi$ in this unbranched case.)

Many classical Riemann mapping behaviors show up as well in the discrete setting. We might note just one example. What explains the large size of the corner circles in Figure 13.1? Figure 13.2 displays the local situations in the domain and range packings at a corner vertex. A classical function F would behave like a square root $(z - z_0)^{1/2}$ at z_0, mapping the (infinitesimally) flat edge at z_0 to a square corner; this implies that $|F'|$ behaves like $|z - z_0|^{-1/2}$ at z_0 – in other words, it blows up at z_0. The large growth in the corner circle of the image reflects $f^{\#}$'s attempt to keep up.

13.3. Discrete Finite Blaschke Products

The next functions we consider are, in some sense, the antithesis of univalent. A *proper* map g of $\mathbb{D}$ to itself is one with the property for any sequence $\{z_j\}$ of points of $\mathbb{D}$ converging to $\mathbb{T}$, the sequence $\{g(z_j)\}$ also converges to $\mathbb{T}$. If g is a light interior mapping, then topological arguments alone imply that it has some constant finite valence n and total branch order $n - 1$.

The analytic proper self-maps of $\mathbb{D}$ are precisely the classical *finite Blaschke products* B. If $a_1, \cdots, a_n \in \mathbb{D}$ are its n zeros (with possible repetitions), then for some real θ, $B(z)$ has the product representation

$$B(z) = e^{i\theta} \prod_{j=1}^{n} \frac{z - a_j}{1 - \overline{a}_j z}, \quad |z| \leq 1. \tag{65}$$

Each factor is a Möbius transformation of the disc and accounts for one zero. Each factor maps $\mathbb{T}$ to itself and one can show that B wraps $\mathbb{T}$ around itself n times in a positively oriented direction.

The formula does us little good in seaching for a discrete analogue, but properness gives us a geometric opening to exploit.

Definition 13.5. Suppose K is a combinatorial disc. A discrete analytic function $b : \mathcal{P}_K \longrightarrow P$ is said to be a **discrete** (finite) **Blaschke product** if it is a proper mapping of $\mathbb{D}$. In other words, P lies in $\mathbb{D}$ and for any sequence $\{z_j\}$ of interior points of $\mathrm{carr}(\mathcal{P}_K)$ converging to $\mathbb{T}$, the sequence of image points $\{b(z_j)\}$ in $\mathrm{carr}(P)$ converges to $\mathbb{T}$.

The existence of such a P implies that K is hyperbolic. However, we restrict ourselves to the finite case, where the requirement that b be proper simply means that all boundary circles of the image packing are horocycles. The existence and essential uniqueness of these packings follows from Theorem 11.6 by prescribing infinite hyperbolic boundary labels. The terminology and notation we use in this situation will be justified in Lemma 13.8.

Definition 13.6. Given a combinatorial closed disc K with branch structure β, a circle packing with branch set β whose boundary circles are horocycles is called a **branched maximal circle packing** for K and β. The notation is $\mathcal{P}_{K,\beta} \longleftrightarrow K(\mathcal{R}_{K,\beta})$.

In particular, then, maps $b : \mathcal{P}_K \longrightarrow \mathcal{P}_{K,\beta}$ are seen to be discrete Blaschke products. (The automorphisms of $\mathbb{D}$ also qualify since $\mathcal{P}_K = \mathcal{P}_{K,\emptyset}$.) If you return to the description accompanying the Blaschke example of Figure 2.6 in the Menagerie, you will see how closely the behavior of these discrete versions parallels that noted above for their classical models.

Lemma 13.7. *Let K be a combinatorial closed disc and $b : \mathcal{P}_K \longrightarrow \mathcal{P}_{K,\beta}$ a discrete Blaschke product with branch set β. Then there exist a quasiconformal homeomorphism $h : \mathbb{D} \longrightarrow \mathbb{D}$ and a classical Blaschke product B with branch set $h(\beta)$ such that $b \equiv B \circ h$ on $\mathrm{carr}(\mathcal{P}_K)$.*

Proof. To apply Stoïlow's Theorem (11.2), we need a temporary modification $\tilde{b}$ of b similar to that used in Section 6.3 (see Figure 6.4). Namely, each edge $e \in \partial K$ corresponds to hyperbolic geodesics in the boundaries of $\mathrm{carr}(\mathcal{P}_K)$ and $\mathrm{carr}(\mathcal{P}_{K,\beta})$. Extend b continuously across e so that the missing half-plane in the domain is mapped homeomorphically onto the missing half-plane in the range. Do this for all boundary edges e and the extended map $\tilde{b}$ is seen to be a proper mapping of $\mathbb{D}$. Stoïlow's Theorem now yields $\tilde{b} = B \circ \tilde{h}$ for a homeomorphism $\tilde{h}$ of $\mathbb{D}$ and an analytic function B; since B must be proper, it is necessarily a classical finite Blaschke product. Define $h \equiv \tilde{h}|_{\mathrm{carr}(\mathcal{P}_K)}$, so that $b \equiv B \circ h$. Our extension $\tilde{b}$ introduces no new branch points, so $br(B) = h(\beta)$. Since b can easily be chosen to be smooth on each face in $\mathrm{carr}(\mathcal{P}_K)$, the same therefore holds for h, so h is quasiconformal on each of finitely many faces, implying that it is κ-quasiconformal on carr($\mathcal{P}_K$) for some κ. (See Appendix A for background on quasiconformal issues.) $\square$

13.3.1. Constructions

Building discrete finite Backslash products for given K and branch set β involves computing the maximal packing $\mathcal{P}_{K,\beta}$. This is a "domain" construction, since it starts with K, and it is computationally trivial: in `CirclePack` one simply adds 2π to the target angle sum for each $v \in \beta$, sets all boundary labels to infinity, repacks, and then lays out the resulting circle packing. Even a complex with several thousand circles will generally pack in a matter of seconds.

It is another thing entirely to create examples with prescribed *branch values*, that is, prescribed *images* of the branch points in the range. We can use the *slitting* and *geometric pasting* processes described in Chap. 12 to construct the *image packing P* directly. After the fact P is seen to be $\mathcal{P}_{K,\beta}$ for the resulting K and β. Computing the (unbranched) maximal packing $\mathcal{P}_K$ then completes construction of $b : \mathcal{P}_K \longrightarrow \mathcal{P}_{K,\beta}$. Let me describe the four-fold example of Figure 13.3.

We begin with the maximal packing Q shown in the inset box; for visual simplicity I use one with hexagonal combinatorics, though nearly any maximal packing could serve. Choose three interior circles and connect their centers to the boundary with simple, nonintersecting, interior edge-paths; the centers for these circles will be the branch values. We are going to slit Q open along each of the edge-paths; recall that this involves cloning to get circles on each side of each slit.

For the construction, create four identical copies, $\tilde{Q}_j$, $j = 1, \ldots, 4$, of Q as the "sheets" for the eventual image surface. Let $P_1 = \tilde{Q}_1$. Choose the left edge of some slit on $\tilde{Q}_2$ and geometrically paste it to the right edge of the corresponding slit on P_1 to get the new packing P_2. Choose the left edge of a slit on $\tilde{Q}_3$ and geometrically paste it to the right edge of a corresponding slit on P_2 to create P_3. Do the same with the last sheet $\tilde{Q}_4$ to get P_4. These pastings are indicated in Figure 13.3, and one can see that all four sheets are now attached.

P_4 has unpasted left and right edges, which we must clean up. You can check as you walk around the boundary of P_4 that every unpasted left edge is contiguous, in the boundary, to an unpasted right edge for the corresponding slit. Geometrically paste each of these pairs of contiguous edges together to arrive at the final packing P. In most instances we are merely resealing unused slits, as though they had never been opened on their particular sheets. In three cases, however we are completing the pastings around branch vertices, and these vertices compose the branch set β. It is important to note that these cleanup pastings identify *contiguous* edges in the boundary of P_4, for this means that all our pastings attach simply connected surfaces along simply connected edge segments, so Van Kampen's Theorem ensures that the complex K for the finished packing P is simply connected. Since all the slit edges have now been identified, the only boundary circles remaining for P are horocycles, so P is a maximal branched packing $\mathcal{P}_{K,\beta}$. Computing the maximal packing $\mathcal{P}_K$, we get our four-fold discrete Blaschke product $b : \mathcal{P}_K \longrightarrow P$.

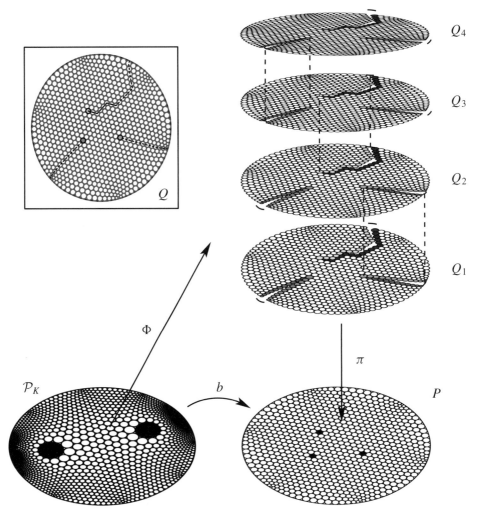

Figure 13.3. Image packing construction.

In the figure you may treat Φ as a one-to-one map of $\mathcal{P}_K$ to the image surface and π as the projection to P in the disc, so $b \equiv \pi \circ \Phi$. The projected packing looks exactly like Q, but of course, each circle represents four (counting multiplicities at branch circles). I have normalized $\mathcal{P}_K$ so that its circle at the origin is mapped by Φ to the circle at the origin on sheet $\widetilde{Q}_2$, implying $b(0) = 0$.

13.3.2. Remarks

Observe in this construction that once the circles of the prospective branch values are chosen, the cuts to these circles themselves are immaterial; they vanish after pasting. However, the pasting *decisions* – order and choice of slit to use – are important. The interested reader may want to check that there are eight distinct (up to Möbius transformations) image surfaces that can be constructed from the four slit sheets of Figure 13.3. (Careful: this "8" is not 2^3.)

Constructions of Blaschke products in the classical and discrete setting complement one another. Classical Blaschke products B are most naturally defined by their n zeros *via* formula (65). This is not an option in the discrete setting. The discrete constructions, on the other hand depend on the branch points and values, which are more directly geometric features. The parallel to our geometric construction would be extremely challenging in the classical case.

In both Figure 2.6 and Figure 13.3 you can see the hyperbolic contraction principle in high gear: branch values are driven in toward the origin in the range of the former, while branch points are pushed *out* toward the unit circle in the domain of the latter. Undoubtedly you have noticed the relative sizes (euclidean) of the horocycles in domain versus range. More on this later.

13.4. Discrete Disc Algebra Functions

We change direction now by bringing boundary behavior into the mixture, taking as our guide the *disc algebra* $\mathcal{A}$, the algebra of functions analytic on $\mathbb{D}$ and extending continuously to $\partial\mathbb{D}$. I formulate and illustrate two types of boundary value problems, the first in the domain, and the second in the range. You will note that "boundary value" has slightly different connotations in the discrete setting, since we are prescribing labels, not locations.

13.4.1. Boundary Derivative

Our domain approach uses the ratio function to set the boundary radii. The motivation is this classical theorem: *Let* $\Omega \subset \mathbb{C}$ *be a Jordan domain,* λ *a continuous positive function on* $\partial\Omega$, *and* β *a finite subset of* Ω *(counting repetitions). There exists an analytic function* $F : \Omega \longrightarrow \mathbb{C}$ *with* $\mathrm{br}(F) \equiv \beta$ *whose derivative* F' *extends continuously to* $\overline{\Omega}$ *with* $|F'| \equiv \lambda$ *on* $\partial\Omega$. F *is unique up to translations and rotations.*

The discrete version is a special case of Theorem 11.6. To give us a benchmark for comparison we will mimic a known classical function F on a given region Ω. In particular, set $F(z) = 4z^3 - (3 - 2i)z^2 + (1 + i)z$ and let Ω be the region on the left in Figure 13.4. Define $\lambda \equiv |F'|$ on $\partial\Omega$ and note the two branch points $\{i/6, (1 - i)/2\}$ of F. These serve as our preassigned data.

The domain packing $P \longleftrightarrow K(R_1)$ is cookie-cut in the shape of Ω from the infinite *ball-bearing* pattern. The image packing $Q \longleftrightarrow K(R_2)$ is created to approximate F as follows. Designate the two shaded circles in P closest to the branch points of F as the prescribed (simple) branch vertices β. We expect the ratio function $f^{\#}$ to approximate

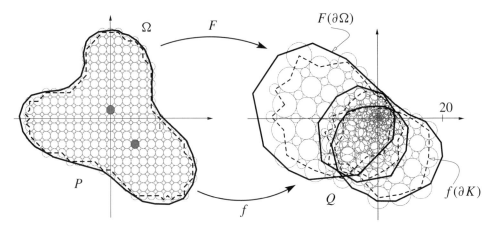

Figure 13.4. Classical and discrete.

$|F'|$ on the boundary, so for each $w \in \partial K$ we define the label $R_2(w)$ by

$$R_2(w) = |F'(z_w)| \cdot R_1(w), \quad z_w = \text{center}(C_w) \text{ in } P.$$

A branched packing label R_2 is guaranteed by Theorem 11.6, and laying out Q with normalizations based on F yields the discrete analytic function $f : P \longrightarrow Q$ shown in the figure. The image boundaries $f(\partial K)$ and $F(\partial \Omega)$ are shown for comparison. (*Note*: coordinate scales differ between domain and range.)

13.4.2. Prescribed Image

Our second disc algebra example begins in the range. Given a continuous closed curve γ with a finite number of transverse double points, there are known combinatorial conditions which are necessary and sufficient for existence of $F \in \mathcal{A}$ with $\gamma = F(\partial \mathbb{T})$ and which determine required branching. Explicit construction of F is another matter; modified Schwarz–Christoffel methods apply in limited circumstances, but this is by-and-large an unsolved problem. The discrete construction, however, is a straightforward application of our slitting and geometric pasting methods. For convenience we will again use hexagonal construction material; the construction is shown in Figure 13.5.

The winding numbers of γ around various points imply that we need four "sheets" and three branch values. We have some latitude with the branch locations and their cuts; see the choices I have made in the inset. The four sheets S_j are distinct, but they are cut from identical hexagonal packings – it is important that circle locations match from one sheet to another so that we can use *geometric* pasting. Note in Figure 13.5 that I have made only those slits needed for the indicated pastings; the reader can verify that the resulting packing P' is simply connected. Let K' be the associated complex; $\mathcal{P}_{K'}$ is our domain. Again, Ψ is a one-to-one mapping of $\mathcal{P}_{K'}$ to the "image surface" that, when followed by the projection, gives the discrete disc algebra function $f : \mathcal{P}_{K'} \longrightarrow P'$ as $f \equiv \pi \circ \Psi$.

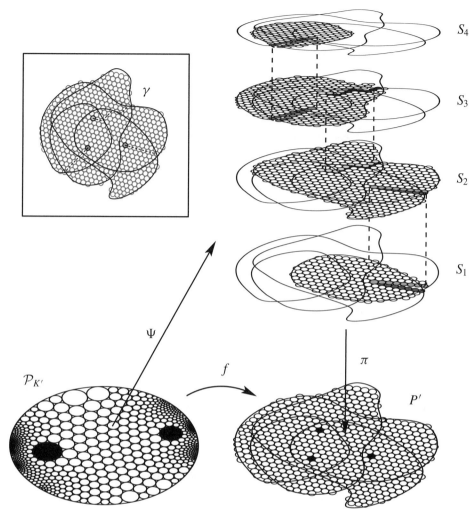

Figure 13.5. Approximating a disc algebra function.

13.5. Discrete Function Theory

Blaschke products are key players in classical function theory on the disc. Here are four of the reasons in discrete form. The first parallels a classical but not widely known 1947 result of Nehari and adds to our monotonicity results (Corollary 11.9); the proof goes exactly as the proof in the unbranched case (Lemma 8.3) and justifies the term "maximal branched packing." (This applies equally well to infinite K.) The companion distortion result subsumes Lemma 13.3 and will be important in the sequel.

Lemma 13.8 (Discrete Branched Schwarz Lemma). *Let K be a combinatorial closed disc and assume $b : \mathcal{P}_K \longrightarrow \mathcal{P}_{K,\beta}$ is a discrete finite Blaschke product with $b(0) = 0$.*

Suppose $f : \mathcal{P}_K \longrightarrow P$ is a discrete analytic self-map of $\mathbb{D}$ with $\beta \subseteq \mathrm{br}(f)$ and $f(0) = 0$. Then $f^{\#}(0) \leq b^{\#}(0)$, and equality implies $\mathrm{br}(f) = \beta$ and $f \equiv \lambda b$ for some unimodular constant λ.

Lemma 13.9 (Discrete Branched Distortion Lemma). *Let K be a combinatorial closed disc and assume $b : \mathcal{P}_K \longrightarrow \mathcal{P}_{K,\beta}$ is a discrete finite Blaschke product with $b(0) = 0$. Suppose $f : \mathcal{P}_K \longrightarrow P$ is a discrete analytic function with $\mathrm{br}(f) \subseteq \beta$ and $f(0) = 0$ and suppose further that for some $t > 0$ every boundary circle of P intersects the complement of the disc $\{|z| < t\}$. Then the euclidean ratio functions satisfy $f^{\#}(0) \geq t b^{\#}(0)$; equality implies $\mathrm{br}(f) = \beta$ and $f \equiv \lambda b$ for some constant λ with $|\lambda| = t$.*

Proof. It suffices to consider the case $t = 1$. With this, the argument is a simple generalization to accommodate branching of the argument used in the proof of hyperbolic uniqueness in Section 8.3.2. Let $\beta' = \mathrm{br}(f)$. Define the label S for K by

$$R(v) = \begin{cases} \mathrm{radius}(c_v) & c_v \subset \mathbb{D} \\ \infty & \text{otherwise.} \end{cases}$$

As before, this can be modified to become a legitimate hyperbolic label for K, and the modified label S will be in the family

$$\Phi = \{R : \theta_R(v) \leq 2\pi(\beta_v + 1) \text{ for every interior vertex } v\}.$$

In particular, S is a β'-superpacking label and (because $\beta' \subseteq \beta$) a β-superpacking label, with infinite boundary labels. But the branched maximal label, $\mathcal{R}_{K,\beta}$, is the infimum of precisely such β-superpacking labels (see the proof of Theorem 11.6). In other words, $S(v_1) \geq \mathcal{R}_{K,\beta}(v_1)$, implying $f^{\#}(0) > 1 = t$ for the hyperbolic ratio function; since $f(0) = 0$, this applies also to the euclidean ratio function. Equality could occur only if no adjustments to S were needed and $\beta' = \beta$; that is, if all the boundary circles of P are horocycles and P is a rotation of $\mathcal{P}_{K,\beta}$. $\square$

The next result explains something you might have picked up in various of our example packings – a particularly good instance is the discrete Blaschke product of Figure 13.3: For self-maps of $\mathbb{D}$ which fix the origin, the boundary horocycles in image packings seem larger (with euclidean eyes) than their counterpart horocycles in the domain.

Here is the classical formulation. Suppose $F : \mathbb{D} \longrightarrow \mathbb{D}$ is analytic. It is well known that F has a *radial limit* $F^*(\zeta)$ at almost every point (with respect to Lebesgue measure) $\zeta \in \mathbb{T}$; that is, $\lim_{r \uparrow 1} F(r\zeta) = F^*(\zeta)$ exists for almost all ζ. Furthermore, should $F^*(\zeta)$ belong to $\mathbb{T}$, then

$$F'(\zeta) = \lim_{r \uparrow 1} \frac{F^*(\zeta) - F(r\zeta)}{r - 1} \tag{66}$$

exists and either is real or diverges to $+\infty$. In the former case the limit is called the *angular derivative* of F at ζ. Now to our point: *If $F(0) = 0$ and $F'(\zeta)$ exists, then $F'(\zeta) \geq 1$, with* *equality if and only if F is univalent and a rotation.* This lemma gives the parallel discrete result.

Lemma 13.10 (**Discrete Angular Derivatives**). *Let K be a combinatorial closed disc and $f : \mathcal{P}_K \longrightarrow P$ a discrete self-map of $\mathbb{D}$ with $f(0) = 0$, and suppose w is a boundary vertex of K whose circle c_w in P is a horocycle. If C_w is the corresponding horocycle in $\mathcal{P}_K$, then their euclidean radii satisfy* radius $(c_w)/$radius $(C_w) \geq 1$, *with equality if and only if f is univalent and a rotation.*

The proof requires a small extension of the fixed point index of Definition 8.10: the curve σ need not always be Jordan, f need not always be a homeomorphism. If λ and σ intersect in at most two points and $f : \gamma \longrightarrow \sigma$ is fixed point–free and orientation-preserving (does not back up), then $\eta(f; \gamma) = \mathfrak{w}(f(z) - z; \gamma)$ is nonnegative.

Proof. Without loss of generality, assume C_w and c_w are both tangent to the unit circle at the point 1 and $P \neq \mathcal{P}_K$. For $\epsilon > 0$ small, let ψ_ϵ denote the Möbius transformation $\psi_\epsilon : z \mapsto (1 - \epsilon)(z + \epsilon)$ and write P_ϵ for the circle packing $\psi_\epsilon(P)$. Since $\psi_\epsilon(z)$ is a contraction toward the point $1 - \epsilon$, P_ϵ lies in the disc $D = \psi(\mathbb{D})$, which has compact closure in $\mathbb{D}$.

Suppose radius $(c_w)/$radius $(C_w) \leq 1$, so $c_w \subseteq \text{int}(C_w)$. By the Schwarz Lemma, the α-circles at the origin satisfy $c_1 \subseteq \text{int}(C_1)$. For ϵ sufficiently small, both these facts persist. Fix such an ϵ and run our winding number argument as in Section 8.4. Add auxiliary interstices to $\mathcal{P}_K$ to get a simple closed augmented boundary Γ lying outside $\mathbb{D}$; add auxiliary interstices to P_ϵ to get augmented boundary Σ inside $\mathbb{D}$.

As before, we may define a cadre $\mathcal{F}_{\mathcal{P}_K, P_\epsilon} = \{f_j\}$ of maps. The f_j for triangular elements $\mathfrak{e}_j$ are defined first and are homeomorphisms, as before; the induced maps f_k for circle elements $\mathfrak{e}_k$ will fail to be homeomorphisms when $\mathfrak{e}_k$ is a branch circle of P_ϵ. The cadre induces an orientation-preserving map $F : \Gamma \longrightarrow \Sigma$, and we have

$$\Sigma \subset \mathbb{D} \subset \text{int}(\Gamma) \Longrightarrow \eta(\Gamma, \Sigma) = 1. \tag{67}$$

On the other hand, cancellation works as always and we have

$$\eta(F; \Gamma) = \sum_j \eta(f_j; \mathfrak{e}_j).$$

All the indices are nonnegative; however, the ones associated with both c_1 and c_w are 1, yielding a contradiction with (67). $\quad\square$

Similar reasoning allows us to formulate a discrete analogue of a classical result related to those on harmonic measure in Section 8.2.1. I leave the details to the reader.

Lemma 13.11 (**Discrete Löwner's Lemma**). *Let K be a combinatorial closed disc and $f : \mathcal{P}_K \longrightarrow P$ a self-map of $\mathbb{D}$ with $f(0) = 0$. Suppose $\langle w_1, \ldots, w_n \rangle$ is a positively oriented boundary edge-path of K, with $\zeta_1, \ldots, \zeta_n$ the centers on $\mathbb{T}$ for the associated* *circles of* $\mathcal{P}_K$. *If* $|f(\zeta_j)| = 1$, $j = 1, \ldots, n$, *then the angular measure of the arc* $[\zeta_1, \zeta_n] \subset \mathbb{T}$ *is less than or equal to the angular measure of the arc* $[f(\zeta_1), f(\zeta_n)] \subset \mathbb{T}$. *Equality implies* f *is a rotation.*

Note how well these last two results complement one another: angular derivative $f^\#(\zeta) > 1$ reflects radial *expansion* toward the origin, while in the tangential (i.e., orthogonal) direction, Löwner's result gives tangential *expansion*. This homogeneous stretching is what you would expect if $f^\#(\zeta)$ were an actual complex derivative. Note, too, that all these results have *invariant* formulations that remove the condition $f(0) = 0$. In Löwner's Lemma, for example, one replaces angular (i.e., Lebesgue) measures of arcs by appropriate *harmonic* measures.

There is additional function theory concealed in our examples. Did you note a resemblance between Figure 13.3 and Figure 13.5? Suppose the curve γ in the disc algebra construction happens to lie in $\mathbb{D}$ (as can be arranged *via* simple scaling). Imagine that the sheets we used in the construction of f were cut from copies of the maximal packing Q used for b. After arranging the branch points, cuts, and pasting pattern for f, it might occur to us to simply extend those cuts out to the unit circle and do the pastings with the *full* copies of Q. That is precisely what is shown in Figure 13.3 (because the pattern of pastings is the same).

In hindsight, you can reverse this reasoning: constructing b first, you can afterward just cut away the excess circles on the various sheets to get f. In particular, the complex K' associated with f is a simplicial subcomplex of the complex K for b. That means that one can obtain the function f via a composition, as shown in Figure 13.6. On the left is $\mathcal{P}_{K'}$. The map φ embeds $\mathcal{P}_{K'}$ in $\mathcal{P}_K$ – its image circles are shaded. Now b maps $\mathcal{P}_K$ to $\mathbb{D}$ on the right, and the images of those shaded circles are those of the packing P' as embedded in P. It is natural to write $f \equiv b \circ \varphi$ in this setting.

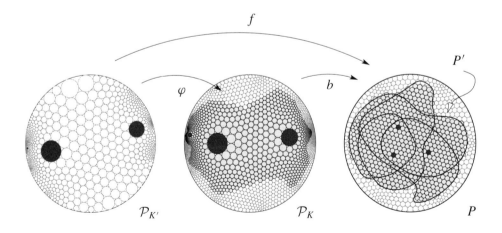

Figure 13.6. Discrete subordination.

This must seem terribly artificial to many readers, and is in fact a setup. However, *classical* disc algebra functions such as F (that is, associated with curves γ lying in $\mathbb{D}$) are always *subordinate* to finite Blaschke products; that is, $F(z) = B(\Phi(z))$, $\forall z \in \mathbb{D}$, where B is a classical Blaschke product, and Φ is a univalent analytic self-map of $\mathbb{D}$ with $\Phi(0) = 0$, the so-called *subordinating* function. Subordination is a ubiquitous classical technique. The chain rule gives

$$F'(0) = B'(\Phi(0))\Phi'(0) = B'(0)\Phi'(0).$$

By the Schwarz Lemma, $|\Phi'(0)| \leq 1$, so $|F'(0)| \leq |B'(0)|$. In light of the Discrete Schwarz Lemma for φ, you can track precisely the same inequalities with circle radii in Figure 13.6.

What I am contending here is that the geometry illustrated in Figures 13.3, 13.5, and 13.6 is right on the money. It appears artificial only because we have had to make the combinatorics of our maps compatible. Indeed, explicit *composition* of discrete analytic functions, as with $f \equiv b \circ \varphi$, is rare only because of such discretization obstacles.

13.6. Infinite Combinatorics

With the small stable of examples we have described, our discrete theory is as yet but a pale shadow of its classical model. To get anything like the classical richness, we must push ahead with the theory for combinatorial *open* discs, that is, for infinite hyperbolic complexes. Everything becomes tougher; theory, computation, construction, interpretation – there are plenty of open questions. I want to give the reader a few additional examples and suggest directions for development.

13.6.1. Discrete Univalent Functions

There is a potential goldmine of examples from the following result of Z-X. He and Oded Schramm:

Theorem 13.12 (Inverse Riemann Mapping Theorem). *Let K be a hyperbolic combinatorial open disc. Given any simply connected domain $\Omega \subset \mathbb{C}$, not the whole plane, there exists a univalent circle packing P for K with* carr$(P) = \Omega$. *That is, there exists a univalent discrete analytic function f for K with range Ω.*

This borders on the unbelievable. Let K be the heptagonal complex $K^{[7]}$, for example, and let Ω be the most horribly pathological Jordan region you can imagine. There exists a univalent circle packing for K filling Ω! Figure 13.7 suggests what we might see if Ω were the unit square. Even this simplest of examples is but speculative suggestion. The fact is that this beautiful theorem is nonconstructive; even for Ω a square and K finite, I know of no algorithm for the computations.

Figure 13.7. Squaring $K^{[7]}$ (simulated).

13.6.2. Discrete Blaschke Products

Definition 13.5 intentionally leaves open the door for discrete finite Blaschke products based on infinite complexes.

Theorem 13.13. *Suppose K is a combinatorial disc that is hyperbolic and of bounded degree and suppose β is a branch structure for K of valence $n - 1$. Then there is an essentially unique maximal branched packing $\mathcal{P}_{K,\beta}$ for K and β and the discrete analytic function $b : \mathcal{P}_K \longrightarrow \mathcal{P}_{K,\beta}$ is a discrete Blaschke product of valence n. In particular, $b \equiv B \circ h$, where h is a quasiconformal homeomorphism of $\mathbb{D}$ and B is a classical Blaschke product with branch set $h(\beta)$.*

For K finite, this is just Lemma 13.7. The infinite case involves an exhaustion $K_j \uparrow K$ by finite subcomplexes, the corresponding decompositions $b_j \equiv B_j \circ h_j$ from the finite case, the Ring Lemma in the Large (coming in Section 14.2.1) for uniform quasiconformality of the h_j, and finally, standard normal families arguments to extract convergent subsequences from the families $\{B_j\}$ and $\{h_j\}$. We will see similar arguments in Part IV of the book, so I will not pursue the details here. However, let me leave the reader with this question: is there a direct circle packing approach which might avoid the use of quasiconformality? one which could remove the bounded degree hypothesis?

Open Question 13.14. *Suppose K is a combinatorial open disc which is hyperbolic but has unbounded degree and let β be is a finite branch structure for K. Does there exist a corresponding maximal circle packing $\mathcal{P}_{K,\beta}$? Does it lie in $\mathbb{D}$? Is it proper?*

One approach would be a downward Perron method like that of Theorem 11.6. The issue is whether the process avoids degeneracy. If you settle this question, move on to the issue of infinite valence for β, where the field is entirely open.

13.6.3. Boundary Behavior

Certain of the results we have mentioned earlier, such as the discrete Schwarz and branched Schwarz lemmas, also apply to infinite K. Others are likely to extend as well once they are appropriately formulated. Classical function theory on the disc revolves largely around "boundary values" – how F(z) behaves as $|z| \uparrow 1$. That is fine in the discrete setting when K has an actual boundary, but what is the approach when K is a combinatorial open disc?

The most natural first target would be a discrete analogue of a famous result of Fatou: *If F is a bounded analytic function on $\mathbb{D}$, then F has a nontangential limit, denoted $F^*(\zeta)$, at almost every $\zeta \in \mathbb{T}$ (with respect to Lebesgue measure).* In words that I will not try to make precise, F has an extension almost everywhere to a boundary function F^* on $\mathbb{T}$. A companion target would be Lindelöf's theorem.

Open Question 13.15. *Suppose $f : \mathcal{P}_K \longrightarrow P$ is a bounded discrete analytic function, meaning that P is a bounded packing in $\mathbb{C}$.*

Discrete Fatou. *Suppose a sequence $\{Z_j\}$ of circle centers from $\mathcal{P}_K$ converges nontangentially to a point $\zeta \in \mathbb{T}$. Does $\{f(Z_j)\}$ converge in $\mathbb{C}$? for almost every ζ?*

Discrete Lindelöf. *If sequences $\{Z_j\}$ and $\{Z'_j\}$ of centers in $\mathcal{P}_K$ converge to the same point $\zeta \in \mathbb{T}$ and sequences $\{f(Z_j)\}$ and $\{f(Z'_j)\}$ converge to values $w, w' \in \mathbb{C}$, respectively, are w and w' necessarily equal?*

Probabilistic methods provide a promising *entree* to these questions. Here is a sample; the reader may have to come back here after picking up some definitions in Chap. 18.

Theorem 13.16 (Benjamini and Schramm). *Let K be a combinatorial open disc that is hyperbolic and of bounded degree and let P be a univalent circle packing for K in the plane. Almost every simple random walk on the circles of P will converge to some point ξ of the boundary of* carr(P).

For example, if $f : \mathcal{P}_K \longrightarrow P$ is a discrete univalent function, then almost every simple random walk will lead in $\mathcal{P}_K$ to a point $\zeta \in \mathbb{T}$ and in P to a point ξ in the boundary of carr(P). It is tempting to define $f^*(\zeta) = \xi$. But, of course, we would have to settle the Discrete Lindelöf question and a few other issues before we could rightly claim a *boundary function* f^*. Moreover, we would hope to extend the results to more general nonunivalent cases.

Once we understand boundary behavior, the next step will be to *control* it. An opening challenge here might be to extend the disc algebra construction of Section 13.4.1 to infinite K. Immediately we face the question of how to impose boundary conditions when we have only an "ideal" boundary.

The reader might note some distinctions arising between what we can *prove* and what we can actually *do* in a practical sense. Putting aside the Inverse Riemann Mapping Theorem, many of these topics seem approachable, and despite the infinities involved they may be amenable to experimentation. In any case, there is a huge amount of classical theory waiting to be tackled here, and much of it is profoundly geometric. Let me end with a few of the buzzwords that might evoke questions for those familiar with the classical studies: *Hardy*, *Bloch*, *Bergmann*, *bounded mean oscillation*, and *Dirichlet* spaces, *Blaschke* condition, *Pick* interpolation, *outer* functions, *Bloch* and *Landau* constants, and (my favorites) the *inner* functions.

13.7. Experimental Challenges

To end on an experimental note, I offer three challenges, each concerned with some fundamental classical constant. It is quite natural to wonder whether analogous constants exist – in fact, whether roughly the *same* constants work – in the discrete theory.

13.7.1. The Class S

Among the most studied classical functions are those of "class **S**," the univalent functions F on $\mathbb{D}$ (also known as *schlicht*) normalized by $F(0) = 0$ and $F'(0) = 1$. These are univalent functions having power series representations

$$F(z) = z + \sum_{j=2}^{\infty} a_n z^n = z + a_2 z^2 + a_3 z^3 + \cdots + a_n z^n + \cdots.$$

According to the famous *Bieberbach Conjecture*, now the deBranges Theorem: *If $F \in$ **S**, then the coefficients in its power series satisfy $|a_n| \leq n$ for $n \geq 2$. A single case of equality implies equality for all n and F is the Koebe function $k(z) = z/(1 - z)^2$.* It was at the 1985 Purdue conference celebrating deBranges's proof that Thurston introduced circle packing to analysts.

Power series are not in our ken, but since every $F \in \mathcal{S}$ is identified uniquely with the simply connected domain $F(\mathbb{D}) \subset \mathbb{C}$, there is a direct geometric link to these functions. Using this we can define a fairly close discrete parallel.

Definition 13.17. The *discrete* **schlicht** class $\mathbf{S}_{\text{dis}}$ is the class of univalent discrete analytic functions $f : \mathcal{P}_K \longrightarrow P$ on the unit disc, normalized by $f(0) = 0$ and $f^{\#}(0) = 1$

(euclidean ratio function). In other words, $P \subset \mathbb{C}$ is univalent and the circles at the origin in $\mathcal{P}_K$ and P are identical.

The most famous result for class **S** is Koebe's 1/4-Theorem: *If* $F \in \mathbf{S}$, *then the image region* $F(\mathbb{D})$ *contains the disc* $D(0, 1/4)$. Z-X. He and Oded Schramm have shown the failure of the discrete parallel: *There is no open disc at the origin which is guaranteed to lie in the carrier of the image packing for every* $f \in \mathbf{S}_{\mathrm{dis}}$. However, their counterexamples rely on packings having ever larger degrees, while experiments with packings having hexagonal combinatorics suggest something close to the Koebe 1/4.

Open Question 13.18. *Suppose* K *is a combinatorial disc and* $f : \mathcal{P}_K \longrightarrow P$ *is a discrete analytic function of class* $\mathbf{S}_{\mathrm{dis}}$. *Does there exist a constant* $s_d > 0$, *depending only on* $d = \deg(K)$, *so that* $\{|w| < s_d\} \subset \operatorname{carr}(P)$? *If affirmative, is* s_6 *roughly 1/4?*

It is unfortunate that I cannot show you a movie, but Figure 13.8 illustrates a sequence of experiments for $d = 6$. Here is the protocol: the domain packing is $\mathcal{P}_K$ in (a). The circle C_1 is at the origin and I have added a dashed circle having 4.0 times its radius. You are permitted to manipulate the image packing P within the following constraints: c_1 is at the origin, $\mathbb{D} \subset \operatorname{carr}(P)$, and $z = -1$ is in the boundary of $\operatorname{carr}(P)$. **Challenge:** *Can you make* c_1 *larger than the dashed circle?* If you succeed with P univalent, then you have a counterexample to the discrete Koebe 1/4.

Perhaps you see the strategy in Figures 13.8(b)–(e); try to force c_1 to grow by growing the boundary circles. However, since some boundary vertices are pinned near $z = -1$, you need to make their radii relatively *small*. Make them too small and you lose univalence, which you balance by making some neighbors smaller as well. If you just keep working to maintain the right balance between these competing demands, you can, in fact, get c_1 to grow, as it does in each successive image here.

The final packing, Figure 13.8(f), is "carrier-univalent" but no longer univalent. I included this packing to help us reconnect to the classical result. The extremal function for the Koebe 1/4 Theorem is the Koebe function $k(z)$ mentioned above, which maps $\mathbb{D}$ univalently to the plane slit from $-1/4$ to $-\infty$ along the negative real axis. The discrete analytic function $f : \mathcal{P}_K \longrightarrow P$ for P of Figure 13.8(f) is trying mightily to mimic $4k(z)$. As you might guess, if you disregard univalence and let the trend in these pictures continue so the two edges of the carrier start overlapping, you can get c_1 to exceed the dashed circle.

13.7.2. Dieudonné–Schwarz

Our second example concerns a classical extension of the Schwarz Lemma due to Dieudonné: *If* $F : \mathbb{D} \longrightarrow \mathbb{D}$ *is analytic with* $F(0) = 0$, *then*

$$|F'(z)| \leq \begin{cases} 1, & |z| \leq \sqrt{2} - 1 \\ \dfrac{(1 + |z|^2)^2}{4|z|(1 - |z|^2)}, & \sqrt{2} - 1 < |z| < 1. \end{cases}$$

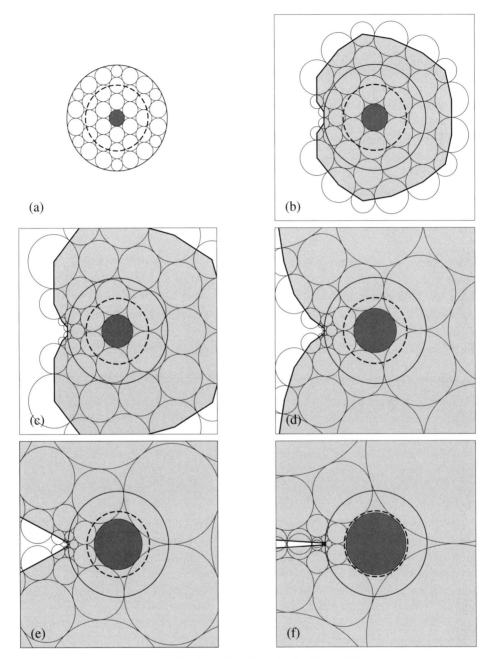

Figure 13.8. Experimenting with the Koebe 1/4.

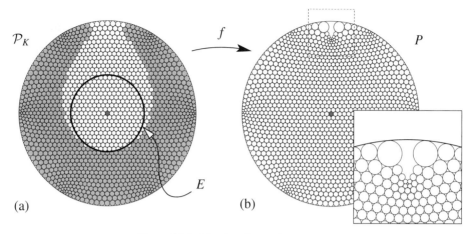

Figure 13.9. Dieudonné experiments.

The discrete test setup is shown in Figure 13.9. The maximal packing of some K is shown on the left, with a circle E of radius $\sqrt{2}-1$ superimposed, a hyperbolic packing P for K is shown on the right, and $f : \mathcal{P}_K \longrightarrow P$. To mimic $|F'|$, we must work with the *euclidean* ratio function $f^{\#}$. Each of the shaded circles on the left has a larger *euclidean* radius in P than it does in $\mathcal{P}_K$. **Challenge:** *Can you get a shaded circle to penetrate the classical "exclusion zone", that is, to move inside E?* On a computer you can continually modify P and watch the pattern of shaded circles. You may be surprised by the geometry. Figure 13.9 shows the effect in a hexagonal case when we mimic a small "dimple" in the image. I will let the reader contemplate the reasons for the behavior, basically the same in the classical and discrete settings. The Dieudonné inequalities are sharp, with 2-fold Blaschke products as extremals; if you push the example in Figure 13.9 further with this in mind, you can in fact penetrate the exclusion zone – but just barely!

Open Question 13.19. *Consider discrete analytic functions $f : \mathcal{P}_K \longrightarrow P$ on $\mathbb{D}$ with $f(0) = 0$ and euclidean ratio functions $f^{\#}$. Does there exist a universal constant $C > 0$ such that $f^{\#}(Z_v) \leq 1$ whenever circle center Z_v satisfies $|Z_v| \leq C$?*

The two experiments described above are quite entertaining – maybe not video game material, but fun – and the geometry is subtle. I tend to believe that the answers to both Open Questions are "yes" and that constants s_6 and C will be roughly $1/4$ and $\sqrt{2}-1$, respectively. I have to admit, however, that my experiments have been largely with small hexagonal packings; perhaps the reader can see how to choose more challenging combinatorics.

13.7.3. Nehari's Condition

One can often manipulate boundary labels for a given complex K to obtain circle packings P that are univalent by visual inspection – several examples occur in the Menagerie. The methods remain ad hoc, however, and it would be helpful to have more honest techniques, perhaps discrete versions of classical criteria for "starlike" or "convex" functions. A model we might hope to emulate is due to Nehari: if F is analytic on $\mathbb{D}$ then its *Schwarzian derivative* S_F is given by

$$S_F(z) = \left(\frac{F''(z)}{F'(z)}\right)' - \left(\frac{1}{2}\right)\left(\frac{F''(z)}{F'(z)}\right)^2.$$

The *Nehari condition* states that if $|S_F(z)| \leq 2(1 - |z|^2)^{-2}$ for $z \in \mathbb{D}$, then F is univalent.

Open Question 13.20. *Do there exist univalence criteria for packing labels R? That is, for given K, are there conditions on R that will ensure that $P \longleftrightarrow K(R)$ will be univalent?*

This is quite an open-ended question, and I do not have a particular approach to suggest. The Schwarzian is a very geometric quantity, as evidenced by the fact that S_F is unchanged if F is pre-/post-composed with a Möbius transformation. It has already found some uses in circle packing; here would be a good place to start developing its discrete version.

Let me wrap up this chapter by suggesting the appropriate spirit of these investigations should you undertake them. First, most questions will involve a few niggling discretization issues – univalent or just carrier-univalent? euclidean or hyperbolic ratio functions? centers or whole circles? and so forth and so on. Do not let these discretization details obscure the underlying geometry.

Second, the geometric issues often reside with the small packings. In Part IV of the book we will be proving that discrete analytic functions tend to converge to their classical counterparts as their combinatorics become finer. One can foresee existence arguments that run "*... and such-and-such constant exists for all sufficiently large complexes ... and since there are only finitely many small complexes ... there exists some universal constant ...*" I do not find such arguments to be very sporting, since they largely miss the geometric spirit of the hunt.

14

Discrete Entire Functions

In this section K will be a parabolic combinatorial open disc, so the maximal packing P_K fills the euclidean plane. The function $f : P_K \longrightarrow P$ is a discrete *entire* function if $P \subset \mathbb{C}$ or a discrete *meromorphic* function if $P \subset \mathbb{P}$.

The practical difficulties of experimenting with infinite complexes are self-evident and our collection of examples remains rather limited. We will illustrate three basic approaches to constructing examples: (1) exhaust K with finite subcomplexes and study converging sequences among their packings; (2) exploit the combinatorial regularity of certain infinite complexes and fortuitous numerical patterns for their labels; or (3) construct image surfaces and confirm that the resulting complexes are parabolic.

14.1. Liouville and (Barely) Beyond

The hyperbolic/parabolic dichotomy is the great divide in the study of classical analytic functions just as it is for us. Landing on the "parabolic" side brings one to an enduring theme in classical analysis, *value distribution theory*, and to some of the legendary names: Liouville, Picard, Ahlfors, Nevanlinna. We can eke out only a small bit of general theory here, beginning with the basic dichotomy itself: P_K is parabolic because its circle packings can not live in the disc.

Theorem 14.1 (Discrete Liouville Theorem). *There exist no bounded discrete entire functions.*

The classical Liouville Theorem extends also to entire functions of slow growth. For example, if F is entire and $|F'|$ is bounded, then F is an automorphism, $F(z) = az + b$. Of course, we look to the ratio function $f^\#$ for our analogy, and we run into something that we see often in the sequel: the *bounded degree* side condition on K. This result, for instance, follows from probabilistic methods that we introduce in Section 18.4.

Theorem 14.2. *Suppose K is a combinatorial open disc that is parabolic and of bounded degree. If $f : P_K \longrightarrow P$ is a discrete entire function for which $f^\#$ is bounded, then f is an automorphism of the plane.*

181

A striking improvement on Liouville's Theorem in the classical theory is the so-called *Little Picard Theorem*. The discrete analogue is more than likely true for general parabolic complexes of bounded degree, but it has been proven only for the hexagonal complex by Callahan and Rodin.

Theorem 14.3 (Discrete Picard Theorem). *Let* $f : \mathcal{P}_H \longrightarrow P$ *be a discrete entire function for the hexagonal complex H. Then f omits at most one point in the plane.*

We cannot go into the proof, which relies on the deep "five islands theorem" of Ahlfors. (The Callahan and Rodin arguments apply to branched hexagonal packings without change.) Like many other results in this vein, such as the as yet untouched "Great" Picard Theorem, this last theorem would follow from the theory of quasiregular functions if one could show that f was in fact quasiregular. That has not yet been shown, even in the hexagonal case. This highlights a major ingredient that has gone missing: the Rodin and Sullivan Ring Lemma. A petal in a flower can be made arbitrarily small compared to the center if its neighboring petals are allowed to overlap one another, and this can happen even without branching, as a look ahead to Figure 14.4 will show.

14.2. Discrete Polynomials

The first class of entire functions one would hope to see is the (complex) polynomials. Tomasz Dubejko has developed a theory giving us a full spectrum of discrete polynomials – in fact, this remains the only setting in which we have a systematic way to create discrete entire functions. Classical polynomials may be characterized as the *proper* analytic self-mappings of $\mathbb{C}$, and we can exploit this geometric property much as we did earlier with Blaschke products.

Definition 14.4. A **discrete polynomial** is a discrete entire function $f : \mathcal{P}_K \longrightarrow P$ which is a proper self-mapping of $\mathbb{C}$. By standard topological arguments, f must have valence n and total branch order $n - 1$ for some $n \geq 1$, in which case f is said to have **degree** n. Degree 1 polynomials are automorphisms of $\mathbb{C}$.

By Stoïlow's Theorem (11.2), $f \equiv F \circ h$, where $h : \mathbb{C} \longrightarrow \Omega$ is a homeomorphism and $F : \Omega \longrightarrow \mathbb{C}$ is analytic and proper. It is well known that these last conditions imply first that $\Omega = \mathbb{C}$ and then that F is an n-degree polynomial. In other words, $f \equiv F \circ h$ for some classical polynomial F; this certainly justifies our terminology.

14.2.1. Ring Lemma in the Large

To construct polynomials we need distortion control, and as we noted above, flowers can now have (noncontiguous) petals which overlap. In the context of a larger packing, however, this has a topological consequence in that the flower must enclose either a branch point or a part of the carrier boundary. That is why the following *local* result requires *global* finiteness conditions.

Lemma 14.5 (Ring Lemma in the Large). *Given integers $d > 0$ and $m > 0$, there exist constants $N = N(d, m)$ and $\mathfrak{c} = \mathfrak{c}(d, m)$ with the following properties. Suppose K is a combinatorial disc with $\deg(K) \leq d$ and P is a euclidean circle packing for K with valence bounded by m. Then if v_0 is a vertex of K that is at least N combinatorial generations from ∂K and if r_0 is its radius, then the radius r for any neighbor of v_0 satisfies $r \geq \mathfrak{c} \cdot r_0$.*

Proof. Suppose the result fails. Then there exists a sequence $\{K_n\}$ of combinatorial closed discs with degrees bounded by d with associated packings $P_n \longleftrightarrow K_n(R_n)$ (say euclidean) with the property that the α-vertex $v_1^n \in K_n$ is separated by at least n generations from ∂K_n and has neighbor w^n with

$$R_n(w^n) < (1/n) \cdot R_n(v_1^n). \tag{68}$$

The bound d on degree allows us to use combinatorial normal families; by Theorem 11.14 we may assume without loss of generality that the K_n converge to a complex K with α-vertex v_1. By a scaling of the packings P_n and a diagonalization argument we may assume that the labels R_n converge to a function $\widehat{R}$ on K, with $\widehat{R}(v) \in [0, \infty]$ and $\widehat{R}(v_1) = 1$. As we have done before, we partition the vertices into the sets of *open* and *solid* dots,

$$\mathcal{D}_o = \{v \in K : \widehat{R}(v) = 0\}$$
$$\mathcal{D}_s = \{v \in K : \widehat{R}(v) \in (0, \infty]\}.$$

By our normalization, v_1 is a solid dot, and by (68) it has a neighbor w which is an open dot. Let $\mathcal{C}$ be the component of $\mathcal{D}_o$ containing w. The combinatoric situation is suggested by Figure 14.1: namely, the solid dot v_1 lies in a two-way infinite, simple edge-path

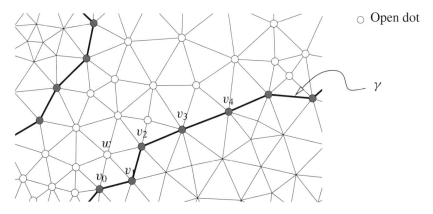

Figure 14.1. Solid dots bordering a component $\mathcal{C}$ of open dots.

$\gamma = \{\ldots, v_0, v_1, v_2, \ldots, v_M, \ldots\}$ of solid dots neighboring $\mathcal{C}$. The justification lies in Lemma 11.7: $\mathcal{C}$ must be infinite to avoid situation (b), and the vertices v_j of γ must be distinct to avoid situation (c).

Define $M = 3m + 1$. Let $\mathcal{V}$ denote the vertices of $\mathcal{C}$ which are neighbors of v_j for $j = 1, \ldots, M$; this is a finite, edge-connected set. Let $\mathfrak{a} = \min\{\widehat{R}(v_1), \widehat{R}(v_2), \ldots, \widehat{R}(v_M)\}$, so $\mathfrak{a} > 0$. Returning to our packings, suppose we translate each P_n so its circles for v_1 and v_2 are centered on the real axis with their tangency point at the origin. It is easy to see that because $R_n(v_1)$ has a finite limit, the fan of its petal vertices from v_2 counterclockwise to v_0 are associated with vertices of $\mathcal{C}$. In particular, their radii go to zero as n grows and it is easy to see that their circles converge to the origin, which is the tangency point of c_1 and c_2 in each of the packings P_n. Moreover, because $\mathcal{V}$ is finite and connected and in $\mathcal{C}$, all its circles likewise converge to the origin.

Summary. As n grows, the M circles for $v_1, \ldots, v_M$ have radii converging to no less than $\mathfrak{a}$ but have neighbors that are converging to the origin. That is to say, we have M large circles huddling ever closer to the origin. Let $\omega = \exp(2\pi i/3)$ and consider the three complex numbers $z_1 = \mathfrak{a}, z_2 = \mathfrak{a}\omega, z_3 = \mathfrak{a}\omega^2$, which are equally distributed about the origin. It is an elementary observation that a circle of radius greater than $\mathfrak{a}$ that is sufficiently close to the origin must contain at least one of the points z_1, z_2, z_3. Since we have M such circles and $M > 3m$, when n is large there must be at least one of the points z_j that lies in m distinct circles of P_n. This contradicts the hypothesis that P_n has valence at most m and we are done. $\qquad\square$

14.2.2. Polynomials: Existence and Uniqueness

Theorem 14.6. *Let K be a combinatorial open disc that is parabolic and of bounded degree and let β be a branch structure for K having cardinality $n - 1$. There exists an essentially unique discrete polynomial f of degree n defined on K for which $\mathrm{br}(f) = \beta$.*

Proof. Follow your experimental instincts here. Choose nested, finite, simply connected complexes exhausting K, $K_j \uparrow K$, so that each K_j contains β and the α-vertex v_1. As we saw in Section 13.3, for each j there is a discrete finite Blaschke product $b_j : \mathcal{P}_{K_j} \longrightarrow P_j$ with branch set β and with v_1 centered at the origin in both domain and range. Treating these packings as euclidean, we may scale each so that its circle for v_1 has euclidean radius 1. Abusing notation, let us keep the same names for the normalized packings. Define f_j as the discrete analytic function $f_j : \mathcal{P}_{K_j} \longrightarrow P_j$.

In the domain the packings $\mathcal{P}_{K_j} \longrightarrow \mathcal{P}_K$ (as in our original construction of $\mathcal{P}_K$). The degrees and valences of the packings P_j are uniformly bounded, so the Ring Lemma in the Large gives us the key condition (4) for normal families, Theorem 11.15. In particular, we may extract a subsequence so that $P_{j_i} \longrightarrow P$ for some euclidean circle packing P for K. Define f as the discrete entire function $f : \mathcal{P}_K \longrightarrow P$ and R as the label for P.

Regarding the properties of f, note first that by construction $\mathrm{br}(f) = \beta$ and $f_j \longrightarrow f$ uniformly on compact subsets of the plane. For j large, the maps f_j will be κ-quasiconformal on each face of $\mathrm{carr}(\mathcal{P}_{K_j})$ for some fixed $\kappa = \kappa(n, d) \geq 1$ depending only on n and $d = \deg(K)$. Let me explain the reasoning, which depends on background in Appendix A. By the Ring Lemma in the Large, there is a constant $\mathfrak{C} = \mathfrak{C}(n, d) > 1$ so that the triples of radii for every face of $\operatorname{carr}(\mathcal{P}_{K_j})$ and $\operatorname{carr}(P_j)$ are $\mathfrak{C}$-bound (see Section A.2.4). Lemma A.9(a) then implies the existence of κ. By Stoïlow's Theorem, $f_j \equiv F_j \circ h_j$ for analytic function F_j and homeomorphism h_j. In particular, then h_j is κ-quasiconformal. By arranging that $h(0) = 0$ and $h(1) = 1$, standard arguments lead one to limit functions F analytic with branch set $h(\beta)$, h κ-quasiconformal, and $f \equiv F \circ h$. Since the F_j are finite valence, F is finite valence, and by Liouville's Theorem h maps onto the plane. Thus F is a classical polynomial. In particular, f is proper, hence a discrete polynomial for K with branch set β.

It remains to prove the essential uniqueness of f. It suffices to prove that the associated label R is unique subject to the normalization $R(v_1) = 1$. This will be a two-step process. Here we only establish the following claim that R is "comparable" to any competing label. When we have developed the necessary probabilistic methods, we will prove in Section 18.4 that "comparable" implies "equal."

Claim. *Suppose $p_1 : \mathcal{P}_K \longrightarrow P_1$ and $p_2 : \mathcal{P}_K \longrightarrow P_2$ are two discrete polynomials for K with branch set β. Suppose $P_1 \longleftrightarrow K(R_1)$ and $P_2 \longleftrightarrow K(R_2)$. Then there exists a constant $\mathfrak{C}$ such that*

$$\frac{R_1(v)}{R_2(v)} \in \left[\frac{1}{\mathfrak{C}}, \mathfrak{C}\right], \quad \forall v \in K. \tag{69}$$

Proof of Claim. Write $p_1 \equiv F_1 \circ h_1$ and $p_2 \equiv F_2 \circ h_2$ for n-degree polynomials F_1, F_2 and κ-quasiconformal homeomorphisms h_1, h_2. Assume that the circles for v_1 are centered at the origin in each of $\mathcal{P}_K$, P_1, P_2 and that $R_1(v_1) = 1 = R_2(v_1)$.

We first need some classical a priori bounds on asymptotic behavior of the p_j. Here E_ρ denotes the circle $\{|z| = \rho\}$. Quasiconformal theory gives a constant $\mathfrak{k}$ (depending only on κ) so that for all sufficiently large ρ, $\max_{E_\rho}|h_j(z)|/\min_{E_\rho}|h_j(z)| \leq \mathfrak{k}$, for $j = 1, 2$. Meanwhile, as polynomials of degree n, the functions F_j behave asymptotically like z^n. Standard arguments combining these facts for $p_j = F_j \circ h_j$ show that there exists ρ' such that

$$\frac{\max_{E_\rho} |p_j(z)|}{\min_{E_\rho} |p_j(z)|} < 2\mathfrak{k}, \quad \text{for } \rho \geq \rho', \quad j = 1, 2. \tag{70}$$

To explain the circle packing geometry, let me introduce a notion I will call "bracketing." A euclidean circle packing Q is said to be *bracketed* by s and t, $0 < s \leq t$, if Q lies in the disc $D(0, t)$, but all its boundary circles intersect the complement of $D(0, s)$. We will write $Q \prec [[s, t]]$; an example is shown on the right in Figure 14.2.

The following lemma is a simple consequence of the Branched Schwarz and Distortion Lemmas of Section 13.5 (though phrased using packings and not functions).

Lemma 14.7. *Suppose L is a combinatorial closed disc and Q is a euclidean circle packing for L with branch set β and with circle c_v centered at the origin. Let $\mathcal{P}_{L,\beta}$ be the*

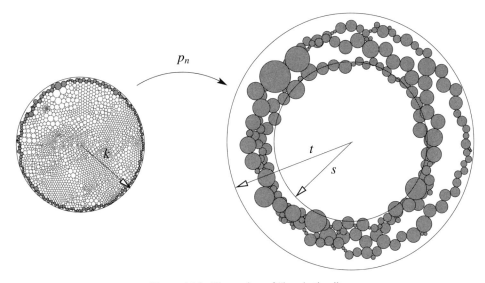

Figure 14.2. The notion of "bracketing."

branched maximal packing with circle C_v centered at the origin. If $Q \prec [[s, t]]$, then its
euclidean radii satisfy $s \leq \text{radius}(c_v)/\text{radius}(C_v) \leq t$.

Our goal is to bracket portions of P_1 and P_2. I am going to temporarily put some
discretization issues aside so that we can focus on the essential geometry. That geometry
is elementary – try not to let the notation obscure that. For integer k, let K_k be the largest
simply connected subcomplex of K so that all the circles of $\mathcal{P}_K$ associated with vertices
of K_k lie in $D(0, k)$ (as illustrated on the left in Figure 14.2). Assume k is sufficiently large
so that $\beta \subset K_k$. Define $P_j^{(k)}$ to be the restriction of P_j to K_k. Let $\mathcal{P}_{K_k,\beta}$ be the maximal
packing for K_k that has the circle for v_1 centered at the origin and let $\mathfrak{r}_k$ be the euclidean
radius for that circle. Define

$$m_{j,k} \equiv \min_{E_k} |p_j|, \qquad M_{j,k} \equiv \max_{E_k} |p_j|.$$

By (70),

$$M_{j,k} \leq 2\mathfrak{k} m_{j,k}, \qquad j = 1, 2. \tag{71}$$

Our choice of K_k (almost) means

$$P_j^{(k)} \prec [[m_{j,k}, M_{j,k}]], \qquad j = 1, 2. \tag{72}$$

Two applications of the lemma for vertex $v = v_1$ and the fact that $R_1(v_1) = 1 = R_2(v_1)$
imply

$$m_{1,k} \leq \frac{1}{\mathfrak{r}_k} \leq M_{1,k}, \qquad m_{2,k} \leq \frac{1}{\mathfrak{r}_k} \leq M_{2,k}.$$

With (71), we get

$$m_{1,k}/2\mathfrak{k} \leq m_{2,k} \leq M_{2,k} \leq 2\mathfrak{k} M_{1,k} \leq 4\mathfrak{k}^2 m_{1,k}.$$

If we define $s_k = m_{1,k}/2\mathfrak{k}$, then $P_j^{(k)} \prec [[s_k, 8\mathfrak{k}^3 s_k]]$, $j = 1, 2$. With another factor of 2 thrown in, we can also bracket translates of these packings. In particular, for given $a \in \mathbb{C}$ let ϕ_a be the translation $\phi_a : z \mapsto z - a$. Then the bracketing above and the fact that $s_k \longrightarrow \infty$ as $k \longrightarrow \infty$ implies that for any fixed $a \in \mathbb{C}$,

$$\phi_a\left(P_j^{(k)}\right) \prec [[s_k/2, 16\mathfrak{k}^3 s_k]], \quad j = 1, 2, \qquad \text{for all sufficiently large } k. \tag{73}$$

This last puts us right where we want to be. Fix an arbitrary vertex $v \in K$ and let a_j be the center for v in P_j, $j = 1, 2$. Fix k so large that $|a_1|, |a_2| \le s_k/2$ and let $\mathfrak{r}_v$ be the euclidean radius for v in the branched maximal packing $\mathcal{P}_{K_k,\beta}$ which puts the circle for v at the origin. Recall that $R_j(v)$ restricted to K_k is the label for $P_j^{(k)}$ and hence also for the translated packing $\phi_{a_j}(P_j^{(k)})$. By (73),

$$s_k/2 \le \frac{R_j(v)}{\mathfrak{r}_v} \le 16\mathfrak{k}^3 s_k, \quad j = 1, 2 \quad \Longrightarrow \quad \frac{R_1(v)}{R_2(v)} \in \left[\frac{1}{32\mathfrak{k}^3}, 32\mathfrak{k}^3\right]. \tag{74}$$

Since v is arbitrary in (74), we would appear to be done with the claim using the constant $\mathfrak{C} = 32\mathfrak{k}^3$.

However, I mentioned some suspended discretization issues. For given k, the circles of K_k in $\mathcal{P}_K$ lie inside the disc $D(0, k)$; the boundary circles do not reach E_k, so some boundary circle c_w of $P_j^{(k)}$ must fail to intersect the complement of $D(0, m_{j,k})$ and the bracketing of (73) may fail. However, c_w must have a *neighboring* circle c_u that intersects $D(0, m_{j,k})$, and this allows us to patch things up. This is easy, but it can hardly be called trivial since this is where the hypotheses of the theorem come in!

Work with p_1. It has valence n, so P_1 has at most n circles that enclose the origin. We assume k is sufficiently large so that these are interior circles of $P_1(k)$. Also, by the Ring Lemma in the Large, there is a bound $\mathfrak{c}$ on the ratio of neighboring radii of circles in P_1. Now, draw a picture with (1) c_u intersecting the complement of, say, $D(0, t)$; (2) c_w tangent to c_u and neither enclosing the origin; and (3) radius (c_u)/radius $(c_w) \in [1/\mathfrak{c}, \mathfrak{c}]$. Conclude that c_w must intersect the complement of $D(0, t/\mathfrak{c})$. The lower part of the bracketing of (74) will hold if we multiply by a factor depending only on the degree of p_1 and the degree of K. A similar argument applies to the upper part of the bracketing and to P_2. We conclude, then, that we can salvage the bracketing of (74) if we replace $32\mathfrak{k}^3$ by a larger constant. This completes the proof of the claim. $\square$

Recall that proof of uniqueness from the claim will be carried out with probabilistic tools in Chap. 18. $\square$

Discrete polynomials are in a sense the "parabolic" versions of discrete finite Blaschke products. In view of the theorem, it is consistent to denote the image packing P by $\mathcal{P}_{K,\beta}$ and describe it as the *maximal branched packing* for K and β. Indeed, existence of $\mathcal{P}_{K,\beta}$ follows from normal families arguments without the need for any bound on the degree of K. Absent control on distortion, however, the issues of properness and uniqueness of the resulting function remain, leaving us with this open question.

Open Question 14.8. *Suppose K is a parabolic combinatorial disc that is not of bounded degree and suppose β is a finite branch structure for K. Does there exist a polynomial for K with branch set β? If it exists, is it essentially unique?*

14.3. Discrete Exponentials

After polynomials, the *exponential* map would probably top everyone's wish list of functions. I know of two openings for its construction, and both are illustrated in the example of Figure 2.7 in Chap. 2.

From the geometric standpoint, $z \mapsto e^z$ is the universal covering map of $\mathbb{C}^* = \mathbb{C}\backslash\{0\}$. Any univalent packing P filling $\mathbb{C}^*$, say with complex K, induces a packing $\widetilde{P}$ of $\mathbb{C}^*$ for the universal covering complex $\widetilde{K}$ of K. Assuming $\widetilde{K}$ is parabolic, the discrete entire function $f : \mathcal{P}_{\widetilde{K}} \longrightarrow \widetilde{P}$ has the mapping properties of the exponential function.

Alternately, with favorable combinatorics, one can leverage the arithmetic properties of the exponential. Our hexagonal constructions are based on this observation of Peter Doyle: *Let $F = \{v : v_1, \ldots, v_6\}$ be a 6-flower. Given any two positive numbers a and b, label F so that the ratios of the petal labels to the central label are, in order, $\{a, b, b/a, 1/a, 1/b, a/b\}$. The angle sum at v for this label is precisely 2π.*

To build an example, let $Q = \mathcal{P}_H$ denote the regular hexagonal packing of $\mathbb{C}$ whose centers form the hexagonal lattice Λ generated by $z_1 = 1$ and $z_2 = (1 + \sqrt{3})/2$: that is, $\Lambda = \{nz_1 + mz_2 : n, m \in \mathbb{Z}\}$. Let $v_{n,m}$ denote the vertex of H whose circle is centered at $nz_1 + mz_2$. Given positive parameters a and b, define label $R = R_{a,b}$ for H by $R(v_{n,m}) = a^n b^m$. Using Doyle's observation, the reader can verify that R is a packing label. Let $P = P_{a,b} \longleftrightarrow H(R_{a,b})$ be its circle packing. When $a = b = 1$, $P = Q$, but in all other cases, P forms an infinite-sheeted packing spiraling about some point ζ in the plane, a so-called *Doyle spiral*. If we assume a normalization placing ζ at the origin and α-circle $v_{0,0}$ at the point 1, then the discrete entire function $f : \mathcal{P}_H \longrightarrow P$ behaves like an exponential map – in fact, for some real λ it will interpolate $\exp(\lambda z)$ at points of the lattice Λ. (To be fair, I must tell you that the spiral in Figure 2.7 is not *typical* among the Doyle spirals. It illustrates what is known as a *coherent* spiral. If you are curious about Doyle's observation and these spirals, be sure to visit Appendix C.)

Discrete exponentials such as f exhibit the mapping properties of their classical models, so we see behavior not encountered with polynomials. In Fig. 2.7, for instance, we observed that f is periodic, and since it omits zero, f demonstrates sharpness of the Discrete Little Picard Theorem. There exist chains of circles going off to infinity in the domain whose images converge to zero in the range, so f is not proper; zero is called an *asymptotic value* for f. The behavior of the image around zero is the archetype for what is known as *logarithmic* branching.

Exponential functions have very clean behavior. More general entire functions, on the other hand, are complicated mixtures of finite and logarithmic branching, asymptotic values, and other geometric features that we cannot yet hope to mimic. We can nonetheless attempt the next logical step.

14.4. The Discrete Error Function

In classical function theory, if $F(z)$ is a locally univalent entire function, then $F'(z)$ will be a nonvanishing entire function, and hence will be the logarithm of a second entire function $G(z)$. That is, F is an indefinite integral, $F(z) = \int^z \exp(G(\zeta)) \, d\zeta$. The simplest example is the exponential itself, with $G(z) = z$. It makes sense that the next function to target would involve $G(z) = z^2$. That, in fact, defines the classical *error function*,

$$\operatorname{erf}(z) = \frac{2}{\sqrt{\pi}} \int_0^z \exp(-\zeta^2) \, d\zeta.$$

Attempts to mimic the error function discretely provided one of the most exciting experiences in my early work on circle packing. My strategy was to leverage knowledge of the derivative, which tells one something about $f^\#$, into knowledge of f *via* integration. Here is the experiment:

- For domain we use a regular hexagonal packing Q having circles of radius ϵ with α-vertex v_1 at the origin, and vertex v_2 on the positive y-axis. Let H_n denote the subcomplex of H of vertices within n generations of v_1.

- Given n, visit each vertex v_j in the boundary of H_n, find the center z_j of its circle in Q, and then compute

$$\rho_j = |\operatorname{erf}'(z_j)|\epsilon = \left|\exp\left(-z_j^2\right)\right| \epsilon.$$

- Solve the boundary value problem using these boundary radii to get an unbranched packing label R_n for H_n. Lay out an associated circle packing P_n with v_1 at the origin and v_2 on the positive y-axis.

- Repeat for successive n to see whether the P_n will converge geometrically to an infinite packing P; equivalently, whether the packing labels R_n will converge to a packing label R on H.

What made this experiment so exciting in practice was that the strategy seemed to be working! Look at Figure 14.3. Frame (a) displays the domain, with the circles of generations $n = 5, 7, 9, 11$, and 13 marked. Frames (b)–(f) show the packings P_n, $n = 5, 7, 9, 11, 13$. The reader will have to imagine sitting at the computer. There are, of course, preliminary dry runs, new code for computing the ρ_j, choices of normalizations and display options, and so forth – the usual sorting out. Then the experiment: in quick succession five pictures, (b)–(f), and the packing grows for you on the screen.

Even in these five pictures one begins to see the geometry of the error function emerging. But you should know that static images give a pale reflection of the actual experiments. As generations are added, the circles begin to overlap (you see a hint of this only on reaching (f)). Static images become hopelessly complex, while at the computer one is seeing a live animation – features can be highlighted, the mouse can be dragged around in the domain lighting up corresponding circles in the range, specific data can be queried, and so forth.

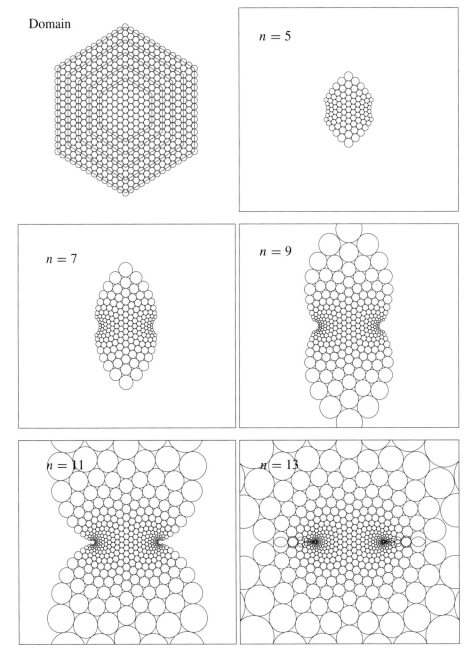

Figure 14.3. Toward a discrete error function.

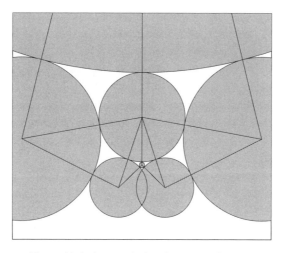

Figure 14.4. A nonunivalent hexagonal flower.

And I ran not 5 but rather 25 stages of this experiment. I was amazed to find that *the labels required no repacking!* Did the radii ρ_j already define a packing label for H?

At this point, I did what many of us would do: I enjoyed myself for a weekend before pushing the experiment further. Hopeful as I had been, I found that I did not have a prescription for an infinite packing: it happened that only at the 26th stage did the angle sum errors became significant enough to trigger repacking.

This hardly means the construction is doomed – but pushing it further requires some new idea. On the positive side, the image packings evolve just as one would hope. If you are not familiar with the geometry of the error function, you can learn it here. You can see logarithmic branching emerging at roughly $w = \pm 1$, the two asymptotic values of erf(z). As each new generation is added, its circles flow out and around ± 1; as the sheets of the image surface are wrapping around one singularity, they are covering the other (the Picard Theorem is not in danger here).

On the negative side, $f^{\#}$ is unbounded, so as n grows the dilatations of various hexagonal flowers in P_n will grow without bound. Figure 14.4 illustrates a mild case. The carrier is univalent, but two (noncontiguous) petals overlap – in fact, this is how the packing manages to wrap around one of its logarithmic branch points without covering it. With such skyrocketing distortions in flowers, is there any chance to discern the behavior and convergence of the labels R_n? Are there some algebraic relationships to exploit within the labels or within some related cross ratios? See what *you* can do; the jury is still out.

Aside from maximal packings and Doyle spirals, there are as yet no known locally univalent discrete entire functions based on any complex K, hexagonal or otherwise. Ultimately, experiments can only suggest whether an approach has possibilities and perhaps give clues to a rigorous construction. At a minimum, however, these experiments mark the discrete error function as a front runner in addressing the following question posed by Peter Doyle.

Open Question 14.9. *Do there exist any locally univalent circle packings for the hexagonal complex H other than regular hexagonal packings and Doyle spirals?*

14.5. The Discrete Sine Function

Injecting branching back into the mix, we can target trigonometric functions. We are all familiar with the behavior of *sine* as a real function, but its geometry as a *complex* mapping may be less well known. It has beautiful symmetries that we exploit via repeated Schwarz doubling.

Begin with the two packings of Figure 14.5, which share the same complex, namely, half of the hexagonal complex H. The packing Q on the left fills the half-infinite rectangular strip $\{z = x + iy : -\pi/2 \le x \le \pi/2, y \ge 0\}$, which has three boundary segments, the left, center, and right edges, $\gamma_l, \gamma_c, \gamma_r$. We have chosen the corners so that γ_c has seven circles. The packing P on the right has dots indicating the circles at ± 1; these correspond to the corners in Q and break the boundary of P into three segments, $\sigma_l, \sigma_c, \sigma_r$.

Construction involves an infinite sequence of simultaneous Schwarz doublings. The doubling process was described in Section 12.7, and is the basis for an important technique, one that we will be seeing again. I formulate it for the sphere, but the technique applies for any combination of domain and range geometries, as long as the doubled packings remain in their respective geometric spaces.

Lemma 14.10 (Schwarz Reflection). *Let $f : Q \longrightarrow P$ be a discrete analytic function between two spherical circle packings with complex K. Suppose there exists an edge-path $\gamma \subset \partial K$ whose circles in Q are orthogonal to a common spherical circle E_Q and whose circles in P are orthogonal to a common spherical circle E_P. Then f extends to a discrete analytic function $\tilde{f} : \tilde{Q} \longrightarrow \tilde{P}$ between the Schwarz doubles of Q and P across γ; that is, the restriction $\tilde{f}|_Q$ is identical to f.*

Apply discrete Schwarz reflection to the packings of Figure 14.5. As suggested in Figure 14.6, doubling Q across γ_c yields the packing Q_1 of a vertical strip, while doubling

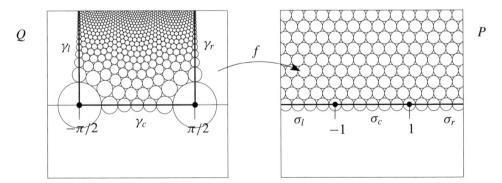

Figure 14.5. Construction material for the sine.

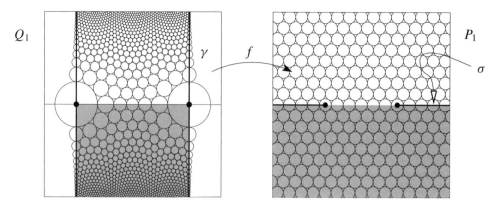

Figure 14.6. Schwarz reflection.

P across $\sigma_c = [-1, 1]$ seems to give the hexagonal packing. However, that is not quite right; P_1 has boundary edges, seen here as the slits into ± 1 from $\pm\infty$, respectively.

It is less evident in Figure 14.6, but we are in position to apply Schwarz reflection once more. If γ is, say, the right edge of Q_1, then its image is the edge σ of P_1 which runs on the positive x-axis from $+\infty$ in to the circle at $+1$, then back out to $+\infty$. Doubling Q_1 across γ to get Q_2 is easy to picture, but doubling P_1 across σ for P_2 is a little trickier. When you have sorted things out, you find that P_2 is basically two copies of P_1 cross-connected along a slit; the result is a double-sheeted packing with the branch circle centered at $+1$.

As you may have guessed, you can Schwarz reflect again and again, ad infinitum. Figure 14.7 suggests the ultimate result $\widehat{Q}$ in the domain. The orientation-reversed copies of Q are shaded. This is clearly univalent, so it is the maximal packing $\mathcal{P}_K$ for its complex K, a fortiori parabolic. Likewise, the companion reflections of P generate the image packing $\widehat{P}$. I have not shown $\widehat{P}$ because it would look just like the hexagonal packing;

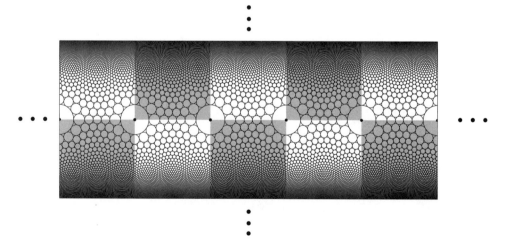

Figure 14.7. A sequence of doublings.

but of course, its complex is K and each circle would represent infinitely many circles of $\widehat{P}$. The function $f : \widehat{Q} \longrightarrow \widehat{P}$ is our *discrete sine* function. Familiar features are evident, for example, its 2π periodicity, oddness, and simple branching at points $\pi/2 + \pi n$. We get the *cosine* function for free by merely translating the domain by $\pi/2$.

14.6. Further Examples?

A major issue in the study of discrete entire functions is clearly the dearth of examples. We have systematic constructions for polynomials and ad hoc constructions for a few key examples. But we have nothing resembling the huge variety one sees among classical entire functions, and we have precious few tools in place. The challenge is clear.

15

Discrete Rational Functions

In this chapter, K is a combinatorial sphere: finite, simply connected, with no boundary. Topological arguments alone imply that *all* circle packings for K must be spherical. In parallel with the classical theory, therefore, functions $f : \mathcal{P}_K \longrightarrow P$ are called *discrete rational* functions.

There is no difficulty in seeing parallels with classical rational maps when you have a discrete example in hand, as we will describe shortly. The challenge lies, however, in creating these mappings in the first place. How does one prove existence of a circle packing with a given branch set? When a branched packing exists, is it unique up to conformal automorphisms of $\mathbb{P}$? What are necessary and sufficient conditions on branch sets?

Unfortunately, our typical experimental approach is stymied by the lack of a circle packing algorithm in spherical geometry. We have precious few methods and all I can offer in this chapter are some ad hoc examples. I hope to whet your appetite, however, for the *existence and uniqueness of circle packings on the sphere is perhaps the most important, most challenging, and most interesting open question in the foundations of circle packing*. There is important work yet to be done here!

15.1. Basic Theory

An open continuous map from the sphere to the sphere will be of constant valence n for some $n \geq 1$. No point will have branching of order greater than $n - 1$, and by the *Riemann–Hurwitz formula* there will be a total of $2(n - 1)$ branch points (counting multiplicities, as usual).

For a classical analytic self-map F of $\mathbb{P}$ these properties imply that F has the form $F(z) = G_1(z)/G_2(z)$. Here $G_1(z)$ and $G_2(z)$ are complex polynomials, and if we assume they have no common factors, then the valence n of F is the maximum of their degrees and is also known as the *degree* of F. Of course it is straightforward to construct F with specified zeros and poles (∞-points) – simply prescribe the zeros of G_1 and G_2, respectively. In principle, one can instead specify the $2(n - 1)$ branch points or alternatively the $2(n - 1)$ branch values and their combinatorial structure, but I am not aware of any concrete implementations.

In the discrete setting, of course, there is no quotient representation, but the topological properties are the same. If we adopt the term *degree* for the valence, then by Stoïlow's Theorem (11.2), a discrete rational function f of degree n will be of the form $F \circ h$, where F is a classical rational function of degree n and $h : \mathbb{P} \longrightarrow \mathbb{P}$ is an orientation-preserving homeomorphism. The full range of classical behavior appears in the discrete setting, though of course there are only countably many discrete rational functions (up to pre-/post-composition by Möbius maps). As for construction, we exercise control only via branching and we have our usual two tracks: (1) given K and branch set β, compute the image packing P or (2) construct an image packing P whose complex is a topological sphere and compute the domain $\mathcal{P}_K$ by the Koebe–Andreev–Thurston Theorem (Proposition 7.1).

So much for generalities. What can we actually *do*? In the next section we discuss a beautiful example that has dropped in our laps; though hardly generic, it shows the general features we would hope for in track (1). The section after that introduces two genuine techniques for track (1), but these are limited to certain branch structures. In the final section, we demonstrate the track (2) technique, which, though it seems rather unexciting now, foreshadows a key method in our later work.

15.2. A Fortuitous Example

Return to Figure 2.3 of the Menagerie for our first example of a discrete rational function $f : \mathcal{P}_K \longrightarrow P$. The complex K has the symmetries of the dodecahedron, and these are central to the construction of P, which relies on repeated reflections of *Schwarz* triangles – the circles get carried along for free. It is a pleasant bit of elementary spherical geometry which I have placed in Appendix H.

For now note that K has 42 vertices, 12 of degree 5, the rest of degree 6. The packing P has simple branching at the vertices of degree 5. We could apply the Riemann–Hurwitz formula to see that f has valence 7, but in the simplicial case it is easy enough to show directly. Recall that the area of a spherical triangle is the sum of its angles minus π. The area of $\mathrm{carr}(\mathcal{P}_K)$ is, of course, 4π; the angle sums of P contribute an additional 2π at each of 12 vertices, a total of 24π, which accounts for another six layers, hence valence 7. This layering, by the way, makes the image of P rather difficult to decipher, so I have isolated in Figure 15.1 one of the 5-flowers in domain and range.

Let V denote the 12 vertices of a regular dodecahedron. Symmetry tells us that we can normalize to center the 5-degree circles at the points of V in both domain and range. Both packings are then invariant under the dodecahedral group, and $f(V) = V$. This gives us enough information to deduce the classical model for f; namely, the rational function F given by $F(z) = \frac{z^2(3z^5-1)}{(z^5+3)}$. In fact, $f = F \circ h$, where the homeomorphism $h : P \longrightarrow P$ fixes the centers of the circles of $\mathcal{P}_K$, so f interpolates F at 42 points of the sphere. One can hardly get more explicit!

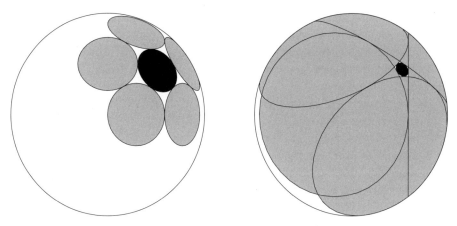

Figure 15.1. A branched spherical 5-flower.

This example is, in fact, *too* good. With a more complete theory, we would not need the luck of an explicit construction – we would (presumably) construct P based solely on K and branch set β. Nonetheless, this example shows what we can aim for in the theory.

15.3. Special Branched Situations

We can prove existence and uniqueness in two special circumstances, both based on constructions of finite Blaschke products in the disc.

15.3.1. Polynomial Branching

Complex polynomials, when extended to map ∞ to ∞, qualify as rational functions on the sphere. They are characterized, up to Möbius transformation, by the fact that half of their branching occurs at a single point. This feature can be exploited in the discrete setting.

Definition 15.1. Given a topological sphere K, a subset β of its vertices (with possible repetitions) is termed a **polynomial branch structure** if the following conditions hold: (1) β contains $2(n-1)$ points for some integer $n > 1$. (2) β contains a vertex v_∞ having branch order $n - 1$. (3) No neighbors of v_∞ lie in β. (4) In the complex K' obtained from K by removing the open star of v_∞, the remaining vertices of β form a branch structure.

Theorem 15.2. *Assume K is a combinatorial sphere. If β is a polynomial branch structure for K, then there exists an essentially unique circle packing P for K in $\mathbb{P}$ with* br $(f) = \beta$. *Conversely, if P is a spherical circle packing for K with half of its branching at a single circle, then its branch structure β is a polynomial branch structure.*

Proof. The arguments that linked packings on the unit disc and the sphere in Chap. 7 can be applied again in the branched setting. Let β' be the branch set β with v_∞ (all occurrences) removed. Let P' be the branched maximal packing $\mathcal{P}_{K',\beta'}$, as in Definition 13.6. Its boundary circles are horocycles wrapping n times about the inside of the unit circle (because β' has cardinality $n - 1$). Project P' stereographically to the sphere; the projected circles lie in the northern hemisphere, with boundary circles tangent to the equator. Adjoining the southern hemisphere as the circle c_∞ for v_∞, we get a spherical circle P packing for K. The boundary circles of P' wrap n times around c_∞, so c_∞ is a branch point of order $n - 1$ in P. Essential uniqueness of P follows from that of P'.

For the converse, there is one side condition to verify. Suppose P is a spherical packing for K whose carrier is an n-fold covering of the sphere and suppose c_∞ is a circle of branch order $n - 1$. The faces in the star of c_∞ account for all n sheets of carr(P) lying over the points of c_∞. In particular, no neighbor could be a branch circle since its petals would necessarily cover some points of c_∞ more than once, giving them valence greater than n. The proof is completed by normalizing to make c_∞ the southern hemisphere and then reversing the previous arguments.   □

15.3.2. Symmetric Branching

A related technique applies under certain strong symmetry conditions. See Section 12.4 for the notion of "doubling."

Theorem 15.3. *Suppose K is a combinatorial closed disc and β is a branch structure for K. Let $\widetilde{K}$ be the combinatorial double of K across its boundary and let $\widetilde{\beta}$ denote the union of β with its image under the doubling reflection. There exists a discrete rational function f with domain $\mathcal{P}_{\widetilde{K}}$ and with $\mathrm{br}(f) = \widetilde{\beta}$.*

Proof. A Blaschke-like product $b : Q \longrightarrow P$ can be created based on K so that the boundary circles of both the univalent domain packing Q and the image packing P are *orthogonal* to, rather than tangent to, the unit circle and so that $\mathrm{br}(b) = \beta$. Schwarz doubling in the unit circle (see Theorem 12.2) results in a rational function $f : \widetilde{Q} \longrightarrow \widetilde{P}$ between the Schwarz doubles of the domain and range, and it is easily confirmed that $\mathrm{br}(f) = \widetilde{\beta}$.   □

By construction, the branch points and the branch values of f are each symmetric with respect to the unit circle. Renormalizing yields a discrete rational function mapping the (extended) real line to itself and the upper and lower half-planes to themselves. With this normalization, the theorem gives discrete versions of a family of rational functions important in differential equations and control theory. Can you prove essential uniqueness of f here? I make no claim, but experience strongly suggests that it should hold.

15.4. Range Constructions

We conclude the chapter by illustrating a range construction. I will be brief, since the methods parallel those used in Figures 13.3 and 14.5–14.7 for the disc and the plane, respectively.

Begin with a univalent circle packing Q of the sphere. We construct a three-sheeted covering of $\mathbb{P}$ using copies $\widetilde{Q}_1$, $\widetilde{Q}_2$, $\widetilde{Q}_3$ of Q as shown in Figure 15.2.

Slit each of $\widetilde{Q}_1$ and $\widetilde{Q}_2$ along a common simple open edge-path γ_1, cloning the circles along the cuts as usual, then geometrically *cross-connect* them as described in Section 12.2 and 12.3, the left side of γ_1 on one copy to its right side on the other. The result is a spherical packing P_1 with simple branching at each endpoint of γ_1. With their cuts and cloned circles, $\widetilde{Q}_1$ and $\widetilde{Q}_2$ are topological closed discs, so when they are pasted together

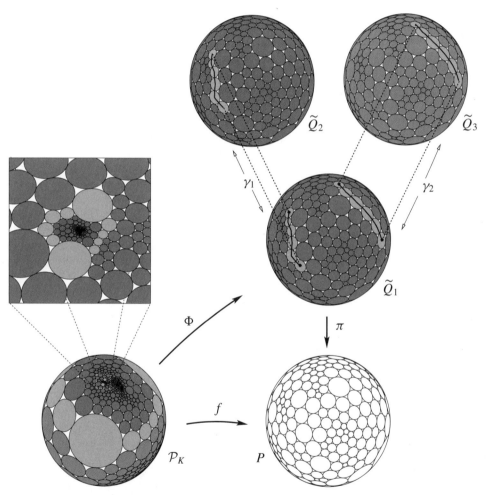

Figure 15.2. Constructing a three-sheeted covering of $\mathbb{P}$.

along their boundaries the complex K_1 for P_1 is a topological sphere by van Kampen's theorem.

Choose a second edge-path γ_2 in Q. Slit each of P_1 and $\widetilde{Q}_3$ along a corresponding chain of circles and cross-connect them to get the final packing P. Again, this introduces two new branch points and the complex K for P is a combinatorial sphere. You may note that there are two copies of γ_2 on P_1; which one we slit and cross-connect with $\widetilde{Q}_3$ is irrelevant here since they are indistinguishable. Were there more cuts, however, pasting decisions would come into play, as would higher order branching should endpoints of cuts coincide.

The three copies of Q in Figure 15.2 are color-coded to show how they fit together in the domain $\mathcal{P}_K$, with gray circles corresponding to the cuts. The normalization makes the red region very small, so a blowup shows the detail. Since the final complex K is a sphere, $f : \mathcal{P}_K \longrightarrow P$ is a discrete rational function. Total branch order is $\mathrm{br}(f) = 4$ and valence is $n = 3$, confirming that $\mathrm{br}(f) = 2(n-1)$.

It is perhaps not widely known that every *classical* rational function can, in principle, be constructed by an analogous process. One pastes spheres together along systems of duplicate cuts, obtaining a topological sphere S with a natural projection $\pi : S \longrightarrow \mathbb{P}$. The conformal structure of $\mathbb{P}$ lifts to S under π. Since a sphere has a unique conformal structure, there must be some univalent conformal map $h : \mathbb{P} \longrightarrow S$. The function $F \equiv \pi \circ h : \mathbb{P} \longrightarrow \mathbb{P}$ is therefore rational with image surface S. One might imagine that this construction can actually be carried out in situations where the branch values and the pattern of the surface pieces (the location and ordering of pastings) are known. However, there is a bump in the classical road: it is typically impossible to realize h concretely, that is, to obtain the conformal structure on S. In the discrete case, this is precisely what the Koebe–Andreev–Thurston Theorem provides, and it is computable!

In concluding this chapter, note how "existence" and "construction" seem to go hand-in-hand, with "uniqueness" close by, also. An algorithm for computing branched circle packings in situ in the sphere would likely be a major breakthrough in theory, as well as in practice.

16

Discrete Analytic Functions on Riemann Surfaces

We move from the simply connected combinatorics of the last few chapters to more general complexes, and hence to Riemann surfaces. Recall that when K is not simply connected, its maximal packing $\mathcal{P}_K$ brings with it the Riemann surface $\mathcal{S}_K$ on which it resides. A mapping $f : \mathcal{P}_K \longrightarrow P$ to a packing P for K in a Riemann surface $\mathcal{S}$ becomes a discrete analytic function between $\mathcal{S}_K$ and $\mathcal{S}$.

We restrict attention primarily to *compact* K (finite, with no boundary), so $\mathcal{S}_K$ and $\mathcal{S}$ will necessarily be compact Riemann surfaces. If P lies in $\mathbb{P}$ we also call $f : \mathcal{P}_K \longrightarrow P$ a *meromorphic* function. The classical theory for analytic functions between compact Riemann surfaces is among the richest in all of mathematics, but do not get your hopes up yet; we are admittedly building a large stage here for what is as yet a small cast of confirmed characters.

As usual, we have two constructive approaches, one starting in the domain, the other in the range. The first is the most intriguing, but is largely uncharted territory. I will take you through a succession of experimental probes – some succeed, some fail, but we learn something with every one. The range constructions give us more control and I will illustrate with systematic techniques for constructing what are known as discrete Belyĭ meromorphic functions. I conclude the chapter with a variety of other examples and instructive nonexamples.

We already have our collective foot in the door with the existence of maximal packings; here is the function theory formulation.

Theorem 16.1 (Discrete Uniformization Theorem). *Given any complex K there exists an associated Riemann surface $\mathcal{S}_K$, a discrete analytic function f from $\mathbb{G}$ (one of $\mathbb{P}$, $\mathbb{C}$, or $\mathbb{D}$) onto $\mathcal{S}_K$, and a discrete subgroup Λ of $\mathrm{Aut}(\mathbb{G})$ under which f is invariant so that $f : \mathbb{G} \longrightarrow \mathcal{S}_K$ is single-valued and one-to-one modulo Λ.*

16.1. Ground Rules

First we must contend with some new subtleties. A circle packing P in a Riemann surface $\mathcal{S}$ is, as usual, a configuration of circles whose pattern of tangencies is encoded in a simplicial complex K which triangulates an oriented topological surface. That raises the first question: What is a circle? and that requires a metric.

Convention. *The metric for circles on a Riemann surface S will be the intrinsic metric, that is, the spherical, euclidean, or hyperbolic metric of constant curvature $+1, 0, -1$, respectively, as appropriate to S. (See Section 3.2.)*

A circle is still the set of points a given distance r from a center $p \in S$, but that can yield some rather odd behavior now: circles tangent to or intersecting themselves, perhaps many times, circles with topological handles in their interiors, and so forth. We need the following

Standing Assumptions. *The circles of a packing P in S have no self-intersections or self-tangencies; they bound topological discs (i.e., have simply connected interiors); circles for neighboring vertices have mutually disjoint interiors and precisely one point of tangency; interstices formed in S by oriented triples (of mutually tangent circles) of P associated with faces of K are topological discs.*

Let me put the reader's mind at ease – these restrictions are not land mines that will blow up with the smallest misstep. They are largely automatic because of the local nature of the packing and layout algorithms; they ensure that a circle packing P in S determines a (possibly branched) immersion of its complex K in S. It is in interpreting patterns that *appear* to be circle packings that one can run afoul of these assumptions; we see some examples in the closing section.

16.2. Discrete Branched Coverings

There is certainly a place for discrete analytic functions within classical covering theory, in the spirit of the Uniformization Theorem above. In particular, if a maximal packing $\mathcal{P}_K$ in a Riemann surface S is invariant under a fixed point–free group $\Lambda \subset \text{Aut}(S)$, then $P' = \mathcal{P}_K/\Lambda$ is a circle packing in the Riemann surface $S' = S/\Lambda$. The map $f : \mathcal{P}_K \longrightarrow P'$ is a discrete analytic covering map from S onto S' (modulo some details regarding the Standing Assumptions; see Section 16.5.6).

I would call these "soft" examples. Our tenet is that *discrete analytic functions are maps between circle packings*. I prefer *concrete* packings to ones which are simply carried along like so much excess baggage in the covering theory. That means we must confront the issue of *branched* packings. Let us shine a flashlight around and see what we are facing.

16.2.1. Three Failures

We will run three experiments based on the 12-vertex, genus-2 complex K we studied at the end of Section 9.4 and in Practicum II. You may recall how easily we computed its maximal packing for Figure 9.5. We set all angle sum targets to 2π; knowing that $\mathcal{S}_K$

must be hyperbolic, we set an arbitrary initial label, repacked with `CirclePack` to get $\mathcal{R}_K$, and then laid the packing out in $\mathbb{D}$ as a fundamental domain for $\mathcal{S}_K$.

Question: What happens if we reset angle sum targets to create branch points?

Experiment 1: Let β be a singleton branch set, say $\beta = \{v_4\}$. Resetting the target angle sum at v_4 to 4π and repacking in hyperbolic geometry yields a branched hyperbolic packing label R_1 for K.

Experiment 2: Add a second branch point, say v_{11}, so that $\beta = \{v_4, v_{11}\}$. Resetting the target to 4π at v_4, v_{11} and attempting to repack in hyperbolic geometry leads to a degenerate label, all the radii decrease to zero. We switch to euclidean geometry and try again. Success! We get a euclidean packing label R_2 with branch set β.

Experiment 3: Add two more branch points, say $\beta = \{v_4, v_{11}, v_1, v_7\}$. Both hyperbolic and euclidean repacking attempts degenerate and we are left with spherical geometry. Unfortunately, this is where our experimental prowess fails (miserably). We do not know whether the spherical label R_3 exists; if R_3 were to exist, we would not know how to compute it; and if computed, we would not know whether it was essentially unique. Life is tough on the sphere.

Some success, some failure, some abject failure. In searching for explanations, we might start with a type of counting we have seen before. Recall (see Section 11.3) that if F, E, and V are the numbers of faces, edges, and vertices of K, then

$$\chi(K) = -F/2 + V = 2 - 2g \quad \Longrightarrow \quad F - 2V = 4g - 4.$$

Let b denote the cardinality of β. For a hyperbolic label R, the total Θ of the angles in all faces can be computed in two distinct ways, sum by face or sum by vertex. This implies $\Theta = 2(V + b)\pi = F\pi - \text{Area}(K(R))$, and we conclude that

$$\text{Area}(K(R)) = (F - 2V - 2b)\pi \quad \Longrightarrow \quad \text{Area}(K(R)) = (4g - 4 - 2b)\pi.$$

With $g = 2$, this is fine in Experiment 1 where $b = 1$; but in Experiment 2 where $b = 2$, this gives area zero, explaining the degeneracy in that hyperbolic computation. In euclidean geometry, on the other hand, $\Theta = F\pi$, so

$$2(V + b)\pi = F\pi \quad \Longrightarrow \quad F - 2V = 2b \quad \Longrightarrow \quad b = 2g - 2,$$

and the algorithm works for the euclidean label R_2. As for Experiment 3, if spherical label R_3 exists, then counting yields

$$\text{Area}(K(R_3)) = \Theta - F\pi = 32\pi - 28\pi = 4\pi, \tag{75}$$

so there remains some hope.

Having a label is not sufficient, however; the next question is whether that label is associated with an honest-to-goodness circle packing P in some Riemann surface $\mathcal{S}'$. Our next test should be the *Riemann–Hurwitz formula.* If P were to exist, then $f : \mathrm{carr}(\mathcal{P}_K) \longrightarrow$ $\mathrm{carr}(P)$ would be a light-interior mapping between compact surfaces $\mathcal{S} = \mathcal{S}_K$ and $\mathcal{S}'$. The relation among the valence n and total branch order b of f and the genera g, g' of $\mathcal{S}$, $\mathcal{S}'$ is

$$2(g - 1) = 2n(g' - 1) + b. \tag{76}$$

Note, among other things, that b is even. This means that Experiment 1 was doomed from the beginning, despite the ease in finding the label. This is perhaps the first example of the book that illustrates why we refer to the *packing* condition for labels as a *local* condition – there can exist global obstructions to realizing a packing label in an actual circle packing. (For those familiar with the concept, Experiment 1 is the parallel to one of Thurston's "bad" *orbifolds.*) We can also dispense with our hapless Experiment 3: by either Eq. (76) or (75), f would have valence 1, contradicting the branching of the label.

Our remaining hope for success lies with the euclidean label R_2 of Experiment 2. It seems to pass muster with both the euler characteristic and Riemann–Hurwitz formula, which tells us it should pack a torus. Figure 16.1 shows our attempt to lay out a packing in $\mathbb{C}$ for packing label R_2. This is an important, though rather jumbled, image.

To sort out this image, recall from Practicum II what `CirclePack` does. Its construction of a drawing order for K is tantamount to cutting K open to form a simply connected complex L and then cloning the vertices along the cut to identify boundary edge-paths of L which are associated with the same vertices in K – we use the term "side-pairings." (So a combinatorial pasting of edges of ∂L under all these side-pairings would reconstitute K.) A *packing* label R for K (either branched or unbranched) induces a packing label for L. Since L is simply connected, an associated circle packing Q can be laid out unambiguously by the Monodromy Theorem. Furthermore, by simple geometry each combinatorial side-pairing of L can be realized geometrically via a conformal automorphism ϕ of $\mathbb{G}$; that is, ϕ will carry the circles of one side precisely onto the circles of the paired side.

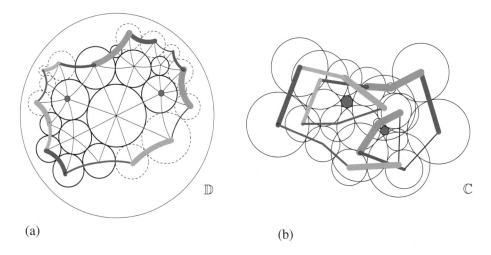

(a)   (b)

Figure 16.1. Layouts for Experiment 2; color-coded side-pairings.

The packing on the left of Figure 16.1 shows how well this works when Q takes its radii from the unbranched maximal label $\mathcal{R}_K$; this duplicates the image we first saw in Figure 9.5. The side-pairing automorphisms generate a discrete group Λ (of the disc in this instance), the images of Q under Λ tile $\mathbb{D}$ and together form the univalent maximal packing $\widetilde{\mathcal{P}}$, and $\mathbb{D}/\Lambda$ defines the Riemann surface $\mathcal{S}_K$, which then supports the maximal packing $\mathcal{P}_K = \widetilde{\mathcal{P}}/\Lambda$. However, if you go back to the proof of Proposition 9.1, you will note that it was actually $\widetilde{\mathcal{P}}$ that came first there: we deduced the properties of Λ from the essential uniqueness of $\widetilde{\mathcal{P}}$.

Now, let us give Q the radii from the euclidean packing label R_2. The layout on the right of Figure 16.1 shows Q in this case: this packing is perfectly coherent and unambiguous as a packing for L, though it is visually complicated due to the presence of the two branch circles (shaded). The side-pairings are marked to match those of the maximal packing. It is in these side-pairings that the experiment fails. It is clear that the group they generate is *not* discrete; discrete subgroups of Aut($\mathbb{C}$) contain only translations, yet the blue side-pairing, for example, is clearly not a translation.

Conclusions. *Experiments 1 and 3 were doomed from the start by counting considerations. Experiment 2 should have provided an image packing of a torus; the reasons for its ultimate failure merit further study.*

16.2.2. Rigged Success

"Merit further study" is a major understatement. This is the crunch point for the discrete theory – we are encountering core rigidity issues in Riemann surfaces. To keep our spirits up we should see at least one successful experiment, even if we have to cheat: the game plan is to use a range construction with plenty of symmetry.

Experiment 4: Two identical copies of a conformal torus $\mathcal{T}$ can be slit and cross-connected along a common arc to create a genus 2 surface $\mathcal{S}$. The natural projection of each piece to $\mathcal{T}$ induces a 2-valent analytic projection $F : \mathcal{S} \longrightarrow \mathcal{T}$ which has two simple branch points.

As a classical construction, this is perfectly rigorous, although it provides no concrete information on the surface $\mathcal{S}$ itself. The following discrete construction, on the other hand, gives us explicit access to $\mathcal{S}$ – that does not mean it is easy to interpret! Figure 16.2(a) shows a hexagonal circle packing Q of a conformal torus $\mathcal{T}$ (with the usual side-pairings). Two identical copies are slit as indicated and then cross-connected using these slits. This is in fact a *geometric* pasting (see Section 12.6) no repacking is needed and the result is a euclidean circle packing P with two simple branch circles (at the two ends of the slit). The packing P again lies in $\mathcal{T}$; in fact, Figure 16.2(a) could serve as an image of P, with each circle (other than the branch circles) representing two circles of P. The Euler characteristic confirms that the complex K for P has genus 2. Figure 16.2(b) displays $\mathcal{P}_K$, which forms the fundamental domain for the genus 2 surface $\mathcal{S} = \mathcal{S}_K$.

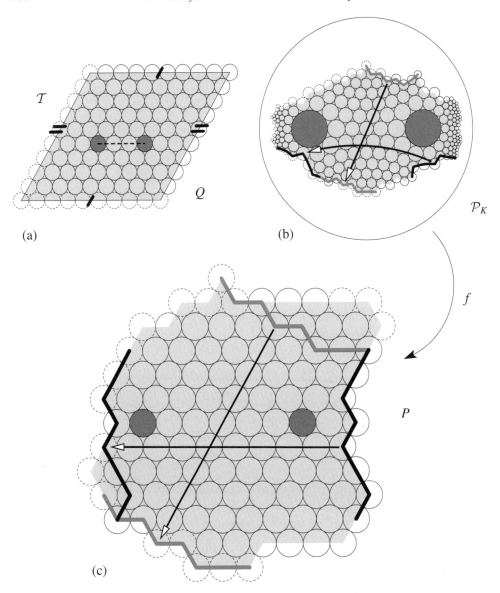

Figure 16.2. Experiment 4.

Of course, we rigged this example to guarantee a concrete branched packing P. But let us try to recapture our experimental spirit: imagine starting with K and β only, with no insider information on P. One would proceed as follows: set target angle sums to 4π for vertices of β, set arbitrary initial euclidean labels, repack to get a branched packing label R, and then see if the circles can be consistently laid out.

As expected, the computed label will be constant. However, Figure 16.2(c) shows the experimental layout – and this is where things become difficult to interpret visually. Does this, in fact, represent a circle packing P for K? I have used the same drawing order in both (b) and (c), but most circles of (c) represent two circles of the image packing. To confirm that the layout is consistent, we need to check the side-pairings, which are rather numerous due to the vagaries of automatic layout (see Practicum II). In fact, there are nine side-pairings, but I have chosen just two for display and have highlighted them in both (b) and (c). Observe that the side-pairings in (c) are translations belonging to the covering group of $\mathcal{T}$. This is true of the other seven side-pairings as well, so (c) does represent a branched circle packing on $\mathcal{T}$. The function $f : \mathcal{P}_K \longrightarrow P$ is a discrete branched covering map from $\mathcal{S}$ to $\mathcal{T}$.

Conclusion. *In Experiment 4 there exists a euclidean packing label R_4 with branch set β. Moreover, there is a circle packing P associated with this label living in a conformal torus.*

16.2.3. The Function Theory

For compact Riemann surfaces, the study of geometry is equivalent to the study of analytic functions. Our experiments have hardly scratched the surface (so to speak) in this extremely rich topic. We can, however, glimpse the central function theory core by contrasting the success of Experiment 4 with the failure of Experiment 2.

Although we succeeded in computing a euclidean branched packing label R_2 in Experiment 2, the notion of *holonomy* will help us confirm that no corresponding circle packing can exist. Indeed, it is difficult to imagine a better setting for introducing holonomy if you have not seen it before.

Using the (euclidean) label R_2, lay out a base face $f_0 \in K$ as a triangle $T_0 \in \mathbb{C}$. Let $\gamma = \{f_0, f_1, \cdots, f_n\}$ be a closed chain of faces in K. Starting with T_0, lay out successive faces of γ, ending with triangle T_γ for f_n. Since $f_n = f_0$, the triangles T_0 and T_γ are similar, and therefore there exists $\phi = \phi_\gamma \in \text{Aut}(\mathbb{C})$ such that $\phi_\gamma(T_0) = T_\gamma$. The reader can verify (in fact, see in action) the following facts: (1) By monodromy, ϕ_γ depends only on the homotopy class $[\gamma]$ of γ. (2) In particular, if γ is null homotopic, then ϕ_γ is the identity. (3) If $\gamma = \gamma_1 \cdot \gamma_2$ (concatenation), then $\phi_\gamma = \phi_{\gamma_2} \circ \phi_{\gamma_1}$. From these the reader can confirm that the map $\Phi = \Phi_{K,R} : \pi_1(K) \longrightarrow \text{Aut}(\mathbb{G})$ defined by $\Phi(\gamma) = \phi_\gamma$ is a group homomorphism.

Definition 16.2. Given a complex K and a packing label R for K in the geometry of $\mathbb{G}$ (one of $\mathbb{P}$, $\mathbb{C}$, or $\mathbb{D}$), the group $\Phi(\pi_1(K))$ is called the holonomy group for $K(R)$ and is unique up to conjugation (i.e., up to choice of a base triangle).

If P is a circle packing for $K(R)$ in Riemann surface $\mathcal{S}$, then the holonomy group for $K(R)$ must be isomorphic to a subgroup of the covering group $\Lambda \subset \text{Aut}(\mathbb{G})$ for $\mathcal{S}$. We asserted that this fails for label R_2 in Experiment 2 based on Figure 16.1(b). That static image is difficult to interpret when you cannot manipulate it on the computer. Perhaps Figure 16.3 can help. On the left is a closed chain γ in the maximal packing. On the right

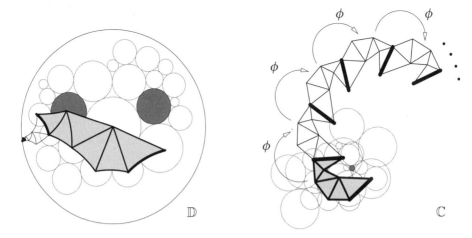

Figure 16.3. Holonomy for $K(R_2)$.

we see the corresponding chain laid out in $\mathbb{C}$ using label R_2. (The circles corresponding to the branch vertices are shaded in these images.) I have shown the faces that result when we have the computer continue the layout through multiple passes about γ. The rigid motion ϕ_γ is clearly not a translation. Note that we do not expect *trivial* holonomy; rather the question is whether the holonomy belongs to a fixed point–free discrete subgroup of $\text{Aut}(\mathbb{C})$. Here the answer is clearly no.

Holonomy of this type is the obstruction to the existence of analytic functions on Riemann surfaces whether you are in the classical or the discrete setting. However, here we face both the geometry and "discretization" effects. An attempt to preassign branch points would undoubtedly fail in the classical case just as it does in Experiment 2. However, in the classical case one can shift the branching to arbitrary locations, whereas our discrete branching is restricted to the vertices. So we face this added "discretization" effect: *the geometry of a situation requires a branch point at location x; unfortunately, there is no vertex at x to carry that branching; using a nearby vertex introduces nontrivial holonomy; the construction fails.* It appears likely that a generic complex K of positive genus will have *no branched labels* whose holonomy groups are discrete. An interesting potential work-around is to deploy some type of *fractional branching*. While a branch point represents an extra angle sum target of 2π concentrated at a vertex, perhaps that 2π could be shared among a set of nearby neighbors in some way that counteracts unwanted holonomy.

We have to leave this topic hanging. That is too bad, since there is a flood of natural questions, and the live experiments are particularly challenging – I am sure there is a holonomy video game lurking somewhere here. It is also unfortunate that we do not yet have the wherewithal to experiment further on the sphere along the lines of Experiment 3, for the classical theory begins most naturally with meromorphic functions – differential forms, "divisors," the Riemann–Roch Theorem, Weierstrass points, and so forth. Classically, *exact holomorphic differentials* are those which can be integrated to give global meromorphic functions. But recall that our labels are related to derivatives via ratio functions and development along chains of faces is akin to definite integrals along curves. In this regard, a spherical packing label on K which has no actual circle packing corresponds loosely to a holomorphic differential on S which is not exact. So although I will admit that our discrete theory has a long way to go, there are some tantalizing clues to pursue.

16.3. Discrete Belyĭ Functions

A less forbidding route into the richness of meromorphic functions is via direct construction of their image packings. I illustrate here a systematic approach that sets the stage for a more extensive theory in Part IV.

The construction is close to that used in the last section of the previous chapter – indeed, we may well get rational functions here, depending on the combinatorics that are prescribed. We start with the packing Q of the sphere, just eight circles, which is shown in miniature in Figure 16.4. For our purposes, we cut along the real axis to get two hemispheres, the upper and lower half planes, $\mathbb{H}^+$ and $\mathbb{H}^-$. That provides three "edges" on the real axis for pasting: $[0, 1]$, $[1, \infty]$, and $[\infty, 0]$. In Fig. 16.4 I have shaded the lower half-plane and colored the three edges as black, red, and blue, respectively, to aid in construction.

Each half-plane is a single hexagonal flower which will be treated as an integral unit here; we build spherical packings P by pasting copies of these units. Each pasting operation attaches a copy of $\mathbb{H}^+$ to a copy of $\mathbb{H}^-$ using the three circles along one of the three designated segments – this is a *geometric* pasting, as described in Section 12.6. If we use finitely many half-planes and arrange that each of the three edges of each half-plane is ultimately pasted to some copy of the other half-plane, then the complex K for P will be compact. Computing the maximal packing for K picks out a Riemann surface S_K, and

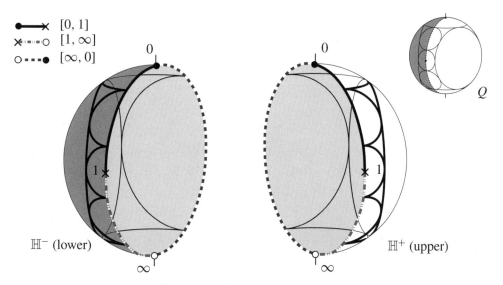

Figure 16.4. The decorated packing Q.

$b : \mathcal{P}_K \longrightarrow P$ is our target discrete meromorphic function on $\mathcal{S}_K$. Note that branching can occur only at circles in P centered at the ends of the pasting slits, i.e., those centered at 0, 1, or ∞. This construction leads to what are known as discrete Belyĭ pairs.

Definition 16.3. A **discrete Belyĭ map** is a discrete meromorphic function $b : \mathcal{P}_K \longrightarrow P$ from a maximal packing in $\mathfrak{s} = \mathcal{S}_K$ to a circle packing P in the sphere whose branch values lie in $\{0, 1, \infty\}$; the pair $(\mathfrak{s}, b)$ is termed a **discrete Belyĭ pair**.

If you have difficulty in seeing precisely how these constructions go, here is a book-keeping tool. A compact, oriented, simplicial 2-complex S will be called *tripartite* if its vertices are partitioned into three subsets so that every face has one vertex from each subset. In illustrations, I mark the vertices with •, ×, and ∘; I shade the faces with orientation × → • → ∘, while I leave unshaded those with orientation • → × → ∘. There will be an even number of faces, half shaded, half unshaded; shaded faces share edges only with unshaded faces and vice versa.

We use a tripartite complex S to create a packing P by associating a copy of $\mathbb{H}^+$ with each unshaded face of S and a copy of $\mathbb{H}^-$ with each shaded face. The points 0, 1, ∞ are identified with •, ×, and ∘, respectively. If faces t_1 and t_2 of S share an edge [•, ×], say, then the associated copies of $\mathbb{H}^+$ and $\mathbb{H}^-$ are geometrically pasted using the three circles along the edge [0, 1]; similarly for edges [×, ∘] and [∘, •]. That is all there is to it – start with tripartite S, do the pastings, and you get an image packing P! (You might note that the complex K for P is in fact a barycentric subdivision of S.)

Tripartite complexes are easy to build, mere child's play. Figure 16.5 displays the maximal packings for specific examples in genus 0, 1, and 2. The hexagonal flowers associated with $\mathbb{H}^-$, $\mathbb{H}^+$ are shaded/unshaded in the carriers; these are (modulo the covering group) embeddings of the associated tripartite complexes S.

Note by way of a converse that if copies of $\mathbb{H}^+$ and $\mathbb{H}^-$ have been pasted to get a compact circle packing P, then the pattern of pastings determines a tripartite complex S. In other

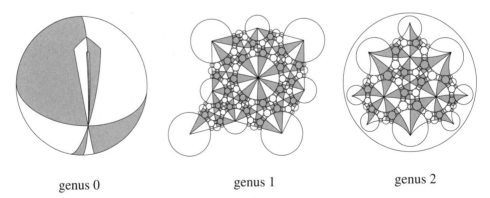

genus 0   genus 1   genus 2

Figure 16.5. Packings for genus 0, 1, and 2 tripartite complexes.

words, there is essentially a one-to-one correspondence between tripartite complexes and discrete Belyĭ pairs. We return to this topic in Chap. 23.

16.4. Further Examples

Let us wrap up with two examples famous from the classical literature. The first is the Klein surface, a genus 3 surface having an automorphism group of maximal possible order (namely, $168 = 84(3 - 1)$). The circle packing for a fundamental domain of this surface is shown in Figure 1.4(a) of the Menagerie. When you add the shading associated with a Belyĭ meromorphic function, you get Figure 1.7(a), which duplicates an image made famous over a century ago by Fricke and Klein (1897–1912). Their draftsman could create that image because of its wealth of internal symmetries – the triangles are *geodesic* triangles. The Picard surface of Figure 2.9 on the other hand, would have been problematic before circle packing since its triangles are (despite their appearance) not geodesic.

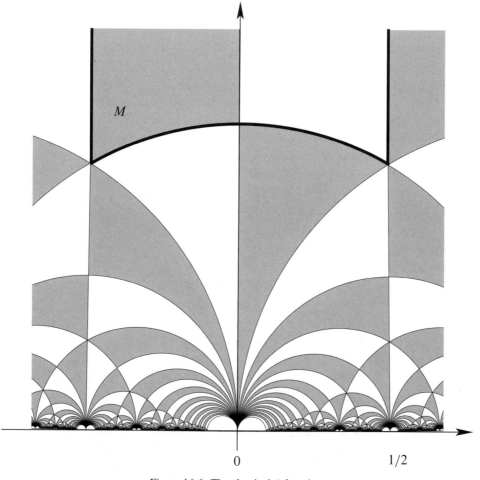

Figure 16.6. The classical j-function.

The *elliptic modular function*, also known as the *j-function*, has a pedigree exceeding even the Klein surface: the triangulated image of the domain in Figure 16.6 was apparently first sketched by Gauss. This is the half-plane model of the hyperbolic plane. The circles underlying the pattern are not shown and the empty gaps along the axis occur because we have halted the construction at 20,000 circles. The discrete *j*-function itself is constructed in Belyĭ fashion (though I dispensed with the barycentric subdivision): the unshaded faces are mapped to the upper half plane of $\mathbb{P}$, the shaded, to the lower half plane.

The *j*-function is a branched covering map from the half plane to the plane; the two faces of the outlined region, labeled *M*, form a fundamental region. The points *z* in the domain where four faces come together are precisely the points with $j(z) = 0$, simple branch points; those where six faces meet are points with $j(z) = 1$, double branch points – all this is familiar from the Belyĭ situation. Things change character over ∞, however. The value ∞ is not assumed by *j* on the interior. The "circles" associated with ∞ are horocycles (not shown); each is covered by infinitely many faces coming together at its tangency point $\zeta \in \mathbb{R} \cup \{\infty\}$, and their images under *j* form an infinite covering of a punctured disc at ∞. This mapping behavior is generally known as *logarithmic branching*.

The *j*-function and the related *modular function* are extremely important in function theory and analytic number theory. The discrete version of *j* can be defined as a point function by using Schwarz reflection on the triangular faces, and in this case is identical to the classical *j*-function. In particular, for example, it is invariant under $\mathrm{PSL}(2, \mathbb{Z})$. Likewise, our discrete Klein example interpolates its classical version. At least in the presence of symmetry, then, one gets much more than analogies out of the discrete constructions. As a next step it might be quite interesting to develop circle packing constructions for the so-called *modular surfaces*.

We close with Fig 16.7, which displays infinite tripartite complexes that arise from triangle groups – I include them more for beauty than theory.

Figure 16.7. Tripartite complexes from triangle groups.

16.5. Alternate Notions of "Circle Packing"

By now the definition of "circle packing" we are using in the book probably seems quite natural. Perhaps it is time to comment on notions that stretch the definition. Several are illustrated in Figure 16.8.

16.5.1. Ambient Metrics

By convention, circles on a Riemann surface are defined using the *intrinsic* metric, making the surface itself an important part of the context. Figure 16.8(a) shows two circle packings for the same complex K, both lying in the punctured plane $\mathbb{C}^* = \mathbb{C}\backslash\{0\}$. The first is euclidean; that is, we are using the *ambient* metric which $\mathbb{C}^*$ inherits as a subset of $\mathbb{C}$. The second packing is the maximal packing $\mathcal{P}_K$; these are circles in the *intrinsic* metric of $\mathbb{C}^*$ (arclength element $ds = |dz/z|$).

16.5.2. Hyperbolic Cusps

By convention, a hyperbolic label R must be finite at an interior vertex v. In fact, the packing algorithm has no scruples about this, and one can often set $R(v) = \infty$ and repack successfully. When laid out, the circle for v is a horocycle. Its petal circles stretch part way around its tangency point ζ, as shown in Figure 16.8(b), and they close up modulo a parabolic Möbius transformation of $\mathbb{D}$ with fixed point ζ. The local geometry is that of a puncture, what is known as a *hyperbolic cusp* at ζ. The example here models a torus with a puncture; the dark lines indicate the two side-pairings.

16.5.3. Apollonian Packings

By convention, circles in our packings have only finitely many neighbors. However, there is a familiar pattern which violates this: the *Apollonian* packing of Figure 16.8(c) apparently reaches from Apollonius of Perga ($\sim$262 B.C.) to the latest computer screensavers. It begins with two tangent horocycles, then a (largest possible) circle is added to each interstice, then to each new interstice, and so forth ad infinitum. There is considerable work over the years on such packings related, for example, to the geometry of the complement of the union of the circles and their interiors – the *Apollonian gasket* – and very recently to the fascinating number theory. (My thanks to Allan Wilks and collaborators for this image; the numbers represent circle "bends," that is, reciprocal radii.)

16.5.4. Nothing-but-Horocycles

Figure 16.8(d) displays a "packing" in $\mathbb{D}$ in which *all* the circles are horocycles, all the faces are *ideal hyperbolic triangles*. This triangulation is well known as the Farey triangulation – the tangency points of the horocycles are precisely the points of $\mathbb{T}$ whose arguments are rational multiples of π. One can carry out a Belyĭ -type construction, mapping each ideal triangle to the upper or lower half plane, with vertices mapped to $\{0, 1, \infty\}$. This gives a map $f : \mathbb{D} \longrightarrow \mathbb{P}$, but now with logarithmic versus finite branching over 0, 1, and ∞. This

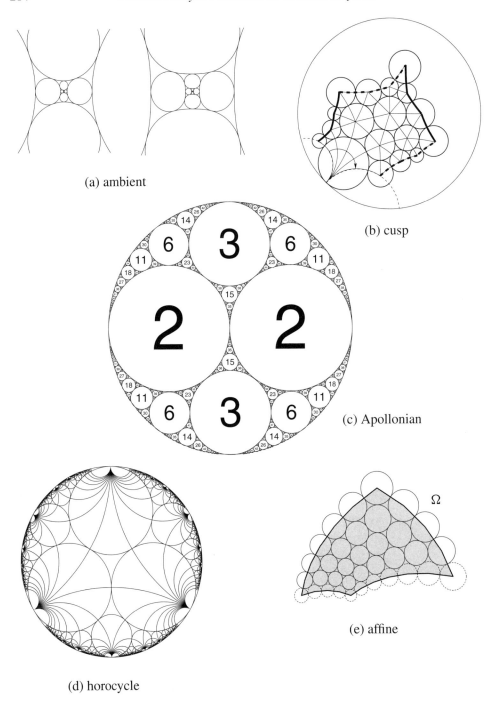

(a) ambient

(b) cusp

(c) Apollonian

(d) horocycle

(e) affine

Figure 16.8. Alternative notions of circle packing.

function f is in essence the *modular function* mentioned earlier, the universal covering map of $\mathbb{C}\backslash\{0, 1\}$ made famous by Picard.

16.5.5. Affine Packings

Euclidean circles may be characterized as the curves of "constant curvature" in $\mathbb{C}$, a property that is preserved under affine mappings. In this sense we can interpret Figure 16.8(e) as a "circle" packing on an *affine torus*. In particular, opposite sides of the carrier Ω may be identified *via* automorphisms $\phi : z \mapsto Az$ and $\psi : z \mapsto Bz$ to form a topological torus T. Since ϕ and ψ are analytic and affine, T inherits conformal and affine structures from Ω. But ϕ and ψ are not isometries, so T does not inherit a metric and these objects are *not* circles on T in the usual metric sense. (Compare the pattern to the Doyle spirals of Appendix C.)

16.5.6. Multiple Tangencies

The "Standing Assumptions" adopted for packings on surfaces preclude self- and multiple tangencies. Figure 16.9 shows what can happen when these assumptions are relaxed.

Any parallelogram $\mathfrak{p}$ in the plane determines a conformal torus $\mathbb{C}/\Lambda$, where Λ is the free group of automorphisms generated by the side-pairing maps for $\mathfrak{p}$. I have drawn four parallelograms $\mathfrak{p}_j$ in Figure 16.9. These four are similar, so their tori are all conformally

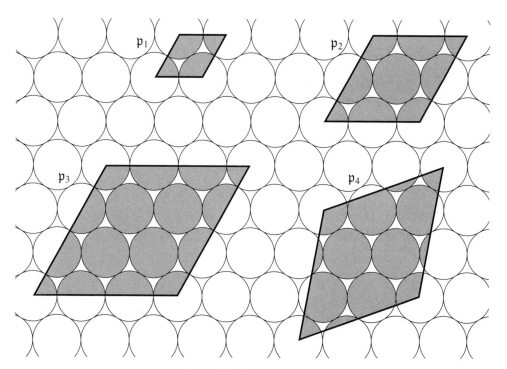

Figure 16.9. Torus specimens.

equivalent to a common conformal torus $\mathcal{T}$. These particular parallelograms where chosen because their side-pairing maps leave the packing $\mathcal{P}_H$ invariant, implying that the circles of $\mathcal{P}_H$ represent circles in $\mathcal{T}$ in each instance. However, for $\mathfrak{p}_1$ this means there is a single circle on $\mathcal{T}$ which is tangent to itself at three distinct points, while with $\mathfrak{p}_2$ there are four circles on $\mathcal{T}$, each tangent to each of the others at two points. Other authors have used such configurations in studies of Riemann surfaces and call them "circle packings," but for us, these do not qualify since they do not triangulate $\mathcal{T}$.

In contrast, the circles for parallelograms $\mathfrak{p}_3$ and $\mathfrak{p}_4$ represent two maximal circle packings on $\mathcal{T}$. The latter is the packing for the triangulation having the smallest possible number of vertices, namely, 7. Note that we can define a map from either $\mathfrak{p}_3$ or $\mathfrak{p}_4$ to, say, $\mathfrak{p}_1$ by mapping all circles of the former to the one circle of the latter, defining a map of $\mathcal{T}$ to itself. However, we would not consider this to be a discrete conformal mapping simply because the image is not a circle packing according to our Standing Assumptions.

16.5.7. Sphere Packing (NOT!)

I frequently get two types of questions concerning "sphere packing." The first results from a confusion of terms: *circle packing* is *not* 2D sphere packing. The term *sphere packing* refers to an important topic with a very long tradition, and many of its 2D images can look like – and in rare instances, can be – circle packings. But the fundamentals of the two topics are entirely at odds. A typical sphere packing question: *How many mutually disjoint discs of radius 1 will fit inside a square of edge length 10?* Thus, *size* is fixed, *pattern* is variable. In circle packing, *pattern* is fixed, *size* is variable. As yet there is little significant cross-talk between the two topics.

The second question concerns the extension of our type of circle packing to 3D (or higher dimensions), that is, to configurations of spheres which are tangent in specified patterns. There has been no progress whatever in this direction, and I think there is a classical explanation: a theorem of Liouville tells us that the only conformal mappings in three or more dimensions are the Möbius transformations. It appears that any reasonable notion of "sphere packing" gives configurations that are equally rigid – the flexibility we enjoy in 2D just evaporates in higher dimensions.

17

Discrete Conformal Structure

Many geometric phenomena one sees with analytic functions can be viewed more directly through the lens of *conformal geometry*, the geometry associated with Riemann surfaces. We would like to appropriate that view to the discrete setting and to identify some properties that rightly attach to *complexes* rather than to their *packings*.

Let us speak informally to get things started; then with the help of a key example I can lay out the discrete formulation. A Riemann surface is a surface in which there is a consistent way to measure angles between curves. A conformal mapping between Riemann surfaces, $f : S_1 \longrightarrow S_2$, taken by tradition to be univalent, is a mapping that preserves these angles. This local behavior itself cannot be discretized, but maps are *conformal* when they are *analytic* and that gives us a convenient starting point.

Definition 17.1. A **discrete conformal mapping** is a discrete analytic function $f : P_1 \longrightarrow P_2$ between univalent circle packings P_1 and P_2 that have a common complex K.

I emphasize to the reader: *discrete "conformal" maps are **not** conformal in the strict classical sense; angles are **not** preserved.* Does that mean all is lost? No! There are many global and ensemble properties associated with conformality that survive discretization quite faithfully, thank you. Below we discuss discrete versions of extremal length, the "type" problem, packability of surfaces, and in the next chapter, probability.

17.1. A Key Example

We have already seen many conformal maps in our work – the univalent functions on $\mathbb{D}$ explored in Chap. 13, for instance. It will be helpful, however, to have a simple, very concrete example which is more in the spirit of this chapter. Here is the "ten-triangle-toy." The complex K has just 10 faces; it is hand-drawn in Figure 17.1 to emphasize its purely abstract nature. We can create a surface S by pasting 10 unit-sided, euclidean, equilateral triangles together in the pattern of K, as suggested in the figure. I call this a "toy" surface because you can actually build it with cardboard triangles or with the plastic equilateral pieces in various polyhedral construction sets (though this fact is irrelevant to our work).

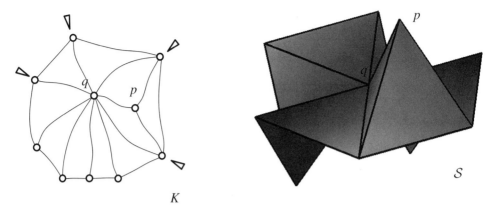

Figure 17.1. The "ten-triangle-toy" surface.

The surface S has a piecewise euclidean metric, which it inherits from its equilateral faces, so one can talk about circles in S. (We will also find in Chap. 22 that it inherits a conformal structure, making it a Riemann surface.) Place a circle of radius $1/2$ at each of the 10 vertices in S. The result is the in situ circle packing Q for K that is drawn on S in Figure 17.2. You may have noted the four "corner" vertices that I marked on ∂K, making it into a combinatorial rectangle. The associated rectangular packing P is computed as described in 13.2. The simplicial map $f : Q \longrightarrow P$ between S, as the carrier of the in situ circle packing Q, and the carrier of the flattened circle packing P in $\mathbb{C}$ is an example of a discrete conformal mapping. (This example will be revisited when we get to Chap. 22.)

(**Note.** I will be using the phrase "in situ" for various packings in the sequel, as I have with Q here. This is perfectly appropriate: these are legitimate circles in the piecewise euclidean metric on S and they are tangent to one another in the pattern of K.)

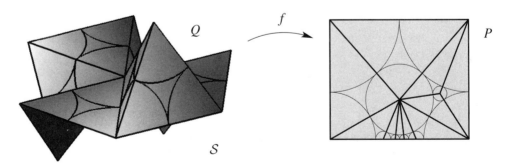

Figure 17.2. The discrete conformal map $f : Q \longrightarrow P$.

17.2. Discrete Formulation

Geometry is a marvelously amorphous notion for Riemann surfaces. Everything is defined locally in the charts of an *atlas*, with compatibility enforced by requiring that the transition maps be analytic. Features that are "geometric" must be local and must be preserved by these transition maps. The principal such features are the angles formed by intersecting curves; hence the adjective *conformal* for *angle-preserving*.

As for the "surface" itself, it is just smoke and mirrors – there is no *there* there. Rather, the surface resides somehow in the collection of transition maps. It can be made concrete via mappings to one geometric setting or another consistent with the conformal structure, that is, analytic. But those geometric properties that are independent of the particular realization should actually be attached to the "web of consistencies" in the transition maps.

In the discrete setting, think of that web as the combinatorics of K. Though K might be realized in one or another geometric setting via circle packings, the features, that are (roughly) invariant should be attached to K itself. In that spirit, conformal mappings $f : Q \longrightarrow P$ will be associated with this generic commutative diagram:

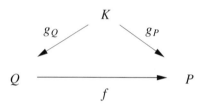

Here g_Q and g_P are simplicial homeomorphisms from K to carr(Q) and carr(P), respectively, and f is the composition $f \equiv g_P \circ g_Q^{-1}$. Note that neither g_Q nor g_P is itself geometric – K has no intrinsic geometry – yet they combine to give a conformal mapping.

Before getting to the main work of the chapter, let me pose a rather circular definition that might nonetheless help your intuition down the road.

Definition 17.2. An abstract (circle packing) complex K is termed a **discrete Riemann surface** and the combinatorics of K provide its **discrete conformal structure**.

We could become formal here – discrete charts, circle-packing transition maps, and so forth – but we will not need discrete Riemann surfaces in the sequel. *All our Riemann surfaces remain classical.* And as for the notion of *discrete* conformal structure, you might just tuck this away until Part IV. After you see the approximation results, you might come to this view for classical surfaces:

classical conformal structure $\longleftrightarrow$ *"infinitesimal triangulation."*

17.3. Extremal Length

The moduli of rectangles and ring domains described in Section A.2.2 of Appendix A is related to a more general notion called *extremal length*. Falling under the rubric "length–area," extremal lengths are among the most powerful conformal invariants.

For the classical notion, let Ω be a Jordan domain in the plane. By a *weight* on Ω we mean a measurable function $\rho \geq 0$. If γ is a rectifiable curve in Ω and Γ is a family of such rectifiable curves, then ρ-lengths and ρ-area are defined by

$$l_\gamma(\rho) = \int_\gamma \rho \, |dz|, \qquad L(\rho) = \inf_{\gamma \in \Gamma} l_\gamma(\rho) \qquad A(\rho) = \iint_\Omega \rho^2 \, dx \, dy.$$

We restrict attention to the class $\mathbb{W}$ of weights ρ with $0 < A(\rho) < \infty$. The *extremal length* of the family Γ in Ω is defined by the homogeneous expression

$$\lambda(\Gamma) = \sup_{\rho \in \mathbb{W}} \frac{(L(\rho))^2}{A(\rho)}. \tag{77}$$

An *extremal weight* ρ is one achieving the supremum.

One of the beauties of extremal length (and the motivation for Beurling and Ahlfors to define it) is its conformal invariance: If $F : \Omega \longrightarrow \widetilde{\Omega}$ is a conformal mapping and $\widetilde{\Gamma}$ is the family of images of Γ under F, then $\lambda(\Gamma) = \lambda(\widetilde{\Gamma})$. Invariance is a consequence of change of variables: in particular, if $w = F(z)$, then computations with weight $\tilde{\rho}(w)$ in $\widetilde{\Omega}$ are mirrored by computations with weight $\rho(z) = |F'(z)| \tilde{\rho}(F(z))$ in Ω. The analogous statement holds in the other direction using F^{-1}. As a consequence, the suprema in (77) are identical in the two situations.

17.3.1. Rectangles

We put general considerations aside now and concentrate on extremal lengths for rectangles. Figure 17.3 illustrates a concrete situation. Ω is a Jordan domain, and there are two marked boundary arcs α, β, which define $\Omega = \langle \Omega; \alpha, \beta \rangle$ as a *conformal rectangle*. By convention, Γ is taken as the family of rectifiable curves $\gamma \subset \Omega$ that begin in α and end in β. In this situation the reciprocal of extremal length, $(\lambda(\Gamma))^{-1}$, is called the *(conformal) modulus* of Ω and is written $\mathrm{Mod}(\Omega)$.

Figure 17.3 suggests how we actually *compute* $\lambda(\Gamma)$; based on the classical result quoted before Theorem 13.4, there exists a (essentially unique) conformal map $F : \Omega \longrightarrow \mathfrak{R}$ to a euclidean rectangle $\mathfrak{R}$ that maps α, β to the two ends of $\mathfrak{R}$. If $\widetilde{\Gamma}$ is the family of rectifiable curves in $\mathfrak{R}$ connecting $F(\alpha)$ to $F(\beta)$, then $\lambda(\widetilde{\Gamma}) = \lambda(\Gamma)$ by conformal invariance. It is not difficult to prove that $\rho \equiv 1$ is an extremal metric in $\mathfrak{R}$ and hence that $\lambda(\widetilde{\Gamma}) = L^2/A$, where L is the length of $\mathfrak{R}$ and A is its area. Of course, $A = LH$, where H is height, so

$$\mathrm{Mod}(\Omega) = \frac{1}{\lambda(\Gamma)} = \frac{1}{\lambda(\widetilde{\Gamma})} = H/L = \mathrm{Mod}(\mathfrak{R}),$$

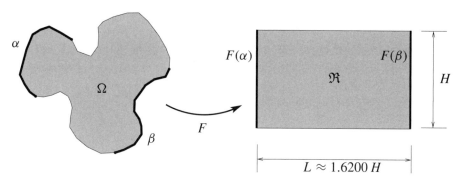

Figure 17.3. Classical extremal length.

the *aspect ratio* of $\mathfrak{R}$. The numerical value of $\text{Mod}(\Omega)$ indicates whether Ω is conformally "long and thin" (small modulus) or "short and fat" (large modulus). Note also that $\rho(z) = |F'(z)|$ is an extremal weight in Ω.

17.3.2. Discrete

Perhaps if you look at the discrete situation depicted in Fig. 17.4, you can guess our discrete analogue. Here Q plays the role of Ω and the corresponding rectangular packing P guaranteed by Theorem 13.4 plays the role of $\mathfrak{R}$. K is marked as a combinatorial rectangle with ends α, β.

Definition 17.3. Let Q be a circle packing whose complex is a combinatorial rectangle $\langle K; \alpha, \beta \rangle$. The **CP-modulus (circle packing modulus)** of K is the number $\text{Mod}^{cp}(K) = \text{Mod}^{cp}(\langle K; \alpha, \beta \rangle) = h/l$, the aspect ratio of the carrier for the rectangular circle packing P of $\langle K; \alpha, \beta \rangle$.

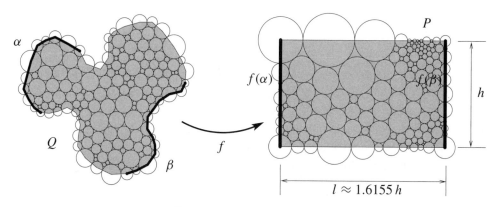

Figure 17.4. CP extremal length.

It may appear that we have thrown out the baby and kept the bathwater here. The quantity $\text{Mod}^{cP}(K)$ has seemingly nothing to do with the packing Q – its value is a discrete-conformal invariant by fiat. We can perhaps get the analogy back on track by addressing the two obvious questions: (1) Where are the weights? (2) Where is the geometry?

(1) *Weights*. There already exists in the literature a notion of discrete extremal length for graphs due to R. J. Duffin and modified by Jim Cannon. The definitions for our combinatorial rectangle $\langle K; \alpha, \beta \rangle$ go like this. Let Γ be the edge-paths γ of K from α to β and let ρ be a discrete density, which we take to mean a set of *vertex weights* $\rho(v) \geq 0$. Define $l_\gamma(\rho)$ as the sum of the weights of the vertices along γ, $L(\rho)$ as the infimum of these lengths over Γ, $A(\rho)$ as the area $\sum_{v \in K} \rho(v)^2$, and w as the collection of weights with positive, finite area. The *graph modulus*, $\text{Mod}^{\text{gr}}(K) = \text{Mod}^{\text{gr}}(\langle K; \alpha, \beta \rangle)$, is defined as before by reciprocal extremal length:

$$\text{Mod}^{\text{gr}}(K) = \frac{1}{\lambda^{\text{gr}}(\Gamma)} = \left(\sup_{\rho \in w} \frac{(L(\rho))^2}{A(\rho)} \right)^{-1}. \tag{78}$$

It is known that there exists an extremal weight ρ^e, unique up to scaling.

Suppose now that P is any euclidean packing for K, say with packing label R. Define the associated *packing weight* ρ_R by $\rho_R(v) = R(v)$, $\forall v$. Note that $l_\gamma(\rho_R)$ is within a factor of 2 of the actual euclidean length of γ in carr(P), while $A(\rho_R)$ is roughly the area of carr(P). Define $\text{Mod}^p(K)$ as in (78) but with w replaced by w^p, the set of packing weights.

Open Question 17.4. *Let R_0 be the packing label for the rectangular packing of $\langle K; \alpha, \beta \rangle$, the packing P of Figure 17.4. Is $\rho_0 = \rho_{R_0}$ the extremal in w^p? That is, does $\text{Mod}^p(K)$ equal $A(\rho_0)/(L(\rho_0))^2$?*

An affirmative answer would have some pleasant consequences. For example, given the packing P and $f : Q \longrightarrow P$ in Figure 17.4, let $\rho = \rho_R$ be the packing density associated with the packing label R for Q. Then the extremal density would be the product $f^\#(v) \cdot \rho(v)$ – recall that this ratio function scales the weights for R to those for R_0. Since $f^\#$ models $|F'|$, this would shadow the classical case. (*Aside:* The classical map F can (in theory) be obtained as the antiderivative of the extremal density on Ω. Could one likewise obtain the packing label for Q by solving an extremal problem?)

There is a bit of a fly in the ointment here, since we restrict to *packing* weights w^p. If we remove that side condition it turns out that we are still in the ballpark. Let ρ_0 be the packing weight for the rectangular packing P, as above, and let ρ^e be the extremal among *all* vertex weights w. Let $\mathfrak{r}$ denote the rectangular carrier of P, with length l and height h. Since ρ_0 consists of radii in P, rough comparisons of euclidean areas give

$$\frac{\pi}{4} A(\rho_0) \leq \text{Area}(\mathfrak{r}) \leq 4A(\rho_0). \tag{79}$$

Paths connecting the ends of τ are clearly at least l in length, so $L(\rho_0) \geq l$; with the inequalities above and the extremality of ρ^e, we get

$$\frac{(L(\rho^e))^2}{A(\rho^e)} \geq \frac{(L(\rho_0))^2}{A(\rho_0)} \geq \left(\frac{\pi}{4}\right) \frac{l^2}{\text{Area}(\tau)} = \frac{\pi}{4} \left(\frac{l}{h}\right). \tag{80}$$

Next we appeal to a duality that occurs with extremal lengths. Let Γ^* be the conjugate family of curves, described abstractly as the family of those curves γ^* that intersect every curve in Γ. In our case this simply means those curves connecting the complementary edges, call them α^* and β^*, of our quadrilateral. The extremal situations we have considered for Γ are simultaneously extremal for Γ^* by symmetry. Letting $H(\cdot)$ (for *height*) play the role of the length $L(\cdot)$ in this dual situation, one can verify that ρ^e is again extremal, and moreover that

$$\sup_{\rho \in w} \frac{(H(\rho))^2}{A(\rho)} = \frac{(H(\rho^e))^2}{A(\rho^e)} = \frac{A(\rho^e)}{(L(\rho^e))^2}. \tag{81}$$

Repeating inequalities of (80) for the dual problem,

$$\frac{(H(\rho^e))^2}{A(\rho^e)} \geq \frac{(H(\rho_0))^2}{A(\rho_0)} \geq \left(\frac{\pi}{4}\right) \frac{h^2}{\text{Area}(\tau)} = \left(\frac{\pi}{4}\right)(h/l)$$

$$\implies \frac{(L(\rho^e))^2}{A(\rho^e)} \leq \left(\frac{4}{\pi}\right)\left(\frac{l}{h}\right).$$

With (81), the definition $\text{Mod}^{cP}(K) = h/l$, and the fact that ρ_0 is a candidate for the supremum $\text{Mod}^p(K)$, we conclude that

$$\text{Mod}^{gr}(K) \sim \text{Mod}^{cP}(K) \sim \text{Mod}^p(K).$$

That is, *all our discrete notions for modulus of a combinatorial rectangle are comparable with a universal constant* $\mathfrak{c} > 1$, meaning that each is bounded by $\mathfrak{c}$ times each of the others. Since $\text{Mod}^{cP}(K)$ is readily computed, we gain a handle on all these quantities.

(2) *Geometry.* If Ω denotes the conformal rectangle represented by $\text{carr}(P)$ in Fig. 17.4, then we have two competing notions for its modulus: $\text{Mod}(\Omega)$ is associated with the classical conformal map $F : \Omega \longrightarrow \mathfrak{R}$ and $\text{Mod}^{cP}(K)$ is associated with the discrete conformal map $f : Q \longrightarrow P$. Technically, f is a point mapping from $\Omega = \text{carr}(Q)$ to $\tau = \text{carr}(P)$, and it will be common in Part IV to define f to be piecewise affine, mapping faces of $\text{carr}(Q)$ affinely to the corresponding faces of $\text{carr}(P)$. We then have quasiconformal mapping arguments that can quantify the relationship of $\mathfrak{R}$ to τ. The following consequence will be used in a key spot later.

Lemma 17.5. *Let* $K = \langle K; \alpha, \beta \rangle$ *be a combinatorial rectangle. Suppose* Q *is a univalent, euclidean circle packing for* K *and* $\Omega = \text{carr}(Q)$. *Then* $\text{Mod}(\Omega)$ *and* $\text{Mod}^{\text{CP}}(K)$ *are comparable by a constant* $\mathfrak{c}$ *which depends only on the degree of* K.

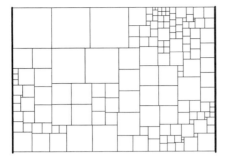 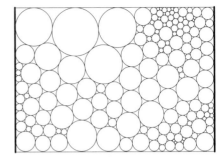

Figure 17.5. "Squared" and "circled" rectangles for K.

Put differently, the aspect ratios of the image rectangles $\Re = F(\Omega)$ and $\mathfrak{r} = f(\Omega)$ are comparable.

I do not want to leave this without sharing the related "packings" in Figure 17.5. On the left is a *squared rectangle* for K. By definition this has a square $s(v)$ for each vertex $v \in K$, with the squares for neighboring vertices sharing (some part of) an edge and the squares for vertices of α and β having edges along the rectangle's ends. Oded Schramm has proven that $\mathrm{Mod}^{gr}(K)$ *is realized precisely by the aspect ratio of this "square packing"*! On the right is an alternative type of "rectangular" packing for K, which perhaps mimics the squared rectangle more closely than does the rectangular packing in Figure 17.4. One might argue that this would be preferable in the definition of $\mathrm{Mod}^{cp}(K)$. Before you conclude that, however, check out the numerical values given below for our examples.

Note. Here are the approximate values of the various extremal lengths for Ω and K used in these examples: $\mathrm{Mod}(\Omega) \approx 1/1.6200$ (thanks to Tobin Driscoll for computing this with his *Matlab* Schwarz–Christoffel Toolbox). $\mathrm{Mod}^{gr}(K) \approx 1/1.4607$ (thanks to J. Cannon, W. Floyd, and W. Parry for the software that generated the squared rectangle). The rectangular packing in Figure 17.4 yields $\mathrm{Mod}^{cp}(K) \approx 1/1.6155$. The alternate style of rectangular circle packing in Figure 17.5 has aspect ratio $\approx 1/1.4215$.

17.3.3. Annuli

We will shortly use another common extremal length, that between the boundary components of a ring domain (annulus) $\mathcal{A}$. If Γ is the family of curves in $\mathcal{A}$ connecting its two boundary components, then the reciprocal of extremal length, $(\lambda(\Gamma))^{-1}$, is precisely the *(conformal) modulus* of $\mathcal{A}$, $\mathrm{Mod}(\mathcal{A})$. This is described in Appendix A and relies on the fact that $\mathcal{A}$ is conformally equivalent to some *round* annulus $A(r, R) = \{z : r < |z| < R\}$. We can again use the maps to standard domains as our approach to the discrete version.

If K is a topological annulus, then there is an essentially unique univalent packing filling some round annulus $A' = A(r', R')$, where $0 \leq r' < R' \leq \infty$. We define the *CP-modulus* of K by $\mathrm{Mod}^{cp}(K) = \mathrm{Mod}(A') = 2\pi / \log(R'/r')$. If this is positive and finite, it reflects whether K is intrinsically a fat (small modulus) or a thin annulus (large modulus). When $\mathrm{Mod}^{cp}(K) = 0$ (in which case K is necessarily infinite), then K packs either

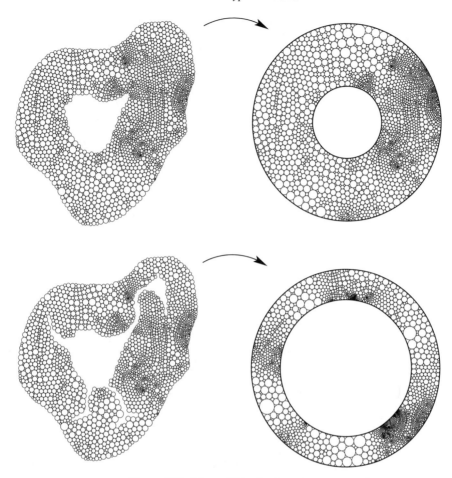

Figure 17.6. Monotonicity for nested annuli.

the punctured disc or the punctured plane, depending on whether K is hyperbolic or parabolic, respectively. In particular, the CP-moduli of ring domains can distinguish *holes* from *punctures*. (*Note*: There is a choice in defining $\text{Mod}^{CP}(K)$ between using packings in the *intrinsic* metric that comes with the maximal packing for K or the *ambient* euclidean metric in a round euclidean annulus; these will be different in general. I have chosen the latter for ease of computation.)

Even putting numerical values aside, discrete extremal length provides important geometric intuition. A small challenge for the reader: *Suppose K is an annulus and subcomplex $L \subset K$ is an annulus whose boundary curves are not null-homotopic in K. Show that $\text{Mod}^{CP}(L) \geq \text{Mod}^{CP}(K)$.* See Figure 17.6 for an example.

17.4. The Type Problem

We first glimpsed the rigid connection between *combinatorics* and *geometry* in the parabolic/hyperbolic dichotomy of Section 8.1.2. In deference to the parallel dichotomy between $\mathbb{C}$ and $\mathbb{D}$ as Riemann surfaces, the attempt to discriminate has come to be called the "type" problem.

Type Problem 17.6. *Let K be a combinatorial open disc. Give criteria to distinguish whether K is parabolic or hyperbolic.*

Let me start with two bookends to the topic. Already in the foundational paper of Rodin and Sullivan there is this very practical one-way criterion based on length–area arguments.

Criterion 1. *For $k > 1$, let n_k denote the number of vertices of K that are in the kth generation from a fixed vertex v_1. If $\sum_k 1/n_k = \infty$, then K is parabolic.*

In contrast, there is a powerful two-way criterion established by Z-X. He and Oded Schramm using graph extremal length. Let Γ be the family of edge-paths of K from v_1 to infinity (i.e., eventually leaving any given finite subcomplex). Consider the collection w of weights ρ assigned to the vertices and having Area $(\rho) = \sum_v \rho(v)^2$ which is finite and positive. Using the notation of the previous section, define the extremal length by

$$\lambda^\infty(K) = \left(\sup_{\rho \in w} \frac{(L(\rho))^2}{A(\rho)} \right).$$

Criterion 2. *Let K be a combinatorial open disc. Then K is hyperbolic if and only if $\lambda^\infty(K) > 0$.*

We learn something from both these criteria. The first suggests that slow "boundary" growth implies being parabolic, allowing one to confirm, for example, that the hexagonal complex $K^{[6]} = H$ is indeed parabolic. Yet this criterion tells us nothing about the heptagonal complex $K^{[7]}$. The second criterion makes it clear that type depends on asymptotic properties of K. For instance, one can show that complexes which are combinatorially identical outside some finite subsets will have the same type, and that refining a complex pushes it toward the hyperbolic side.

Although this second criterion is both necessary and sufficient, one should not declare the Type Problem solved – far from it. The *power* of a criterion is often directly at odds with its *practicality*. When facing concrete situations, it is best to have a variety of criteria to draw upon. Here is a small sampling of criteria between our two bookends.

Archetypes for the type problem are the constant degree complexes $K^{[n]}$. Flowers of small degree are inclined to positive curvature, those of degree 6 try to be "flat," while degree 7 and above tend to negative curvature. We can easily dispense with these and some closely related cases. Consider radii $\{r_0; r_1, \ldots, r_k\}$ for a univalent, closed k-flower, $k \geq 3$. If the petal radii were identical, their common value would be $\lambda_k r_0$ for some $\lambda_k > 0$, and monotonicity would imply

$$\min(r_1, \ldots, r_k) \leq \lambda_k r_0 \quad \text{and} \quad \max(r_1, \ldots, r_k) \geq \lambda_k r_0. \tag{82}$$

Explicit calculation gives

$$\lambda_k = \frac{\sin(\pi/k)}{1 - \sin(\pi/k)} \begin{cases} > 1, & k \leq 5, \\ = 1, & k = 6, \\ < 1, & k \geq 7. \end{cases} \tag{83}$$

Criterion 3. *Suppose K is infinite. If* $\deg(K) = 6$, *then K is parabolic, while if* $\deg(v) \geq 7$ *for all v, then K is hyperbolic. (Note: if* $\deg(K) < 6$, *then K is necessarily finite.)*

Proof. Let P denote the maximal packing for K in $\mathbb{C}$ or $\mathbb{D}$, and in either case denote by R its *euclidean* packing label.

Suppose, now, that every vertex of K has degree at least 7. By (83) and (82), one can construct an infinite simple edge-path $\gamma = \{v_1, v_2, v_3, \ldots\}$ whose euclidean radii decrease by a factor of at least $\lambda_7 < 1$ at each step. This infinite path will thus have finite length in $\mathbb{C}$, implying that P is not locally finite, and hence does not fill the plane; K must be hyperbolic.

On the other hand, suppose every vertex of K has degree at most 6. One then constructs γ so that its (euclidean) radii are nondecreasing. Then P has infinite euclidean area, so K must be parabolic. (A small detail: Can γ be chosen to be simple?) □

Things become more interesting when local tendencies clash. Let us start by knocking 6 and 7 directly against one another. As a first experiment, build K by starting with a half-plane of degree 6 vertices (cut $K^{[6]}$ along one of its hexagonal axes) and then grow it by adding infinitely many successive layers of degree 7 vertices. Figure 17.7(a) suggests the maximal packing: the tendency of the 7-degree region toward negative curvature wins and the result is hyperbolic. Of course, a picture is not a proof (even in this book!), but there are various formal arguments. The reader might, for example, leverage the combinatorial symmetry in K along the "seam" between the two halves into a Möbius shift ϕ leaving $\mathcal{P}_K$ invariant. If $\mathcal{P}_K$ packs $\mathbb{C}$, argue for a path γ of finite length.

The second experiment begins with a hexagonal seed. Generations are added one at a time in successive rings, alternating in some fashion between new rings of degree 6 and new rings of degree 7 vertices. The 7-rings push K toward the hyperbolic camp, while 6-rings try to make it flat. It is surprising to find how few 7-rings it takes to dominate the 6-rings; even Figure 17.7(b) does not do them justice. Indeed, suppose, as in this image, that each 7-ring is a single generation, and let h_j denote the number of 6-degree generations between the jth and $(j+1)$th 7-ring. Ryan Siders has proven that even if h_j grows as rapidly as the j-fold composition of the exponential, K is nevertheless hyperbolic. On the other hand, if h_j grows as the $2j$-fold composition of the exponential, then the 7-rings aren't enough and K is parabolic.

There are a number of type conditions based on isoperimetric consideration, as with Criterion 1 above and yet others which rely on some notion of *average* degree, the critical value invariably being 6. Let me illustrate the latter with a very useful criterion of He and Schramm.

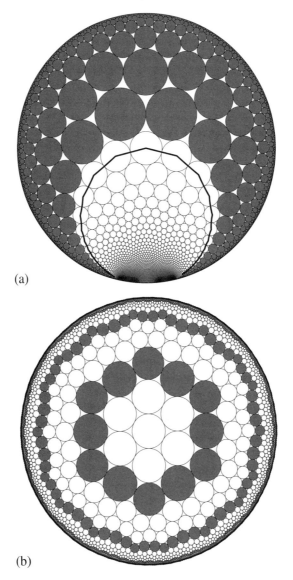

(a)

(b)

Figure 17.7. Mixtures of degree 6 and 7 (shaded); circles: 7 wins! (a) Half-and-half; (b) Rings.

Criterion 4. *Define the "lower average degree" of K as*

$$\mathrm{lav}(K) = \sup_{V_0} \inf_V \left\{ \left(\frac{1}{|V|} \right) \sum_{v \in V} \deg(v) \ : \ V \supset V_0, \ |V| < \infty \right\},$$

where the supremum is over all nonempty finite sets V_0 of vertices of K and $|V|$ denotes the cardinality of V. If $\mathrm{lav}(K) > 6$, then K is hyperbolic.

Criteria that recast the type problem in some equivalent formulation risk the substitution of one mystery for another. On occasion this brings valuable new insight, however, and random walks are a case in point. A *simple random walk* on K is modeled by a random walker on the vertex set of K who, with each tick of the clock, moves from his/her current vertex with equal probability to one of the neighboring vertices. The random walk is called *recurrent* if the random walker starting at v_1 will, with probability 1, return to v_1; otherwise, the random walk is *transient*. I give fuller definitions and the proof for this criterion in the next chapter.

Criterion 5. *Let K be a combinatorial open disc of bounded degree. Then K is parabolic if and only if the simple random walk on K is recurrent.*

17.5. Packable Surfaces

Distinguishing among complexes K that are *not* simply connected is the type problem on steroids! The classical situation is amazingly nuanced and revolves around so-called *moduli* and *Teichmüller* spaces. Though we cannot linger on the topic here, I would like to glimpse some of the issues.

According to the Discrete Uniformization Theorem, every complex K determines a Riemann surface S_K, the surface on which P_K lives. When K is simply connected, S_K is conformally equivalent to one of $\mathbb{P}$, $\mathbb{C}$, or $\mathbb{D}$ – when K is infinite, this leads to the "type" problem of the last section. However, when K is not simply connected, S_K has one of uncountably many possible conformal structures. *Which structure does the combinatorics of K choose?* This is the expanded type problem.

Definition 17.7. A Riemann surface S is called **packable** if it is conformally equivalent to the Riemann surface S_K for some complex K.

The issues are not so much about some particular complex K – in practical terms the best one can hope to do is approximate the covering group (see Section 9.4). The issues, instead, are about the general landscape. What can one say about the packable Riemann surfaces among all Riemann surfaces? There are many open questions and a few preliminary answers. Let us start with tori, the most well understood case.

17.5.1. Tori

In Chap. 9 we observed that every conformal torus T is associated with a lattice group $\Lambda \subset$ Aut($\mathbb{C}$). One can choose independent generators $\phi_1 : z \mapsto z + w_1$ and $\phi_2 : z \mapsto z + w_2$ of Λ so that the complex number $\tau = w_1/w_2$ has positive imaginary part. Tori T and T' are conformally equivalent if and only if $\tau' = \phi(\tau)$ for some $\phi \in$ PSL(2, $\mathbb{Z}$); that is,

$$\tau' = \frac{a\tau + b}{c\tau + d}, \quad \text{for some } a, b, c, d \in \mathbb{Z}, \ ad - bc = 1. \tag{84}$$

The equivalence class of τ, modulo PSL(2, $\mathbb{Z}$), is called the *(conformal) modulus* of T and is denoted Mod(T). A concrete representation is available. Return to Fig. 16.6 and the set M defined by the dark line, a union of two ideal triangles with a vertex at ∞. The j-function is one to one on M (up to identifications of certain boundary segments) and (84) is equivalent to $j(\tau) = j(\tau')$. In essence, M provides a picture of all possible conformally distinct tori; M (with edge identifications) is a complex manifold known as the *moduli space* X_1 for the 1-torus. (As an example, all the tori in Figure 16.8 have modulus $(1/2)(1 + \sqrt{3}i)$.)

One deduces from above that there are uncountably many (conformally distinct) tori. But there are only countably many triangulations K of tori, hence only countably many tori $\mathcal{T}$ are packable. Which ones? It is quite easy to show that there are packable tori arbitrarily "close" to any torus. Suppose $\tau \in M \cap \mathbb{Q}(i)$ (i.e., τ has rational real and imaginary parts). We can cut a parallelogram from the infinite ball-bearing packing of Figure 1.1 having side lengths w_1, w_2 so that $w_1/w_2 = \tau$; identifying opposite sides as usual gives a circle packing in a torus $\mathcal{T}$ with $\mathrm{Mod}(\mathcal{T}) = \tau$.

Gareth McCaughan has proven that if $\mathcal{T}$ is any packable torus, then $\mathrm{Mod}(\mathcal{T})$ is an algebraic number. Whether the converse holds remains open; it is difficult to imagine methods for showing that a given torus is *not* packable. For what pairs K, K' is it the case that $\mathcal{S}_K \simeq \mathcal{S}_{K'}$? How does $\mathrm{Mod}(\mathcal{S}_K)$ change when the combinatorics of K change? for instance, when one applies the "Whitehead moves" described in Appendix E? Can one infer information about $\mathrm{Mod}(\mathcal{S}_K)$ directly from the combinatorics of K? The study of *elliptic curves* is a fertile area, and this discrete approach offers natural open questions and great experimental possibilities.

17.5.2. More General Surfaces

The principal features of the genus 1 case extend to more general Riemann surfaces. I will restrict these brief comments to Riemann surfaces of *finite topological type*. Let g, m, and n be nonnegative integers with $6g + 3m + 2n > 6$. A Riemann surface $\mathcal{S}$ has (topological) type (g, m, n) if it is conformally equivalent to a compact surface of genus g with m topological discs and n points removed. There are infinitely many possible conformal structures, and again there is an associate moduli space, denoted $X_{g,m,n}$, which is a complex manifold whose points are in one-to-one correspondence with the possible conformal structures. As with $X_1 = X_{1,0,0}$ for tori, we can use this to judge how close two structures are.

Theorem 17.8. *The set of points of $X_{g,m,n}$ that are associated with packable surfaces is dense.*

Let me illustrate an approach that comes out of work of Robert Brooks. Assume for convenience that $\mathcal{S}$ is compact ($m = n = 0$). Start by sprinkling a bunch of small, disjoint circles about $\mathcal{S}$ and then letting their radii grow until they lock together in some tangency pattern. Look at the various interstices. With amazing luck, they might all be triangular (three-sided, bounded by three circular arcs); the circles in this case form a circle packing and $\mathcal{S}$ itself is packable. More likely, there are some n-sided interstices for $n \geq 4$. Any

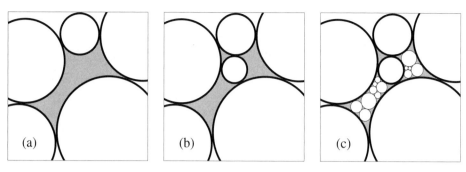

Figure 17.8. Cleaving a five-sided interstice.

interstice having $n \geq 5$ sides, such as that of Figure 17.8(a), can be broken in two by adding a circle that is tangent to opposite sides, as in (b); each of the new interstices will have no more than $n - 1$ sides.

A finite number of such additions will leave you with a configuration of circles whose interstices are all either triangular or quadrilateral. Now we appeal to a result on *Brooks quadrilaterals* described in Appendix D. It tells us that the circles forming a quadrilateral interstice can always be jiggled by an arbitrarily small amount so that their new interstice can be filled with a circle packing, as the two interstices of Fig. 17.8(c) are filled. The procedure, then, is to jiggle all the quadrilateral interstices in $\mathcal{S}$ and fill them with packings, leaving only triangular interstices. The adjustments can be chosen to give an arbitrarily small kick to the conformal structure of $\mathcal{S}$. In other words, our packing lives in a new surface $\mathcal{S}$ – a packable surface – close to the original surface $\mathcal{S}$ in $X_{g,0,0}$.

When $\mathcal{S}$ is compact there are only countably many triangulations, and hence only countably many packable surfaces; density in $T_{g,0,0}$ is the most one expects. In contrast, however, Brock Williams has recently shown that for open surfaces, $m > 0$ or $n > 0$, *every* surface is packable. If the Brooks's quadrilaterals are *shock absorbers* for making small adjustments to the conformal structure on the inside, the open "ends" of a surface are *surge protectors*, compensating to keep the global conformal structure unchanged.

A couple of notes in closing: (1) There is a rather mysterious connection that McCaughan has established between compact packable surfaces and number theory. If $\mathcal{S}$ is compact and packable, then there exists an associated cover group Λ in Aut($\mathbb{D}$) or Aut($\mathbb{C}$), as appropriate, so that each Möbius transformation $\phi \in \Lambda$ has coefficients that are algebraic. (We will see another connection with algebraic numbers in Section 23.1.). (2) A further detail about surface classifications: We have defined a non-simply-connected K to be *parabolic* if its universal covering complex $\tilde{K}$ is parabolic. In classical function theory, however, the terminology is used differently. An open Riemann surface is said to be *parabolic* if it does not support a Green's function. For instance, $\mathbb{C}\backslash\{0, 1\}$ is parabolic, even though its universal covering surface is the unit disc. Discrete function theory is not yet developed to the point that we can make such fine distinctions. Stay tuned, however; at the end of the next chapter there is a hint of a Green's function.

18

Random Walks on Circle Packings

I had the good fortune to be reading the beautiful book "Random Walks and Electrical Networks" by Doyle and Snell as I began working seriously on circle packing, and its themes resonated immediately. In this chapter I show that graph theory, random walks, and electrical networks do in fact fit quite naturally into our topic. I provide enough technical background to prove the type Criterion 5 of Section 17.4 and the uniqueness claim for polynomials in Theorem 14.6.

Potential and probability theory are well-known fellow travelers with analytic function theory, but the link here is still quite remarkable. Whether or not you persevere through the technical sections, please visit the final section of the chapter: you may find that random walks help pin down some of the intuition you may have developed about the dynamics of circle packing.

18.1. Random Walks and Electrical Networks

One can treat the 1-skeleton (i.e., the vertices and edges) of a complex K as a graph. The metaphor for a random walk posits a walker on the nodes of that graph; the walker starts at v_1 at time zero and, at each tick of the clock, moves along an edge from his/her current node v to a neighboring node u with some specified *transition probability* $p(v, u)$. More formally, a *random walk* on a graph is a discrete-time Markov process $\mathfrak{W}$ whose state space is the set of nodes of the graph. A standard example is the *simple random walk*, which we denote by $\mathfrak{W}_s$; if v has degree n, then the walker will move to u with probability $p(v, u) = 1/n$ for $u \sim v$ ($\sim$ indicating neighboring).

Much in the study of random walks has to do with long-term behavior. In the case of infinite graphs, a central question is: *Will the walker almost surely (i.e., with probability 1) return to his/her starting point?* If the answer is yes, then the random walk is said to be *recurrent*; otherwise, it is *transient*. Although there is no *geometry* involved in this statement, there are familiar linkages: It is a famous result of Polya that a regular lattice is recurrent in dimension 2, but transient in dimensions 3 and above. For example, the hexagonal lattice is recurrent; a simple random walker on $K^{[6]}$ will almost surely return to his/her starting point. We cannot change dimension, but we can change combinatorics. In $K^{[7]}$ (see Figure 1.9) the number of vertices in each generation about v_1 grows so quickly – there

232

are so many paths to follow – that walkers easily get lost and never return to v_1; the simple random walk is transient on $K^{[n]}$ for $n \geq 7$.

What does any of this have to do with electricity? Electrical networks are frequently modeled as graphs with resistive wires for edges and with electrons as the walkers. The transition probabilities reflect electrical properties: each edge $\langle v, u \rangle$ (i.e., wire) is labeled with a positive number $\mathcal{C}(v, u)$ known as its *conductance* (reciprocal to electrical *resistance*). Transition probabilities from v are then defined by

$$p(v, u) = \frac{\mathcal{C}(v, u)}{\sum_{w \sim v} \mathcal{C}(v, w)}, \quad \text{for } u \sim v. \tag{85}$$

Thus, the probability that an electron moves from v to u reflects the fraction of the total conductance from v that is attributable to edge $\langle v, u \rangle$. In an infinite network, the key question becomes whether the *conductance to infinity* is positive: *If a particular vertex v_1 is held at one volt and the network is grounded at infinity, will there be any current flow?* Loosely speaking, it is lost electrons that account for current flow. Thus, positive conductance to infinity corresponds with transience, zero conductance with recurrence.

You may already be picking up some vibes from our topic. Consider Figure 17.7. An electron straying into 7-degree territory in the half-and-half packing has a chance of getting lost, more or less as it would get lost in $K^{[7]}$. And as for rings, the 7-rings have more edges, meaning expanded conductance; enough of these will guarantee a flow of electrons to infinity.

18.2. Technical Background

All our random walks will be of this "electrical network" type, with transition probabilities determined by sets $\mathcal{C} = \{\mathcal{C}(v, u) : u \sim v\}$ of positive edge conductances according to (85) – technically, these are *reversible* random walks. The simple random walk $\mathfrak{W}_s$ is itself reversible; set $\mathcal{C}(v, u) \equiv 1$, $u \sim v$. The (1-skeletons of) complexes K are connected, so our random walks are also *ergodic*, meaning that given any two vertices v_1, v_2 there is a positive probability that a walker starting at v_1 will reach v_2. In particular, whether a random walk is recurrent or transient is independent of the vertex where the walk starts.

We will use various results from the theory of random walks for distinguishing between types, that is, recurrence/transience. The first is a comparison: Suppose $\mathfrak{W}_1$ and $\mathfrak{W}_2$ are reversible random walks on K associated with sets $\mathcal{C}_1$ and $\mathcal{C}_2$ of edge conductances, respectively.

Comparison Principle: *If there exists a constant $\mathfrak{C} > 1$ such that $\mathcal{C}_1(v, u) \leq \mathfrak{C}\mathcal{C}_2(v, u)$ whenever $u \sim v$, and if $\mathfrak{W}_2$ is recurrent, then $\mathfrak{W}_1$ is recurrent.*

The second concerns existence of certain functions on the graph. A real function h on the vertices of K is *superharmonic* (for $\mathfrak{W}$) if its value at any vertex is at least as large as the average of its values on the neighboring vertices, weighted by the transition probabilities; that is,

$$h(v) \geq \sum_{u \sim v} p(v, u)h(u). \tag{86}$$

Noting that $\sum_u p(v, u) = 1$, we may rewrite this as

$$\sum_{u \sim v} p(v, u)(h(u) - h(v)) \leq 0, \quad v \in K. \tag{87}$$

Superharmonic: *If there exists a nonconstant, nonnegative superharmonic function for the random walk $\mathfrak{W}$, then $\mathfrak{W}$ is transient.*

If equality holds in (87) for every v, then h is called a *harmonic function* for $\mathfrak{W}$. (h can be real- or complex-valued.) You might compare notions of "superharmonic" and "harmonic" with "superpacking" and "packing" used for labels in Part II. The terminology is not accidental: both random walks and our circle packing arguments exploit the Perron methods of classical potential theory.

The remaining preliminary concerns use of the Dirichlet norm.

Dirichlet Criterion: *The simple random walk on K is transient if and only if there exist a vertex v_1 and a constant $\mathfrak{C}$ such that*

$$h(v_1)^2 \leq \mathfrak{C} \sum_{u \sim v} (h(u) - h(v))^2, \quad h \in C_0(K), \tag{88}$$

where $C_0(K)$ is the space of real functions h on the vertices of K having finite support (i.e., zero off a finite set of vertices).

18.3. Random Walks and "Type"

Establishing a connection between random walks and circle packing type is a two-step process: one needs the geometry–probability link, and then the probability–circle packing link.

Geometry–Probability: The link here is suggested by the classical connection between potential theory and continuous Markov processes. Imagine replacing our electrical network by a continuous, electrically homogeneous, conductive plate, modeled as a Riemann surface S. The random walk is then replaced by its continuous analogue, known as *Brownian motion*. Among many consequences is the classical type condition of Kakutani.

Theorem 18.1 (Kakutani Theorem). *Let S be a noncompact, simply connected Riemann surface. Fix some compact set $E \subset S$ with nonempty interior and some point $p \in S \backslash E$. S is parabolic if and only if almost every Brownian traveler starting at p eventually hits the set E.*

Criterion 5 is the discrete incarnation of this theorem, with $\{v_1\}$ playing the role of E. The geometric similarity lies with the intuition we developed in the previous chapter, summarized as

Riemann surface $\longleftrightarrow$ **infinitesimal triangulation**.

To complement this, we need the further association

Brownian motion $\longleftrightarrow$ **infinitesimal random walk**.

In other words, we would like to claim that Brownian motion on a Riemann surface is essentially a random walk on an *infinitesimal* complex. But as we have seen before, this only works if the continuous and discrete notions are *geometrically faithful*. This seems highly unlikely if we use the simple random walk – it is a blunt instrument that glosses over the local geometry. To capture that geometry we will introduce a random walk that is "tailored" to the circle packing.

Probability–Circle Packings: Let P be a circle packing for complex K with euclidean label R. For interior edge $\langle v, u \rangle \in K$ consider the tangent circles c_v, c_u, as shown in Figure 18.1, and their two common neighboring circles c_a, c_b.

The *radical center* of the triple $\langle c_v, c_u, c_a \rangle$ of circles refers to the center w_a of the circle (dashed in the figure) that is orthogonal to c_v, c_u, and c_a; likewise for w_b vis-à-vis the triple $\langle c_u, c_v, c_b \rangle$. Let z_u, z_v denote the centers of c_u, c_v. Define the conductance for interior edges $\langle u, v \rangle$ of K as follows:

$$C(v, u) = C(u, v) = \frac{|w_a - w_b|}{|z_u - z_v|}, \quad v \sim u. \tag{89}$$

Note that $C(u, v)$ is just the ratio of the lengths of the two dark segments in the picture, the denominator being $R(u) + R(v)$. (There is a similar expression that applies to boundary edges.)

Definition 18.2. Let P be a euclidean circle packing for complex K. The **tailored random walk** $\mathfrak{W}_P$ for P is the reversible random walk on the 1-skeleton of K associated with the conductances $C(u, v)$ of (89).

My real reason for pursuing the type condition is to introduce this beautiful result of Tomasz Dubejko. Define the *center function* $z = z_P : K \longrightarrow \mathbb{C}$; namely, $z(v)$ is the euclidean center of the circle c_v in P.

Theorem 18.3 (Dubejko's Theorem). *The center function z_P is a harmonic function with respect to the tailored random walk $\mathfrak{W}_P$.*

Proof. Consider the closed combinatorial flower $\{v; v_1, \ldots, v_k\}$ for some interior vertex v. For each face $\langle v, v_{j-1}, v_j \rangle$, $j = 1, \ldots k$, let w_j denote the radical center of the associated triple of circles. (In these expressions, $v_0 = v_k$ and $w_{k+1} = w_1$.) Define

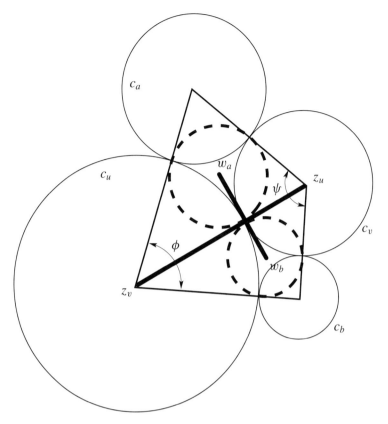

Figure 18.1. Edge conductances.

$\theta_j = \arg(w_j - w_{j+1})$, $j = 1, \cdots, k$. Then

$$0 = \sum_{j=1}^{k}(w_j - w_{j+1}) = \sum_{j=1}^{k}(w_j - w_{j+1})\exp(i\pi/2)$$

$$= \sum_{j=1}^{k}|w_j - w_{j+1}|\exp(i(\theta_j + \pi/2)).$$

As is evident in Figure 18.1, the complex numbers $(w_j - w_{j+1})$ and $z(v_j) - z(v)$ are orthogonal, and in particular, $\exp(i(\theta_j + \pi/2)) = (z(v_j) - z(v))/|z(v_j) - z(v)|$. Putting this in the expression above and recalling the conductances gives

$$0 = \sum_{j=1}^{k}|w_j - w_{j+1}|\frac{(z(v_j) - z(v))}{|z(v_j) - z(v)|} = \sum_{j=1}^{k}\mathcal{C}(v, v_j)(z(v_j) - z(v)).$$

In particular, for every v,

$$\sum_{u \sim v} \mathcal{C}(v, u)(z_P(u) - z_P(v)) = 0$$

$$\iff \sum_{u \sim v} \frac{\mathcal{C}(v, u)}{\sum_{u \sim v} \mathcal{C}(v, u)}(z_P(u) - z_P(v)) = 0$$

$$\iff \sum_{u \sim v} p(v, u)(z_P(u) - z_P(v)) = 0.$$

That is, the center function z_P is harmonic. $\qquad\square$

Let us now return to our discussion of Criterion 5. Suppose K is a combinatorial open disc. The packing $P = \mathcal{P}_K$ fills $\mathbb{C}$ or $\mathbb{D}$, and in either case we will treat it as a euclidean packing. We rely on the tailored random walk $\mathfrak{W}_P$, so let us rework the expression in (89). Suppose R is the euclidean packing label for P. With the help of Herron's formula and using the notation in Fig. 18.1,

$$C(v, u) = \frac{\sqrt{R(u)R(v)}}{R(u) + R(v)} \left(\sqrt{\frac{R(a)}{R(u) + R(v) + R(a)}} + \sqrt{\frac{R(b)}{R(u) + R(v) + R(b)}} \right). \quad (90)$$

Suppose that the simple random walk on K is recurrent. Since the center function $z_P(v)$ is (complex) harmonic, its real part, $\Re z_P$, is also harmonic, implying that $1 + \Re z_P$ is harmonic. If P lies in $\mathbb{D}$, then $1 + \Re z_P$ is a positive, nonconstant harmonic function, and by one of our technical results, $\mathfrak{W}_P$ is transient. On the other hand, note from (90) that $\mathcal{C}(v, u) \leq 2$, so by the Comparison Principle, $\mathfrak{W}_s$ is also transient. We have therefore proven sufficiency in Criterion 5; namely, if $\mathfrak{W}_s$ is recurrent, then K is parabolic. Note that we did not require the bounded degree condition on K.

The proof of necessity does require the bounded degree condition. Gareth McCaughan has proven that if the simple random walk on K is transient, then the inequality in (88) proves an analogous inequality for twice differentiable functions of compact support on carr(P). A classical result then implies that carr(P) is a proper subset of the plane, so K is hyperbolic. The details would take us a little far afield, so I will leave this here and declare our proof complete.

18.4. Completion of a Proof

We now shift attention to the proof of Theorem 14.6, which I suspended until we introduced these probabilistic methods. (See Section 14.2.) Complex K is a parabolic combinatorial open disc of bounded degree; β is a finite branch structure for K. We have established the existence of a packing label R_0 for K with branch set β, normalized so that $R_0(v_1) = 1$. It is only the uniqueness of R_0 that remains to be established.

In the following Θ will denote the angle sum map for R_0, $\Theta(v) = \theta_{R_0}(v)$. We established (see (69)) the existence of a constant $\mathfrak{C}$ such that if R is any label for K with angle sums Θ and $R(v_1) = 1$, then $R(v)/R_0(v) \leq \mathfrak{C}, \forall v \in K$. In other words, any competitor to R_0

lies (after scaling) in this family:

$$\Phi = \{R : R(v_1) = 1 \quad \text{and} \quad \theta_R(v) \geq \Theta(v) \quad \text{and} \quad R(v) \leq \mathfrak{C}R_0(v), \ v \in K\}. \quad (91)$$

To arrive at a contradiction, we assume that Φ contains label R_1 with $\theta_{R_1} \equiv \Theta$ but $R_1 \not\equiv R_0$. We may also assume without loss of generality that $R_1(v_2) > R_0(v_2)$ for some vertex $v_2 \sim v_1$.

Note first that Φ is a Perron family: $R_i, R_j \in \Phi \implies \max\{R_i, R_j\} \in \Phi$. As we have done before, we can verify that $\widehat{R} = \sup_{R \in \Phi} R$ belongs to Φ and that $\theta_{\widehat{R}}(v) = \Theta(v)$ for all v other than v_1. (Recall that if $\theta_{\widehat{R}}(v) > \Theta(v)$ for some v we could increase $\widehat{R}(v)$ and remain in Φ.) Also, since $\widehat{R} \geq R_1$ and $\widehat{R}(v_2) \geq R_1(v_2) > R_0(v_2)$, we have

$$\theta_{\widehat{R}}(v_1) > \Theta(v_1). \quad (92)$$

We use a variational argument. Choose a nested sequence $\{K_n\}$ of combinatorial closed discs exhausting K and containing v_1 and v_2. We are going to develop for each n a parameterized family of labels $R_{n,t}, t \in [0,1]$. Each $R_{n,t}$ is determined by boundary values on $K_n^* = K_n \setminus \{v_1\}$; namely, define

$$R_{n,t}(v_1) = 1, \quad \text{and} \quad R_{n,t}(w) = (1-t)R_0(w) + t\widehat{R}(w) \quad \text{for} \quad w \in \partial K_n. \quad (93)$$

We may use Perron arguments (e.g., Section 11.4) to solve the boundary value problem for the unique label $R_{n,t}$ for K_n satisfying (93) and

$$\theta_t(v) = \theta_{R_{n,t}}(v) = \Theta(v), \quad \text{for } v \in \text{int}(K_n^*). \quad (94)$$

We can deduce various properties of the labels $R_{n,t}$ at interior vertices from their prescribed boundary values. For example, labels $R_{n,t}$ are monotone increasing (in t) on ∂K_n, so by global monotonicity they are monotone increasing at interior vertices. By the Maximum Principle, $R_{n,t}(v)/R_0(v)$ achieves its maximum on ∂K_n, so

$$R_{n,t}(v)/R_0(v) \leq \mathfrak{C} \implies R_{n,t}(v) \in [R_0(v), \mathfrak{C}R_0(v)], \ v \in K_n. \quad (95)$$

Moreover, fixing $t \in [0,1)$ and $\epsilon > 0$ small, we have

$$0 \leq \frac{1}{R_{n,t}(v)}\left(\frac{R_{n,t+\epsilon}(v) - R_{n,t}(v)}{\epsilon}\right) \leq \left(\frac{1}{\epsilon}\right)\left(\frac{R_{n,t+\epsilon}(v)}{R_0(v)} - 1\right)$$

$$\leq \left(\frac{1}{\epsilon}\right)\max_{w \in \partial K_n}\left(\frac{R_{n,t+\epsilon}(w)}{R_0(w)} - 1\right) = \max_{w \in \partial K_n}\frac{1}{R_{n,t}(w)}\left(\frac{R_{n,t+\epsilon}(w) - R_{n,t}(w)}{\epsilon}\right).$$

We now let ϵ go to zero. By (93), the right hand side converges to $R'_{n,t}(w)/R_{n,t}(w)$. (Differentiations are with respect to t.) Since $R_{n,t}(v)$ is monotone, we conclude that $R'_{n,t}(v)$ exists for all but at most a countable set of values $t \in [0,1]$. Avoiding sets of measure zero in what follows, we may define $f_{n,t}(v) = R'_{n,t}(v)/R_{n,t}(v)$ and conclude that

$$0 \leq f_{n,t}(v) = \frac{R'_{n,t}(v)}{R_{n,t}(v)} \leq \max_{w \in \partial K_n}\frac{R'_{n,t}(w)}{R_{n,t}(w)} \leq \mathfrak{C}. \quad (96)$$

Next, we bring in conductances. In particular, let $C_{n,t}(u, v)$ denote the quantity given by the formula of Eq. (90) for label $R_{n,t}$. A straightforward differentiation of the angle sum $\theta_t(v)$ with respect to t gives

$$\theta'_t(v) = \sum_{u \sim v} C_{n,t}(v, u) \left(\frac{R'_{n,t}(u)}{R_{n,t}(u)} - \frac{R'_{n,t}(v)}{R_{n,t}(v)} \right).$$

The angle sums $\theta_{R_{n,t}}(v) \equiv \Theta(v)$ for $v \in \text{int}(K_n^*)$, while $\theta_{R_{n,t}}(v_1)$ is monotone increasing in t, since $R_{n,t}(u)$ is increasing for each $u \sim v_1$. Therefore, this last expression implies that

$$\sum_{u \sim v} C_{n,t}(v, u)(f_{n,t}(u) - f_{n,t}(v)) \begin{cases} = 0, & v \in \text{int}(K_n^*) \\ \geq 0, & v = v_1. \end{cases} \tag{97}$$

We are now ready to go for our contradiction by sending n off to infinity. Let $\mu = \log(\widehat{R}(v_2)/R_0(v_2))$, and note that $\mu > 0$. For each n, the label at v_2 increases from $R_0(v_2)$ to $R_{n,1}(v_2) = \widehat{R}(v_2)$ as t varies from 0 to 1. In particular, for each n,

$$\mu = \log\left(\frac{R_{n,1}(v_2)}{R_0(v_2)} \right) = \int_0^1 d(\log R_{n,t}(v_2)) = \int_0^1 \frac{R'_{n,t}(v_2)}{R_0(v_2)} dt,$$

implying existence of $t_n \in [0, 1]$ such that

$$0 < \mu \leq \frac{R'_{n,t_n}(v_2)}{R_0(v_2)}. \tag{98}$$

Applying a diagonalization argument if necessary, we may assume that

$$f(v) = \lim_{n \to \infty} f_{n,t_n}(v) \quad \text{and} \quad C(v, u) = \lim_{n \to \infty} C_{n,t_n}(v, u) \tag{99}$$

exist for $v \in K$ and $u \sim v$. By (96), $f(v) \in [0, \mathfrak{C}]$, and by (95) and observation, $C(v, u) \in (0, 1]$. Moreover, by the choice of the t_n, f is nonconstant because $0 = f(v_1) < \mu \leq f(v_2)$.

Expression (97) implies that

$$\sum_{u \sim v} C(v, u)(f(u) - f(v)) \geq 0, \quad v \in K. \tag{100}$$

Dividing by the vertex conductance $\sum_{u \sim v} C(v, u)$ to get probabilities and replacing f by the nonconstant and nonnegative function $\tilde{f} = \mathfrak{C} - f$, we have

$$\sum_{u \sim v} p(v, u)(\tilde{f}(u) - \tilde{f}(v)) \leq 0, \quad v \in K. \tag{101}$$

In other words, the random walk $\mathfrak{W}$ on K with the conductances $C(v, u)$ has a nonconstant and nonnegative superharmonic function $\tilde{f}$. That is, $\mathfrak{W}$ is transient. On the other hand, the simple random walk on K is recurrent by the discrete Kakutani condition, since K is of bounded degree and parabolic; the inequalities $C(v, u) \leq 1$ imply by the Comparison Principle that $\mathfrak{W}$ is recurrent.

This contradiction means that Φ cannot contain a competitor to R_0 and completes the proof of uniqueness in Theorem 14.6. (Note that when $\beta = \emptyset$ this provides an alternate proof of the uniqueness of the maximal packing $\mathcal{P}_K$ when K has bounded degree.)

18.5. Geometric Walkers

In the probabilistic model of an electrical network it is clear that the random walkers are the electrons. *Who are the random walkers for tailored random walks?* For insight we hark back to experiments at the very beginning of Part II.

Start with a euclidean circle packing $P \longleftrightarrow K(R)$. Recall the effects of a small increase made in one label, say $R(v)$: the angle sum $\theta_R(v)$ decreases while the angle sums $\theta_R(v_j)$ at the petal vertices $v_1, \ldots, v_k$ increase. Of course, some of the angle shed by v arrives at v_j and must be passed along to maintain the packing condition there, so we adjust $R(v_j)$ and chase the excess to its neighbors; the neighbors do the same, and so forth and so on, ad infinitum. At least in some intuitive sense, then, the effects of our initial adjustment at v will be modeled by the expanding and never-ending movement of "angle" among the vertices of K.

As to that movement, note that in euclidean geometry the angles of triangles add up to π, so angle is never lost: *100% of the angle **leaving** one vertex must be parceled out as angle **arriving** at its neighbors.* This movement can be interpreted as a Markov process: the transition probability from v to v_j is the proportion of a *decrease* in angle at v that becomes an *increase* in angle at v_j. In other words, the random walkers are the quantities of *angle* moving from one vertex to another. (In the hyperbolic setting, there is also some "leakage," angle that becomes part of the area of the vertex's star.)

To make all this work formally, we compute at a differential level, so that all the changes are *infinitesimal*. I will just quote the results and leave the pages of differential calculus to the reader.

Begin with an edge $\langle v, u \rangle$ in the packing as isolated in Figure 18.1. Our interest attaches to the angles labeled ϕ and ψ. By abuse of notation, I will write v, u, a, b for the euclidean radii of these circles, treating v as the variable and the others as fixed parameters. Computation gives

$$\frac{d\psi}{dv} = \frac{\sqrt{u}}{\sqrt{v}(v+u)} \left(\sqrt{\frac{a}{v+u+a}} + \sqrt{\frac{b}{v+u+b}} \right). \tag{102}$$

Moving to a (closed) flower $F = \{v; v_1, \ldots, v_k\}$, let θ_v denote the angle sum at v and let ψ_{v_j} denote the angle at the outer end of $\langle v, v_j \rangle$, $j = 1, \ldots, k$. Adding all the angles of all the faces gives

$$\theta_v + \sum_j \psi_{v_j} = k\pi \implies \frac{d\theta_v}{dv} = -\sum_j \frac{d\psi_{v_j}}{dv}. \tag{103}$$

Now $d\theta_v/dv$ is negative by monotonicity, so the total of (differential) angle that the (differential) increase dv in radius drives to the neighbors is $-d\theta_v/dv$. For a specific neighbor $u = v_j$, the amount arriving at ψ_u is $d\psi_u/dv$, so the transition probability from v to u that we described earlier is

$$p(v, u) = \left(\frac{d\psi_u}{dv}\right) \Big/ \left(-\frac{d\theta_v}{dv}\right) = \left(\frac{d\psi_u}{dv}\right) \Big/ \left(\sum_j \frac{d\psi_{v_j}}{dv}\right). \qquad (104)$$

Before you go ahead bravely and substitute, I suggest you compare Eq. (102) to the conductances of the tailored random walk $\mathfrak{W}_P$ as expressed in (90). Since $d\psi_u/dv = \sqrt{v}\, \mathcal{C}(v, u)$, our transition probabilities become

$$p(v, u) = \frac{\sqrt{v}\, \mathcal{C}(v, u)}{\sum_j \sqrt{v}\, \mathcal{C}(v, v_j)} = \frac{\mathcal{C}(v, u)}{\sum_j \mathcal{C}(v, v_j)}, \quad u \sim v, \qquad (105)$$

which are precisely the transition probabilities of $\mathfrak{W}_P$.

Conclusion. *The tailored random walk $\mathfrak{W}_P$ models the movement of angle among the circles of the circle packing P.*

18.6. New Intuition?

This has been a very technical chapter, but I hope the spirit of a valuable new association has come through:

$$\text{\textit{circle-packing}} \longleftrightarrow \text{\textit{tailored random walk.}}$$

At a minimum, it may give you a way to internalize some of the dynamic properties of circle packing – how disturbances in radii diffuse about the packing like electrons diffusing through an electrical circuit. It may also give much more, however. Let me end with some thought experiments suggesting that random walks have just the right intuitive *feel* for getting at some things we have touched on earlier, namely Green's functions and harmonic measure.

18.6.1. Green's Functions

Consider the uniqueness result we completed earlier. You are given a discrete polynomial $p : \mathcal{P}_K \longrightarrow P$. You perturb P by making a small (i.e., differential) increase in one label, say $R(v_1)$. In our electrical analogy, you have now placed a "charge" $d\theta_{v_1}$ of angle at v_1, namely, the deficit in the angle sum at v_1. We are going to watch as that angle diffuses about K by the process $\mathfrak{W}_P$. Note that although we are watching the angle, it is radii that *drive* the process – when a positive charge arrives at a vertex u, a small increment in the radius at u is triggered to push that charge to the neighbors.

As the charges do their random walks, any charge that returns to v_1 cancels some of its angle sum deficit – current is absorbed at v_1. (We never increment the radius at v_1 because its angle sum is already too low.) But the random walk is recurrent, so every random walker eventually returns to v_1. That is, the initial charge gets entirely reabsorbed at v_1

and the network is again quiet, the packing condition has been reestablished at all the vertices. Now look around and access the radii changes that drove the process. You will find that all the increments are identical; that is, the new label is a scalar multiple of the original.

Consider this same experiment when P has complex K which is transient, however. Some of the initial charge is now lost, never to return to v_1. The deficit in angle sum at v_1 is never entirely canceled, the distribution of radii adjustments is not constant. That residual deficit reflects existence (and perhaps properties?) of a Green's function for the packing.

These are thought experiments, but they provide an alternative view of the dynamics we have seen in other forms. In particular, for non–simply connected complexes they might suggest an internal mechanism for detecting finer distinctions in the "type" of complexes, as Green's functions do for the type of classical surfaces.

18.6.2. Exit Probabilities

As a follow-on question you might well ask: What happens to the missing current in these thought experiments – the lost electrons? This bring us to the notion of exit probabilities. In a transient random walk, almost every walker eventually "falls off the edge" of K. Formally, discrete potential theory defines a *boundary*, $\partial_{\mathfrak{W}} K$, for K and $\mathfrak{W}$, and this boundary has an associated σ-algebra and a probability measure ω. For a measurable set $E \subset \partial_{\mathfrak{W}} K$, $\omega(E)$ represents the probability that a random walker starting at v_1 will "exit" K in the set E. This ω is known as the *harmonic measure* (for K, $\mathfrak{W}$, and v_1). (Compare Definition 8.5.)

Here is a concrete challenge. Consider the maximal packing $\mathcal{P}_K$ for $K = K^{[7]}$ (Figure 1.9 of the Menagerie) with C_1 at the origin. Since the circles all share the same radius, the (hyperbolic) tailored random walk on K is precisely the *simple* random walk. By Theorem 13.16, almost every random walker on K will end at some point of the unit circle $\mathbb{T}$.

Open Question 18.4. *Consider the simple random walk on the circle packing $\mathcal{P}_K$ for $K = K^{[7]}$ starting at the α-circle centered at the origin. Is the harmonic measure induced on $\mathbb{T}$ equal to Lebesgue measure? Is it absolutely continuous with respect to Lebesgue measure? If singular, does it have any point masses?*

Exit probabilities make perfectly good sense on finite complexes K with boundary as well and again appear to be related to harmonic measure. Suppose K is a combinatorial closed disc K. Run a random walk $\mathfrak{W}$ on K, letting the walkers stop (i.e., be *absorbed*) when they reach a vertex in ∂K. Here, $\partial_{\mathfrak{W}} K$ is just ∂K, and the measure ω is the probability distribution reflecting how walkers have exited at the various vertices of ∂K. If $\mathfrak{W}$ happens to be a tailored random walk, $\mathfrak{W}_P$, then ω is certain to be intimately related to the geometry of P. Specifics have yet to be pinned down, but I will show you some actual experiments in Chap. 20 and let you be the judge.

Practicum III

For me, the real hook in Thurston's 1985 talk, beyond the pictures themselves, was the algorithm that produced them. I hope the reader took some time with the "mind game" that opened Chap. 6 or the meta-code of Practicum I. I have personally implemented several packing algorithms by now, studied hundreds of images, both dynamic and static – I have watched many, *MANY* circles pack! Still, I find the packing process itself an enduring mystery.

One constant, in my experience, is that any (sensible) adjustment scheme seems to work – the circles are amazingly single-minded about coming together in a packing! There is some geometric imperative, and the notion that best captures it for me is "curvature." At interior vertex v, with angle sum $\theta_R(v)$, the difference $2\pi - \theta_R(v)$ may be treated as *curvature* concentrated at v. The goal, of course, is a label R that is locally "flat" – curvature zero. Thurston's algorithm begins with label R_0 and then adjusts its entries iteratively to push curvature from vertex to vertex. As to why this works, say R is a (euclidean) label for K. Define

$$\mathcal{E}(R) = \sum \{(\theta_R(v) - 2\pi) : v \in \text{int}(K), \quad (2\pi - \theta_R(v)) < 0\}$$

$$\mathcal{S}(R) = \sum \{(2\pi - \theta_R(v)) : v \in \text{int}(K), \quad (2\pi - \theta_R(v)) > 0\}.$$

Sum $\mathcal{E}(R) \geq 0$ is the total "excess" for angle sums over 2π, while sum $\mathcal{S}(R) \geq 0$ is the total "shortage" for angle sums below 2π. $\mathcal{E}(R) + \mathcal{S}(R)$ measures the distance from R to a packing label.

Now monotonicity kicks in. A sensible adjustment to a label, one that does not overshoot, decreases either $\mathcal{E}$ or $\mathcal{S}$. If it decreases one, the worst it can do is increase the other by an equal amount. Moreover, there are always adjustments that will decrease both, and hence bring you closer to the packing label. Even without any sense of the global situation, just making sensible local adjustments will invariably move toward the packing label.

This, of course, gives tremendous latitude in choosing a strategy – even foregoing the possible uses for nonsensible adjustment. I have tried many, so let me pass along some observations:

(a) A radius adjustment has a very short half-life; it will be made obsolete by subsequent changes – perhaps by the very next change – in other radii.

(b) If vertex v is n combinatorial generations from vertex u, then it will take at least n repack cycles before radius changes at u are felt at v.

(c) Curvature (in the euclidean setting) never dies. Curvature of one sign can cancel that of the other and curvature can reach boundary circles (which do not change). Otherwise, all nonzero curvature is sitting somewhere among the interior vertices.

(d) Two deductions from (c): having few boundary circles tends to slow convergence; having a mixture of positive and negative curvature tends to improve convergence.

(e) A packing algorithm never achieves absolute zero curvature – there is always a little residual "heat" in the packing.

Do not expect too much from your algorithm. Try this little game. Suppose P is a euclidean circle packing with packing label R. Now, modify R at a single interior vertex v, say by making a small increase in $R(v)$. In an ideal world, repacking would require one radius change, namely, resetting $R(v)$. But in the real world? Most algorithms will behave like the random walks of the last chapter: label changes will start diffusing about the packing, first messing up all the circles that already had correct labels, but then slowly, inexorably converging back to R. The correct radius at v is eventually reestablished.

In other words, although the circles intend to come together, they may not know the most direct route. Coupled with the fact that angle sum computations are nonlinear, hence expensive, the observations suggest some things to avoid: (1) Do not bother with highly accurate individual label adjustments. (2) Do not waste time deciding which vertices to adjust next. (3) Do not update the angle sums too often. (What about noniterative methods? R solves a large, sparse, nonlinear system. How do generic direct solvers do? The short answer: horribly!)

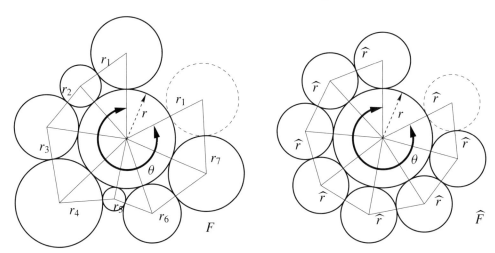

Figure III.1. The uniform neighbor model.

Let me describe the most efficient implementation that I know for the basic iterative approach. Given an interior vertex v we would ideally replace its current label r with that unique label $\bar{r}$ which (with the current labels for the petals) gives angle sum 2π at v. Newton iterations or even simpler bisection methods to estimate $\bar{r}$ are time-consuming. Most recently I have used a "nearest neighbor" model conceived by Charles Collins. It uses a very subtle geometric monotonicity that we might never have guessed had we not seen it experimentally.

The idea is illustrated in Figure III.1. We compare the current k-flower F for v with a "uniform neighbor" model $\widehat{F}$, meaning a k-flower $\widehat{F}$ with label r for v but with petal labels set to a constant $\widehat{r}$ chosen so that the angle sum θ is the same in $\widehat{F}$ as it is in F. The beauty of the model $\widehat{F}$ lies first in its simple computations. Namely, let $\beta = \sin(\theta/2k)$ and $\delta = \sin(\pi/k)$, where k is the number of petals. Then the constant $\widehat{r}$ required for the petal labels and the packing label ρ for v in the resulting uniform flower are given (in the euclidean case) by

$$\hat{r} = \left(\frac{\beta}{1-\beta}\right) r, \qquad \rho = \left(\frac{1-\delta}{\delta}\right) \hat{r}. \tag{106}$$

The strategy is now to replace r not by $\bar{r}$ but instead by ρ. The new monotonicity result I alluded to is this: *The label ρ always lies between r and $\bar{r}$.* Therefore, this strategy is *conservative*: it changes r in the *correct direction* yet *never overshoots*. In the flowers of the figure, for example, k is 7, the initial label is $r = 0.31500$, and the angle sum is $\theta = 1.65000\pi$. Computing the true replacement $\bar{r} = 0.23142$ takes considerable effort. The two computations above, on the other hand, give $\hat{r} = 0.17862$ and $\rho = 0.23306$ nearly for free. And note that ρ gets you more than 98% of the way from r to $\bar{r}$.

It is not simply that we can plug the uniform neighbor computation into our latest code to save compute cycles. The model is so simple that you can streamline the entire algorithm: Given an initial label and threshold ϵ, cycle through the interior vertices in any order – say, based simply on their indexing. On visiting each vertex, replace its label r by the uniform neighbor result ρ (whether r needed changing or not). After some preset number N of passes through the interior vertices, recompute all curvatures in one pass and compare them to ϵ. If you're finished, stop; if not, cycle through another N passes.

What I like most about the uniform neighbor approach is that little bit of geometry behind it. Algorithms benefit when the character of the underlying geometry can show itself. We are not done with repacking yet, and I think there will be some further surprises for you when we get to Practicum IV.

Notes III

The theory of analytic functions is incredibly rich and links to geometry have long been recognized and exploited (sample sources: Ahlfors, 1973; Duren, 1983; Forkas and Kra, 1980; Garnett, 1981; Nevanlinna, 1970). See Appendix A for classical conformal and quasiconformal background and further sources.

Chap. 10. The foundations of circle packing in the Koebe–Andreev–Thurston Theorem and Thurston's Conjecture concerned univalent packings. (*Caution.* In early literature univalence is often assumed without explicit mention.) More comprehensive "discrete" parallels were introduced in Beardon and Stephenson (1990, 1991b); for the experimental approach, see Dubejko and Stephenson (1995b).

Chap. 11. Discrete branching was introduced in Beardon and Stephenson (1990); necessary and sufficient conditions for branch sets (Corollary 11.8) were proved independently in Bowers (1993) and Dubejko (1995). The models for normal families are classical; see, for example, Lehto and Virtanen (1973) and Rudin (1986). For Stoïlow's Theorem see Lehto and Virtanen (1973).

Chap. 12. These construction techniques have classical analogues; e.g., see Stephenson (1988, 1989).

Chap. 13. Discrete disc algebra results are in Carter and Rodin (1992), Dubejko (1997), and Dubejko and Stephenson (1995b); see Marx (1974). For Lemma 13.8, see Nehari (1947); Theorem 13.12, He and Schramm (1995c); Section 13.7.2, Duren (1983, Section 6.3); Theorem 13.16, Benjamini and Schramm (1996). For classical Fatou, Lindelöf, and Löwner theorems, see Nevanlinna (1970). For classical Class **S**, Koebe's 1/4-Theorem, and Nehari's condition, see Duren (1983).

Chap. 14. For the Liouville, Picard, and more general value distribution theory, see Nevanlinna (1970). Discrete polynomials and the important Ring Lemma in the Large, Lemma 14.5, are due to Dubejko (1997c). For the geometry of erf, Nevanlinna (1970, p. 168).

Chap. 15. For polynomial branching and the "fortuitous" example, Bowers and Stephenson (1996).

Chap. 16. Classical Riemann surface theory, e.g., Ahlfors and Sario (1960), Beardon (1984), Farkas and Kra (1980), and Weyl (1955); orbifolds, Thurston; discrete Belyĭ functions, Bowers and Stephenson (2004). Circle packings of surfaces were introduced in Beardon and Stephenson (1991a), Minda and Rodin (1991), Thurston; for affine packings, see Kojima (2003) and Mizushima (2000). For the classical Klein surface (Figure 1.4(a)) and the *j*-function, see Fricke and Klein, (1897–1912). For Apollonian packings see references in Appendix B. See the book *Indra's Pearls* (Mumford, Series and Wright, 2002) for a discussion and gorgeous pictures related to these topics.

Chap. 17. For the "ten-triangle-toy," see Collins, Driscoll, and Stephenson (2003). For extremal distance and squared rectangles, Ahlfors (1973), Cannon (1994), Duffin (1953, 1962), and Schramm (1993); Driscoll's *Matlab* toolbox is Driscoll (1996), and see Driscoll and Trefethen (1998). For the "type" problem, see Beardon and Stephenson (1991c), Bowers (1998), He and Schramm (1995a), McCaughan (1998), Rodin and Sullivan (1987), and Siders (1998); packable surfaces, Bowers and Stephenson (1992, 1993), Brooks (1986), McCaughan (1998), and Williams (2003); matrices with algebraic entries, McCaughan (1996).

Chap. 18. The Doyle and Snell book is (1984); see also Woess (2000). For connections with analyticity, see, e.g., Davis (1979) and Durrett (1984); Kakutani's criterion, (1945). For random walks on circle packings, Benjamini and Schramm (1996), Dubejko (1996–1997, 1997d, 1997e), He (1999), and Stephenson (1990, 1996).

Practicum III. Collins, Driscoll, and Stephenson (2003).

Part IV
Resolution: Approximation

Circle packings were first studied by Paul Koebe in the 1930s in the context of conformal mapping, but the topic quickly dropped from sight. It was independently discovered in the 1970s by William Thurston in the process of constructing certain hyperbolic 3-manifolds. He recognized some special character of rigidity in these circle configurations that was reminiscent of that shown by analytic functions, and was led to conjecture in 1985 that one could use circle packings to approximate classical conformal maps. This is the starting point for Part IV: discrete analytic functions not only *mimic* their classical counterparts, as we have seen in a variety of situations, but actually *approximate* them.

We begin our study with the original Rodin–Sullivan proof of Thurston's conjecture, which remains the cleanest framework for seeing the key ideas. We then extend their proof to remove its hexagonal bias, and in so doing find the flexibility to extend approximation results much further, tapping into the richness of our discrete theory. It turns out that given the slightest chance, circles packings will almost trip over themselves in their rush to converge – all we need to provide is a few geometric hints and some principles of combinatoric refinement. There are two related but distinct strands here. One involves the approximation of analytic functions, and we revisit a few of our earlier examples to illustrate their limit behavior. The other strand, more directly geometric, shows that discrete conformal structures also converge to their classical counterparts.

I want to draw your attention particularly to Chapter 23, where we use both discrete functions and discrete conformal structures in several settings of active research interest, ranging from number theory to conformal tilings to (of all things!) human "brain mapping." These are all settings involving *classically defined* structures for which no numerical approximation methods were available until circle packing arrived on the scene. Perhaps you will be drawn into trying circle packing in your own area of mathematical or scientific interest.

19

Thurston's Conjecture

The setting Thurston established for his 1985 conjecture is illustrated in Figure 19.1. When Ω is a bounded simply connected domain in the plane, the Riemann Mapping Theorem tells us that there exists a conformal mapping F from $\mathbb{D}$ onto Ω. Given arbitrary points w_1, $w_2 \in \Omega$, F may be chosen so that $F^{-1}(w_1) = 0$ and $F^{-1}(w_2) > 0$, and in this case F is unique.

I have illustrated Thurston's idea (for this region) in Figure 2.8. One fills Ω to the extent possible with a hexagonal packing P_ϵ of circles of radius $\epsilon > 0$ and defines the discrete analytic function $f_\epsilon : Q_\epsilon \longrightarrow P_\epsilon$, where Q_ϵ is the maximal packing associated with P_ϵ. For small ϵ there will be distinct vertices v_1, $v_2 \in K_\epsilon$ whose flowers in P_ϵ contain w_1 and w_2, respectively, and one normalizes Q_ϵ to center their circles at the origin and on the positive x-axis, respectively.

In our terminology, each of the functions $f_\epsilon : \mathbb{D} \longrightarrow \Omega$ is a *discrete conformal mapping*. In point of fact, each is the simplicial (i.e., piecewise affine) mapping between euclidean carriers, $f_\epsilon : \text{carr}(Q_\epsilon) \longrightarrow \text{carr}(P_\epsilon)$. Figure 2.8 depicts three such mappings with successively finer hexagonal combinatorics. Here is Thurston's Conjecture about their fate, now the Rodin–Sullivan Theorem:

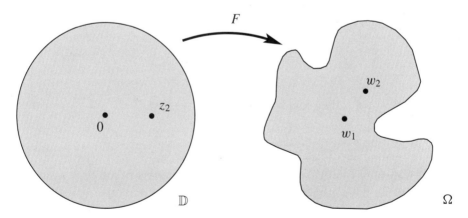

Figure 19.1. A classical conformal mapping F.

Theorem 19.1 (Rodin–Sullivan Theorem). *Let Ω be a bounded and simply connected plane region with distinguished points w_1, w_2. Assume that the classical conformal mapping $F : \mathbb{D} \longrightarrow \Omega$ and the discrete conformal mappings $f_\epsilon : \mathbb{D} \longrightarrow \Omega$ are defined and normalized as described above. Then the mappings f_ϵ converge uniformly on compact subsets of $\mathbb{D}$ to F as $\epsilon \longrightarrow 0$.*

This theorem has been generalized and extended far beyond this original setting, as we will see in the sequel. But the essential ingredients appear so cleanly in Rodin and Sullivan's argument that I will start with their proof.

19.1. Geometric Ingredients

We have already seen the important Ring Lemma in a more precise form, but I restate it here for convenience.

Lemma 19.2 (Ring Lemma). *Given $k \geq 3$, there is a constant $\mathfrak{c}(k) > 0$ such that in any univalent k-flower the ratio between the euclidean radius of a petal and that of the center is bounded below by $\mathfrak{c}(k)$.*

From there we move to two more subtle results. The first is a concrete prototype for extremal length as discussed in Section 17.3.

Lemma 19.3 (Length–Area Lemma). *Let c be a circle in $\mathbb{D}$ and let $S_1, S_2, \cdots, S_m$ be m disjoint chains of circles, each of which separates c (relative to $\mathbb{D}$) either from the unit circle or from the origin and a point of the unit circle. Assume that all the circles involved have mutually disjoint interiors. Denote the combinatorial lengths of the chains by $n_1, n_2, \cdots, n_m$. Then*

$$\text{radius}\,(c) \leq \left(n_1^{-1} + n_2^{-1} + \cdots + n_m^{-1}\right)^{-\frac{1}{2}}.$$

Proof. Suppose S_j consists of circles with euclidean radii ρ_{ji}, $1 \leq i \leq n_j$. By the Schwarz inequality

$$\left(\sum_i \rho_{ji}\right)^2 \leq n_j \sum_i \rho_{ji}^2.$$

Since $l_j = 2 \sum_i \rho_{ji}$ is the geometric length of S_j, we have $l_j^2 n_j^{-1} \leq 4 \sum_i \rho_{ji}^2$, implying

$$\sum_j l_j^2 n_j^{-1} \leq 4 \sum_{ji} \rho_{ji}^2 \leq 4. \tag{107}$$

The last inequality simply reflects that the disjoint interiors of the circles can have area summing to no more than the area of $\mathbb{D}$. From this, $l = \min\{l_1, l_2, \cdots, l_m\}$ satisfies $l^2 \leq 4(n_1^{-1} + n_2^{-1} + \cdots + n_k^{-1})^{-1}$. The separation conditions in the chains guarantee that l is greater than the diameter of c, so this inequality proves the lemma. $\qquad\square$

The Hexagonal Packing Lemma is perhaps the heart of the proof. It depends on uniqueness of hexagonal packings, a fact Sullivan had established in another context. We will see how to interpret and refashion this for nonhexagonal cases later.

Lemma 19.4 (Hexagonal Packing Lemma). *There is a sequence $\{s_n\}$, decreasing to zero, with the following property. Let c_1 be a circle in a univalent euclidean circle packing P and suppose the first n generations of circles about c_1 are combinatorially equivalent to n generations of the regular hexagonal packing around one of its circles. Then for any circle $c \in P$ tangent to c_1,*

$$\left| 1 - \frac{\text{radius}(c)}{\text{radius}(c_1)} \right| \leq s_n. \tag{108}$$

Proof. Assume the result fails. Then there exist an $x \neq 1$ and a sequence of univalent packings $P_n \longleftrightarrow K_n(R_n)$ with neighboring vertices v_1^n, $w^n \in K_n$ such that v_1^n is surrounded by a growing number of generations of hexagonal combinatorics, yet

$$R_n(v_1^n)/R_n(w^n) \longrightarrow x, \text{ as } n \to \infty.$$

By discarding excess circles, we may assume that each P_n is in fact hexagonal, so its complex K_n can be identified with a subcomplex of H, the infinite hexagonal complex, so that vertices v_1^n and w^n are identified with the α-vertex v_1 and a neighbor $w \sim v_1$ in H, respectively.

Since the P_n are univalent, the Ring Lemma gives a bound on the radii of neighboring vertices. This is crucial to the use of normal families (Theorem 11.15): we extract a subsequence P_{n_j} converging to a packing $P \longleftrightarrow H(R)$, with $R_{n_j} \longrightarrow R$. Since the P_{n_j} are univalent, P is univalent. However, we know that all univalent packings for H have labels R that are constant. In particular, $R_n(v_1^n)/R_n(w^n) \longrightarrow R(v_1)/R(w) = 1$. This contradicts the convergence to $x \neq 1$ and establishes the result. $\square$

19.2. Proof of the Rodin–Sullivan Theorem

We now have all the tools in place. Recall that the maps f_ϵ carry (euclidean) triangles t in the carrier of Q_ϵ to the corresponding triangles $f_\epsilon(t)$ in the carrier of P_ϵ. We need some facts about the domains and ranges of these functions, and then about their convergence.

19.2.1. Range Carriers

Each packing P_ϵ is hexagonal, with circles having constant radius ϵ. It is clear that their carriers exhaust Ω as ϵ goes to zero. In fact, given any compact set $E \subset \Omega$, not only is $E \subset \text{carr}(P_\epsilon)$ for small ϵ, but the number of generations of circles between E and $\partial\Omega$ goes to infinity as $\epsilon \to 0$.

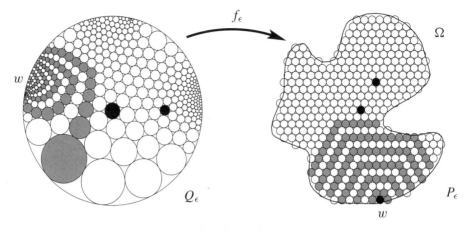

Figure 19.2. Separating chains.

19.2.2. Domain Carriers

Here we have the maximal packings Q_ϵ, with boundary horocycles. Recall, however, that we are treating these as euclidean packings and working with their euclidean carriers. To show that carr(Q_ϵ) exhausts $\mathbb{D}$, it clearly suffices to show that the boundary circles go to zero in euclidean radius – this is where we use the Length–Area Lemma.

Fix ϵ. Let w be a boundary vertex of K_ϵ which is $k + 1$ generations away from v_1. Over in Ω, P_ϵ has at least k pairwise disjoint chains $S_1, S_2, \cdots, S_k$ of circles that separate w from v_1, each chain starting and ending with boundary circles. The combinatorial lengths n_j of S_j (assuming they are numbered according to their generation from w) and the number k of chains satisfy

$$n_j \leq 6j, \qquad k > \frac{\text{dist}(w_1, \partial\Omega)}{3\epsilon}. \tag{109}$$

Moving back to Q_ϵ, let $\sigma_1, \cdots, \sigma_k$ represent the corresponding chains there. These circles have mutually disjoint interiors and each σ_j starts with a horocycle and ends with a horocycle; in particular, each separates the circle C_w for w from a part of the unit circle and from the origin, the center of the circle for v_1. The situation is depicted in Figure 19.2 with alternate chains shaded. By (109) and the Length–Area Lemma,

$$
\begin{aligned}
(\text{radius}(C_w))^2 &\leq \frac{1}{(1/n_1 + 1/n_2 + \cdots + 1/n_k)} \\
&\leq \frac{1}{(1/6 + 1/12 + \cdots + 1/6k)} = \frac{6}{(1 + 1/2 + \cdots + 1/k)}.
\end{aligned}
$$

The right side has the divergent harmonic series in its denominator, implying that the fraction goes to zero, and hence the radius of C_w goes to zero. In fact, by the growth of k in (109), the radii of the C_w go uniformly to zero for all $w \in \partial K_\epsilon$ as $\epsilon \to 0$.

Since the boundary horocycles of Q_ϵ go uniformly to zero in euclidean radius, we conclude that the Q_ϵ have carriers which exhaust $\mathbb{D}$. Moreover, by comparing their labels to the constant labels for P_ϵ and applying the maximum principle, Lemma 11.11, we conclude that the interior radii go uniformly to zero as well.

19.2.3. Quasiconformality

For the distortion properties of maps f_ϵ we rely on general properties of quasiconformal maps; see Appendix A for the basics. The domains of the f_ϵ vary slightly with ϵ, but the behavior of interest manifests itself on compact subsets. It is enough, therefore, to work with an open set D having closure in $\mathbb{D}$ and to take ϵ sufficiently small that D lies in the union of interior faces of $\mathrm{carr}(Q_\epsilon)$.

The faces of $\mathrm{carr}(P_\epsilon)$ are equilateral triangles, those of $\mathrm{carr}(Q_\epsilon)$ are typically not. However, since Q_ϵ is univalent, the Ring Lemma implies that the ratio of (euclidean) radii for neighboring interior circles of Q_ϵ is bounded by constant $\mathfrak{c}(6)$. This says that the interior triangles of $\mathrm{carr}(Q_\epsilon)$ are boundedly close to equilateral. In particular, their triples of radii are $\mathfrak{c}(6)$-bound in the sense of Section 2.4 of Appendix A, and Lemma A.9(a) implies that f_ϵ is κ-quasiconformal on each interior face of $\mathrm{carr}(Q_\epsilon)$ where $\kappa \in [1, \infty)$ depends only on $\mathfrak{c}(6)$. Since face edges are line segments, f_ϵ is κ-quasiconformal across edges, and therefore κ-quasiconformal on all of D. The f_ϵ are also uniformly bounded because their images lie in Ω, so they form a normal family. There exists a sequence $\{\epsilon_j\}$ for which $f_j = f_{\epsilon_j}$ converges, say $f_j \longrightarrow f$.

By normal family results for quasiconformal mappings, we know that f either is a constant or is univalent and κ-quasiconformal. By construction, $f_j(0) = w_1$ and $f_j(x_j) = w_2$ for some $x_j > 0$. Choose $\mathfrak{c}$ so that Ω lies inside the disc $D(w_1, \mathfrak{c})$. Translating P_{ϵ_j} to put w_1 at the origin, scaling by $1/\mathfrak{c}$, and then applying the Discrete Schwarz Lemma, one can check that $x_j > \nu = |w_1 - w_2|/\mathfrak{c} > 0$. On the other hand, pick some arc γ connecting w_1 and w_2 in Ω and choose a ring domain A encircling γ, with $\mathrm{Mod}(A) \in (0, \infty)$, and with closure lying in $\Omega \backslash \gamma$. For small ϵ, A is covered by interior faces of $\mathrm{carr}(P_\epsilon)$. Thus, for j large, the ring domain $A_j = f_j^{-1}(A)$ in the carrier of Q_ϵ separates both $0 = f_j^{-1}(w_1)$ and $x_j = f_j^{-1}(w_2)$ from the unit circle. But the moduli of these ring domains A_j are bounded: $\mathrm{Mod}(A_j) \leq \mathrm{Mod}(A)/\kappa < \infty$. A result of Grötzsch provides us with a lower bound $\mu > 0$ on the distance from x_j to the unit circle. In other words, we now have

$$0 < \nu < x_j < 1 - \mu < 1, \quad \text{for } j \text{ large.}$$

If $x \in (0, 1)$ is an accumulation point for the x_j, then

$$f(x) = \lim_j f_j(x) = w_2 \neq w_1 = \lim_j f_j(0) = f(0).$$

That is, f is not constant, so it must be κ-quasiconformal on D. These arguments apply to any open set D with closure in $\mathbb{D}$, consequently f is defined, univalent, and κ-quasiconformal on all of $\mathbb{D}$. By the Carathéodory Kernel Theorem its range is Ω since the ranges of the f_j exhaust Ω.

Summary. *Given any sequence of ϵ's going to zero, there exists a subsequence $\{\epsilon_j\}$ so that the associated simplicial maps $\{f_{\epsilon_j}\}$ converge uniformly on compact subsets of $\mathbb{D}$ to a κ-quasiconformal homeomorphism $f : \mathbb{D} \longrightarrow \Omega$ satisfying*

$$f^{-1}(w_1) = 0 \quad and \quad f^{-1}(w_2) > 0. \tag{110}$$

19.2.4. Conformality

The last piece of the puzzle requires that we replace quasiconformality by conformality. As above, let $f_j = f_{\epsilon_j}$ and assume $f_j \longrightarrow f$. It suffices once more to work on a fixed open set D whose closure is in $\mathbb{D}$. For each j there exists an integer $n_j > 0$ such that if t is any face of $\text{carr}(Q_{\epsilon_j})$ that intersects D, then the circles of t are at least n_j combinatorial generations deep in Q_{ϵ_j}. Using the numbers s_{n_j} provided by the Hexagonal Packing Lemma, the triples of radii for t are $(1 + s_{n_j})$-bound. Lemma A.9(a) implies existence of a bound κ_j for the dilatation of f_j on t, and extending across the edges, we conclude that f_j must be κ_j-quasiconformal on all of D. As $j \to \infty$, we have

$$n_j \longrightarrow \infty \implies s_{n_j} \longrightarrow 0 \implies (1 + s_{n_j}) \longrightarrow 1.$$

But a 1-bound triple forms an equilateral triangle – t is becoming more equilateral. By Lemma A.9(b), $\kappa_j \longrightarrow 1$, and we conclude that the limit function f is 1-quasiconformal on D. This holds for any D with closure in $\mathbb{D}$, so f is 1-quasiconformal on $\mathbb{D}$.

But "1-quasiconformal" is equivalent to "conformal," and there is only one conformal mapping of $\mathbb{D}$ onto Ω that satisfies (110), namely, the classical conformal mapping that we denoted by F at the beginning of this chapter. Therefore $f \equiv F$. Finally, note that because every sequence of ϵ's going to zero has a subsequence whose mappings $\{f_{\epsilon_j}\}$ converge to this same limit F, one can conclude that $f_\epsilon \longrightarrow F$ as $\epsilon \to 0$. This completes the proof of the Rodin–Sullivan Theorem. □

19.3. Example

Figure 19.3 illustrates once more the discrete conformal maps to our example region Ω. As in the earlier Figure 2.8, f_n denotes f_ϵ for $\epsilon = 1/n$. I have tried to convey a sense of the distortion issues in the proof by color-coding the faces in the image carriers according to their individual dilatations. Light red to dark red codes dilatations from $\kappa = 1$ (no conformal distortion) to $\kappa = 1.1$ (roughly "10%" distortion). Faces that are blue have dilatations exceeding 1.1. One sees clearly that in the interior the maps are improving quite rapidly with refinement. As expected, the worst distortions are near the boundary, particularly at sharp bends or narrow necks.

It is important to remember that a κ-quasiconformal map has *maximal* dilatation κ; in other words, this is a worst-case measure of local distortion. If you are not careful,that κ might taint your opinion of the whole mapping. In fact it is likely that these maps all have large maximal dilatations, but it is behavior on compact sets in the interior that

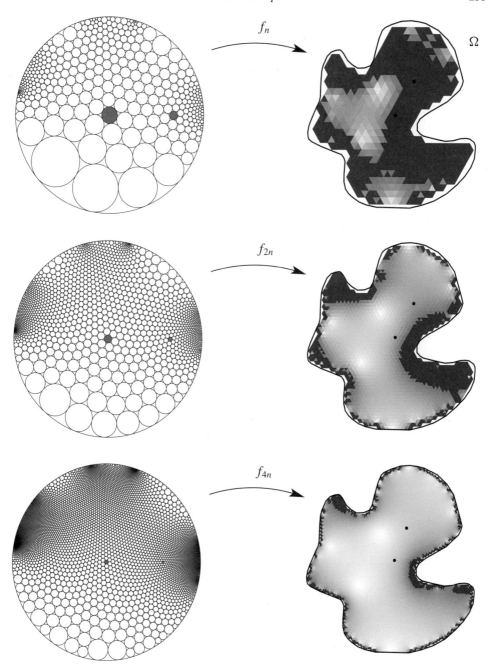

Figure 19.3. Color-coded dilatations.

counts. One should also be aware that real (versus complex) dilatation does not take into account the "direction" of distortion; often the distortions are noncoherent and tend through cancellation to moderate the impact of κ on global features.

These pictures largely confirm what we expected from the proof. However, once you have these pictures in hand, they begin to raise some new questions on their own. Note, for example, the regions of very light red, i.e., very low distortion; there are three notable ones in the interior. You can see them show up even in the coarse stage; they do not correlate with combinatorial depth, and they seem to have very stable locations. Are they telling us something about the geometry of Ω?

20

Extending the Rodin–Sullivan Theorem

The hexagonal combinatorics present in Thurston's Conjecture were a convenience, not a fundamental requirement, and it is easy to become enamored of their special properties. Here we remain in the univalent setting and prove a modest extension of the Rodin–Sullivan Theorem, but one that highlights the two key geometric ingredients.

Theorem 20.1. *Let Ω be a simply connected bounded domain in the plane with distinguished points w_1, w_2 and let $F : \mathbb{D} \longrightarrow \Omega$ be the unique conformal mapping with $F^{-1}(w_1) = 0$ and $F^{-1}(w_2) > 0$. Assume P_n is a sequence of univalent circle packings in Ω which, along with their complexes K_n, satisfy the following conditions:*

(a) Mesh (P_n) (the largest radius) goes to zero as n goes to infinity.

(b) The carriers, carr(P_n), exhaust Ω.

(c) There exists a uniform bound d on deg(K_n).

(d) The sequence $\{K_n\}$ either is a nested sequence which exhausts a parabolic combinatorial disc or is asymptotically parabolic.

For each n, let $f_n : \mathcal{P}_{K_n} \longrightarrow P_n$ be the associated discrete conformal mapping, normalized by $f_n^{-1}(w_1) = 0$, $f_n^{-1}(w_2) > 0$. Then on compact subsets of $\mathbb{D}$, the f_n converge uniformly to F and their ratio functions $f_n^{\#}$ converge uniformly to $|F'|$.

This extension has been eclipsed by a result of Z-X. He and Oded Schramm that removes hypotheses (c) and (d). Their proof, which is wonderfully elegant, would take us too far afield, but I state their result and a startling consequence at the end of the chapter.

Any advantages our extension might have would lie with computational practicality. Hypotheses (c) and (d) hold when the K_n are hexagonal (so this does extend Rodin–Sullivan) or when they are generational complexes of some known parabolic complex of bounded degree, such as the ball-bearing pattern (Figure 1.1(a)) or the universal covering of a torus (Figure 1.3(e)). The more abstract notion of asymptotic parabolicity, defined at the end of Chap. 11, gives considerably more latitude. We will see that it holds, for example, if the K_n have packings in which all circles are comparable in radius with constant independent of n.

As for convergence of the ratio functions, this was anticipated by Rodin and Sullivan and became a key early topic in circle packing. A more general result is now possible using direct geometric methods, so this part of the theorem is split off in a separate section. Following that, I discuss experiments demonstrating boundary approximations which are companions to what we see on the interior; these connections are not yet fully quantified, but suggest that we have much to learn.

20.1. Outline

The proof follows the general lines set out by Rodin and Sullivan. Again, the maps f_n are barycentric maps from the euclidean carrier of the maximal packing $\mathcal{P}_{K_n} = Q_n$ to that of P_n and their normalizations are as before. Since the packings are univalent and of uniformly bounded degree, the Ring Lemma gives us a uniform bound $\kappa = \kappa(d)$ on the quasiconformal dilatations of the maps f_n (and f_n^{-1}) on interior faces.

Our first task is to show that the carriers of Q_n exhaust $\mathbb{D}$, for which it suffices to show that the boundary horocycles go to zero in (euclidean) radius. In the situation as described, suppose w is a boundary vertex of K_n, C_w and c_w are the corresponding circles of Q_n and P_n, respectively, and z_w is the center of c_w. We prove the following:

Lemma 20.2. *There exists a constant $\mathfrak{C} > 0$ such that for n sufficiently large*

$$\operatorname{radius}(C_w) \leq \mathfrak{C} \left(\log \frac{|z_w - w_1|}{\operatorname{radius}(c_w)} \right)^{-\frac{1}{2}}. \tag{111}$$

Note that since $\operatorname{mesh}(P_n) \longrightarrow 0$, boundary circles c_w have radii going to zero and centers going to $\partial\Omega$, so the quantity on the right of the inequality goes to zero. The inequality implies that the carriers of the packings Q_n exhaust $\mathbb{D}$. An additional consequence follows from this and the Ring Lemma: namely, given any compact $E \subset \mathbb{D}$, the combinatorial distance between a circle of Q_n intersecting E and the boundary circles will grow with n.

The maps f_n are κ-quasiconformal, so we are now in position to extract a subsequence converging uniformly on compact subset of $\mathbb{D}$ to a κ-quasiconformal homeomorphism $f : \mathbb{D} \longrightarrow \Omega$ having the desired normalizations. Our last task involves showing that the dilatations of the f_n go to 1 on compacta so that f is actually *conformal*. The key is a reformulation of the constants s_n from the Hexagonal Packing Lemma to remove their hexagonal bias.

Lemma 20.3. *Let $\{K_n\}$ be a sequence of combinatorial closed discs satisfying hypotheses (c) and (d) of the theorem. Then there exists a sequence $\{s_m\}$ of constants, decreasing to zero, with the following property: Suppose that for some n, $v \sim u$ are interior vertices of K_n whose combinatorial distances from ∂K_n are at least m and suppose that $P \longleftrightarrow K_n(R)$*

and $\widetilde{P} \longleftrightarrow K_n(\widetilde{R})$ are two univalent, euclidean circle packings for K_n. Then

$$\left| \frac{\widetilde{R}(u)}{\widetilde{R}(v)} - \frac{R(u)}{R(v)} \right| \leq s_m. \tag{112}$$

Why does this serve in place of the Hexagonal Packing Lemma? For a vertex which is deep within K_n, this says that the ratio of its radius to that of a given neighbor is nearly the same in P_n and Q_n. First, this means that the triples of radii for faces deep in K_n are $\mathfrak{C}$-bound in both packings, giving a uniform bound on the dilatation of f by Lemma A.9(a). But more than that: corresponding faces in domain and range are approaching a common "shape" – it is just that this is no longer necessarily the equilateral shape. The comments following Lemma A.9 explain that as the radii of neighbors approach common ratios, part (b) provides bounds on dilatations that approach 1. The quasiconformal dilatation is being driven to 1 on compacta just as in the Rodin-Sullivan proof. The remaining details also follow that proof. Our job, then, is just to prove the lemmas.

20.2. Proving the Lemmas

These lemmas require some added machinery, but if you watch closely you will see that they are, in their hearts, very direct geometric descendants of the Length–Area and Hexagonal Packing lemmas, respectively.

Proof of Lemma 20.2 We use extremal length, the direct progeny of length–area. The situation is as described in the theorem above; all radii and centers in the proof will be euclidean. The only real difficulty in the arguments involves typical fringe discretization issues. We do not have good quasiconformal bounds for f_n on faces containing boundary circles, so we target instead the interior circles *neighboring* boundary circles.

To that end, define $\widetilde{Q}_n$ and $\widetilde{P}_n$ by removing the boundary circles from Q_n and P_n. Let $\widetilde{K}_n$ denote the complex and let $\widetilde{f}_n : \widetilde{Q}_n \longrightarrow \widetilde{P}_n$ be the restriction of f_n. The interior circles that neighbored the boundary circles now themselves become the boundary circles and $\widetilde{f}_n$ is κ-quasiconformal on all of $\mathrm{carr}(\widetilde{Q}_n)$. We first prove the analogue of inequality (111) for $v \in \partial \widetilde{K}_n$.

The situations in $\mathbb{D}$ and Ω are illustrated in Figure 20.1. Consider $\mathbb{D}$ first. Let $v = v_n \in \partial \widetilde{K}_n$ be chosen so that its circle C_v has center Z_v that is closest to the origin among all boundary circles of $\widetilde{Q}_n$. Let $t_n = |Z_v|$; the disc $D(0, t_n)$ is (essentially) contained in $\mathrm{carr}(\widetilde{Q}_n)$. For later use, write S_v for the star of faces in $\mathrm{carr}(\widetilde{Q}_n)$ containing v.

It is enough to show that radius (C_v) goes to zero as n grows. From there it is easy to argue (e.g., as in the Ring Lemma) that its neighboring horocycle's radius goes to zero and hence that $t_n \longrightarrow 1$, which is our true objective. (*Note:* There are several combinatorial and geometric details that I must leave to the reader. Here is one preliminary: by the Discrete Schwarz Lemma and the moduli of ring domains, there exists $\mathfrak{c}_1 > 0$ such that $t_n \geq \mathfrak{c}_1$ for all large n.)

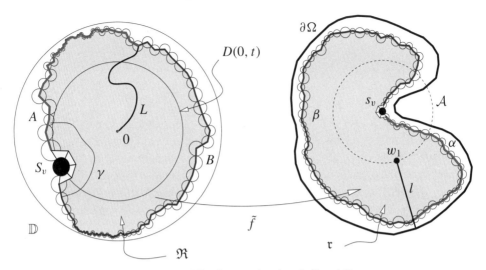

Figure 20.1. The discrete situations in $\mathbb{D}$ and Ω.

Now, move over to Ω. Let s_v be the star of faces in $\mathrm{carr}(\widetilde{P}_n)$ for v (the same vertex $v = v_n$ as above). Draw a straight line l in Ω that starts at w_1, moves directly *away* from s_v, and ends when it first encounters $\partial\Omega$. Let $\mathfrak{r}_n$ be the carrier of $\widetilde{P}_n$ with l and s_v removed, $\mathfrak{r}_n = \mathrm{carr}(\widetilde{P}_n)\backslash(l \cup s_v)$. We treat this as a conformal rectangle $\langle \mathfrak{r}_n; \alpha, \beta \rangle$ by designating as ends the two arcs α, β in the boundary of $\mathrm{carr}(\widetilde{P}_n)$ that connect s_v to the end of l.

Back in $\mathbb{D}$ once more, let $L = \tilde{f}_n^{-1}(l)$ and define $\mathfrak{R}_n = \mathrm{carr}(\widetilde{Q}_n)\backslash(L \cup S_v)$. We also treat this as a conformal rectangle $\langle \mathfrak{R}_n; A, B \rangle$ whose ends A, B are the two arcs in the boundary of $\mathrm{carr}(\widetilde{Q}_n)$ that connect S_v to L. Note that $\tilde{f}_n$ is a homeomorphism that identifies corresponding ends, so $\tilde{f}_n(\langle \mathfrak{R}_n; A, B \rangle) = \langle \mathfrak{r}_n; \alpha, \beta \rangle$.

We can get an upper bound on $\mathrm{Mod}(\mathfrak{R}_n)$ by direct observation. Namely, consider the weight $\rho \equiv 1$ (i.e., consider Lebesgue measure) on $\mathfrak{R}_n$. Every path in $\mathfrak{R}_n$ from A to B (e.g., see γ in the figure) has length at least $\mathrm{radius}\,(C_v)$ and the area of $\mathfrak{R}_n$ is at most π. The modulus is the reciprocal of extremal length, so (77) and the fact that $\tilde{f}_n$ is κ-quasiconformal implies

$$\mathrm{Mod}(\mathfrak{R}_n) \leq \frac{\pi}{\mathrm{radius}\,(C_v)^2} \implies \mathrm{radius}\,(C_v) \leq \sqrt{\frac{\pi}{\kappa\,\mathrm{Mod}(\mathfrak{r}_n)}}. \qquad (113)$$

Over in Ω it is intuitively clear, since the diameter of s_v goes to zero, that $\mathrm{Mod}(\mathfrak{r}_n) \longrightarrow \infty$. We can quantify this using properties of extremal length. Let Γ be the family of curves in $\mathfrak{r}_n$ connecting α and β; the *conjugate* family Γ^* consists of curves in $\mathfrak{r}_n$ connecting s_v to the line segment l. Let $\mathcal{A}$ be the round annulus $\mathcal{A} = \{w : \mathrm{radius}\,(c_v) < |w - z_v| < |w_1 - z_v|\}$. The following inequalities are standard in the study of moduli:

$$\frac{1}{\kappa\,\mathrm{Mod}(\mathfrak{r}_n)} = \frac{1}{\kappa\,\mathrm{Mod}(\Gamma)} = \frac{\mathrm{Mod}(\Gamma^*)}{\kappa} \leq \frac{\mathrm{Mod}(\mathcal{A})}{\kappa} = \frac{2\pi}{\kappa}\left(\log\left(\frac{|w_1 - z_v|}{\mathrm{radius}\,(c_v)}\right)\right)^{-1}.$$

With (113), we conclude that

$$\text{radius}\,(C_v) \le \pi \sqrt{\frac{2}{\kappa}} \left(\log \left(\frac{|w_1 - z_v|}{\text{radius}\,(c_v)} \right) \right)^{-\frac{1}{2}}. \tag{114}$$

Technically, we are not done yet since C_v is not a boundary circle of Q_n, only a neighbor of one, and C_v was selected for having center closest to the origin. But a horocycle in a maximal packing and its neighbors will have comparable radii (depending on d); thus radius (C_v) is comparable to its distance to the unit circle; thence it is comparable to the radii of other boundary circles of $\widetilde{Q}_n$; and thereby it is also comparable to the radii of their neighboring horocycles. Whew! I'll let the reader sort this out. The upshot is that inequality (114) implies the conclusion of the lemma with another constant factor thrown in. $\qquad\square$

Proof of Lemma 20.3 This argument by contradiction is similar to those we have seen before (as for example with the Ring Lemma in the Large). If the result fails, there exists an $x > 0$ and a sequence $m_k \to \infty$ so that the following holds: for each k there is an $n = n_k$, associated packings $P_n \longleftrightarrow K_n(R_n)$ and $\widetilde{P}_n \longleftrightarrow K_n(\widetilde{R}_n)$, and associated vertices $v_n \sim u_n$ that are m_k generations deep in K_n, so that

$$\left| \frac{\widetilde{R}_n(u_n)}{\widetilde{R}_n(v_n)} - \frac{R_n(u_n)}{R_n(v_n)} \right| \ge x.$$

For each n, designate v_n as the α-vertex for K_n. By a euclidean scaling and translation we may center the α-vertex at the origin and make its radius 1 for both P_n and $\widetilde{P}_n$. In particular, our inequality becomes

$$|\widetilde{R}_n(u_n) - R_n(u_n)| \ge x. \tag{115}$$

Applying normal families, two subsequence extractions give a subsequence n_j with $K_{n_j} \longrightarrow K$ for some K and $P_{n_j} \longrightarrow P$ and $\widetilde{P}_{n_j} \longrightarrow \widetilde{P}$ for univalent packings P, $\widetilde{P}$ of K. We have suppressed the dependence on the m_k, but if you trace back you find that the α-vertex of K is infinitely many generations deep, so K has no boundary and is a combinatorial open disc. The family $\{K_n\}$ is by hypothesis nested or asymptotically parabolic. So K is parabolic and P and $\widetilde{P}$ are both maximal packings with the unit circle as α-circle. Since $x > 0$, inequality (115) contradicts the uniqueness of maximal packings. $\qquad\square$

It would be nice, of course, to have a stable of asymptotically parabolic sequences $\{K_n\}$ readily at hand to use with this theorem. However, at this time I can offer only one class possessing any generality.

Lemma 20.4. *Let $\{K_n\}$ be a sequence of combinatorial discs having some subsequence which converges to a combinatorial open disc. Suppose there exists a constant B such that each K_n has some univalent, euclidean packing P_n with the property that the ratio between* *the radii of any two of its circles is bounded above by* $\mathcal{B}$. *Then* $\{K_n\}$ *is asymptotically parabolic.*

Proof. We may suppose, without loss of generality, that $K_n \longrightarrow K$ as $n \to \infty$ with K a combinatorial open disc. Recall that this means "rooted" convergence, so we have chosen α-vertices $v_1^n \in K_n$ and these are identified with the α-vertex $v_1 \in K$. Fix some nested sequence L_n of combinatorial closed discs containing v_1 which exhaust K, $L_n \uparrow K$. By the type dichotomy, it is enough to show that maximal labels $\mathcal{R}_{L_n}(v_1)$ converge to zero.

Verification is via the Length–Area Lemma. Fix integer $m > 0$. In the hypothesized packing P_n we may assume that the circle for v_1^n is normalized to be the unit circle; the remaining circles therefore have radii in $[1/\mathcal{B}, \mathcal{B}]$. This comparability implies three interrelated facts that I will let the reader sort out. First, because there are infinitely many combinatorial generations about v_1 in K, there exists an N so that for $n > N$ the carrier of P_n covers the disc of radius $8m\mathcal{B}$ at the origin,

$$D(0, 8m\mathcal{B}) \subset \text{carr}(P_n), \; n > N. \tag{116}$$

Second, there is a constant $\mathcal{A}$ depending only on $\mathcal{B}$ such that the circles of P_n that intersect $D(0, 8m\mathcal{B})$ are at most $m\mathcal{A}$ generations from v_1^n. Third, there is a constant $\mathfrak{C}$ depending only on $\mathcal{B}$ such that the number of circles of P_n which intersect the circle $|z| = r$ for $r \leq 8m\mathcal{B}$ is bounded by $\mathfrak{C} r$.

Now consider concentric circles $E_j = \{|z| = 8j\mathcal{B}\}$, $j = 1, \ldots, m$. By the first fact, for $n > N$ one can find, for each j, a chain T_j of circles of P_n that roughly follows E_j (e.g., each circle intersects E_j or has a neighbor that does) and that separates the α-circle c_1^n from the boundary circles of P_n. The circles E_j are sufficiently far apart so that the chains $T_1, \ldots, T_m$ are disjoint. From the second fact we deduce that the combinatorial length of T_j is $n_j < 16\mathfrak{C} j \mathcal{B}$. From the third fact, we deduce that the T_j lie within generation $m\mathcal{A}$ of c_1^n.

This is combinatorial information we want to transfer to the hyperbolic maximal packings $\mathcal{P}_{L_n}$. By the convergence $K_n \longrightarrow K$, there exists an $N' > N$ such that for $n > N'$, L_n is a subcomplex of K_n containing at least the first $m\mathcal{A}$ generations of vertices from v_1. In particular, the chains T_j correspond to chains S_j in $\mathcal{P}_{L_n}$ that separate the α-circle from the unit circle. These are disjoint and have combinatorial lengths $n_j \leq cj$ for some constant c depending only on $\mathcal{B}$. By the Length–Area Lemma,

$$\mathcal{R}_{L_n}(v_1) \leq (1/\sqrt{c})\left(1 + \frac{1}{2} + \frac{1}{3} + \cdots + \frac{1}{m}\right)^{-\frac{1}{2}}, \quad n > N'.$$

Since c depends only on $\mathcal{B}$, this goes to zero as m grows, and because the L_n exhaust K, we conclude that K is parabolic. $\qquad\square$

If this seems overly complicated, here is a cautionary example. Assume K_n has n generations of 7-degree vertices about v_1^n, followed by some huge number k_n of generations having hexagonal combinatorics. Let P_n be its maximal packing, rescaled so that the

α-circle is the unit circle. We can choose the k_n to obtain condition (116). It would seem that the packings P_n are looking ever more parabolic as n grows. But no, these complexes converge to the heptagonal complex $K^{[7]}$, which is hyperbolic. It should be noted in a similar vein that a sequence $\{K_n\}$ that is nested and exhausts a parabolic limit K is not necessarily *asymptotically parabolic*. (Hint: it is a matter of the roots.)

20.3. The Ratio Functions

The convergence $f_n^\# \longrightarrow |F'|$ in our theorem is actually a local property. I have split it off here because it holds in much broader circumstances.

Proposition 20.5. *Suppose univalent circle packings Q_n lie in a plane domain Ω and for each n, $f_n : Q_n \longrightarrow P_n$ is a discrete analytic function. Suppose the carriers of the Q_n exhaust Ω, mesh$(Q_n) \longrightarrow 0$, and the functions f_n converge uniformly on compacta of Ω to a nonconstant analytic function $F : \Omega \longrightarrow \mathbb{C}$. Then, on compacta, $f_n^\#$ converges uniformly to $|F'|$ and* br(f_n) *converges as a point set to* br(F).

Proof. This result is a fairly simple consequence of the geometry of complex derivatives along with the Schwarz and Distortion Lemmas, so I will only outline the proof. (Note that the ratio functions $f_n^\#$ are assumed to be extended affinely from circle centers to all of carr(Q_n).)

We must fix $z_1 \in \Omega$, $w_1 = F(z_1)$, and show $f_n^\#(z_1) \longrightarrow |F'(z_1)|$. Assume first that $|F'(z_1)| > 0$. Let $\epsilon > 0$ be given. The definition of the complex derivative implies that for $|z - z_1|$ small, $|F(z) - F(z_1)| \approx |F'(z_1)| \cdot |z - z_1|$. In other words, for t small, F maps the disc $D(z_1, t)$ one-to-one onto a topological disc Δ satisfying

$$D(w_1, (1 - \epsilon)|F'(z_1)| t) \subset \Delta \subset D(w_1, (1 + \epsilon)|F'(z_1)| t). \qquad (117)$$

Now keep t fixed. Let $\widetilde{Q}_n$ be that portion of the domain packing Q_n lying in $D(z_1, t)$, let $\widetilde{P}_n$ denote the corresponding portion of the image packing P_n, and let $L_n \subset K_n$ be their common complex. There is some vertex $v_n \in L_n$ whose circle $C_n \in \widetilde{Q}_n$ is nearest z_1 and hence whose circle $c_n \in \widetilde{P}_n$ is near w_1. We compare each of $\widetilde{Q}_n$ and $\widetilde{P}_n$ to the maximal packing $\mathcal{P}_{L_n}$; let R_n, r_n, and ρ_n be the respective euclidean labels for v_n.

Look at $\widetilde{Q}_n$ first. Since mesh(Q_n) shrinks, the boundary circles of $\widetilde{Q}_n$ approach $\partial D(z_1, t)$ and c_n approaches z_1. Applying the Schwarz and Distortion Lemmas we can conclude that $R_n/\rho_n \longrightarrow t$ as $n \to \infty$. As for $\widetilde{P}_n$, since F is one-to-one on $D(z_1, t)$ and $f_n \longrightarrow F$ uniformly on compact sets, f_n will be one-to-one on $D(z_1, t)$ for large n. The boundary circles of P_n approach the annulus

$$\{w : (1 - \epsilon)|F'(z_1)| t < |w - w_1| < (1 + \epsilon)|F'(z_1)| t \},$$

and its center circle c_n approaches w_1. The Schwarz and Distortion Lemmas now imply that

$$(1 - \epsilon)|F'(z_1)| t \leq r_n/\rho_n \leq (1 + \epsilon)|F'(z_1)| t, \quad \text{for large } n.$$

Dividing our inequalities gives

$$(1 - \epsilon)|F'(z_1)| \leq r_n/R_n \leq (1 + \epsilon)|F'(z_1)|, \quad \text{for large } n.$$

The ratio r_n/R_n is, of course, equal to $f_n^\#(v_n) \approx f_n^\#(z_1)$, and ϵ was arbitrary, so $f_n^\#(z_1) \longrightarrow |F'(z_1)|$, as we claimed.

When $|F'(z_1)| = 0$, F has a branch point at z_1. There is a constant $\mathfrak{C}$ such that for small $t > 0$, $|F(z) - w_1| < \mathfrak{C}|z - z_1|^{3/2}$ for $|z - z_1| = t$. Given any ϵ, choose t so that $\mathfrak{C}t^{1/2} < \epsilon$. Using the notations set above, we now have $\Delta = F(D(z_1, t)) \subset D(w_1, \epsilon t)$. The Schwarz lemma alone implies $r_n/\rho_n \leq \epsilon t$, and hence $f^\#(z_1) = r_n/R_n \leq \epsilon$. Since ϵ was arbitrarily small, $f_n^\#(z_1) \longrightarrow 0 = |F'(z_1)|$. The convergence of $\mathrm{br}(f_n)$ to $\mathrm{br}(F)$ on compacta is a purely topological consequence of the convergence of light interior mappings. $\qquad\square$

20.4. Companion Notions

Anyone observing concrete examples of discrete mappings, such as those in Fig. 19.3, is likely to pick up on various local features that seem to be symptomatic of geometric properties of Ω, though what they are symptoms *of* you may not have a clue. I have already mentioned, for example, the three interior, low-distortion spots that your eye easily picks out in those color images. I do not know yet what to make of them, so let me discuss instead the varying sizes and densities of horocyles one sees in the domains of these Riemann mappings. Here there are some notions that one might investigate by running some classical–discrete comparisons.

A convenient motif for these particular experiments is *harmonic measure*, which we first saw in Section 8.2.1. Classically, the harmonic measure $\omega(\cdot) = \omega(\cdot, w_1, \Omega)$ (relative to $w_1 \in \Omega$) is a probability measure on $\partial\Omega$: for each Borel set $E \subset \partial\Omega$, $\omega(E)$ is the probability that a Brownian traveler starting at w_1 will first encounter the boundary of Ω at a point of E. Thus it measures *exit probabilities*.

In $\mathbb{D}$, the harmonic measure (relative to the origin) is normalized Lebesgue measure on $\partial\mathbb{D}$, which we will denote by $|\cdot|$. It is well known that conformal maps between plane regions preserve harmonic measure; in our case, the classical conformal map $F : \mathbb{D} \longrightarrow \Omega$ maps 0 to w_1 and extends to a homeomorphism of the boundaries, so

$$\omega(E) = \left|F^{-1}(E)\right| \quad \text{for Borel sets } E \subset \partial\Omega. \tag{118}$$

In other words, the boundary behavior of F is encoded in the harmonic measure ω on the image boundary, $\partial\Omega$.

For our experiments I have chosen to work with the situation in the middle panel in Figure 19.3. Let K denote the complex, P the image packing. I will assume that Ω is precisely the polygonal region $\Omega = \mathrm{carr}(P)$ defined by the 189 boundary vertices of K. The maps we work with are nominally $F : \mathbb{D} \longrightarrow \Omega$ and $f : \mathrm{carr}(\mathcal{P}_K) \longrightarrow \mathrm{carr}(P)$, but

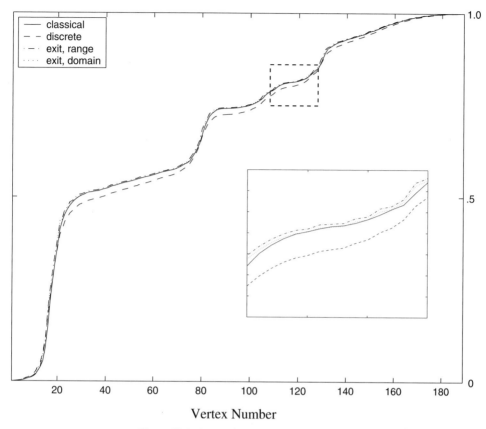

Figure 20.2. Harmonic measure comparisons.

the reader will appreciate that in fact we must use numerical approximations in both instances. In particular, I have used Don Marshall's "zipper" conformal mapping software to approximate F. His software produces a legitimate classical conformal mapping that is within roughly five decimal places of the true F at points of the unit circle. The function f is, of course, approximated using `CirclePack`.

The idea is now to compare F and f as maps from the unit circle to $\partial\Omega$. In Figure 20.2, the solid line (labeled "classical") represents the *cumulative* harmonic measure on $\partial\Omega$ – in other words, the derivative of this curve is the classical harmonic ω measure. (*Note*: The numbers on the abscissa in the graph count vertices around the boundary of Ω; since the circles are all the same radius in P, plotting this way is equivalent to plotting against *arclength* on $\partial\Omega$.)

When it comes to the discrete map f, we have various options. The most direct analogue of the classical plot is given as the dashed line ("discrete"). At each boundary index w, the graph's height is the normalized Lebesgue measure of the arc of the unit circle starting at the tangency point of the horocycle for the first boundary vertex, passing through the tangency points for the intervening vertices, and ending with the tangency point of the horocycle for w itself. In other words, it is the cumulative distribution for the measure mimicking formula (118). It does an admirable job.

An alternate approach is *via* Brownian motion. The random walks we developed in Chap. 18 were touted as discrete analogues, so here is an opportunity to try them out. The dashed–dotted line ("exit, range") plots the cumulative distribution of exit probabilities of the simple random walk on P, starting at the origin. I expected the curve to track the classical harmonic measure, and it did remarkably well.

Encouraged, I did one further experiment, this time with the maximal packing $\mathcal{P}_K$. Recall that once $\mathcal{P}_K$ is converted to a euclidean packing, the *tailored* random walk on $\mathcal{P}_K$, $\mathfrak{W} = \mathfrak{W}_{\mathcal{P}_K}$, involves the movement of "angle" about the packing. Here I have actually run the experiment described in the last paragraph of Chap. 18 (the finite version of the "thought experiment" of Section 18.6.1). Namely, I incremented the (euclidean) radius of v_1, the circle at the origin, and locked in this perturbed value. In essence, this placed a "charge" of angle on the circuit at v_1. I then ran the packing algorithm to get a packing label for $K \setminus \{v_1\}$. The final deficit in angle sum at v_1 is precisely matched by the sum of increases in angle sum at the boundary vertices (and, by monotonicity, these are all increases). I accumulated these increases and normalized to get the fourth plot in Figure 20.2 ("exit, domain").

This last graph is perhaps the most impressive – and the most surprising to this experimenter. In the blowup in Figure 20.2 one sees the precision with which it tracks the random walk results on P, despite the huge differences in circle radii between P and $\mathcal{P}_K$. And the precision with which both these track the classical harmonic measure, despite the higher dilatations near the boundary in Figure 19.3, seems yet more remarkable. These experimental results certainly suggest that the discrete conformality which we have established on the interior has some interesting companions to investigate on the boundary.

20.5. The He–Schramm Theorem

The Rodin–Sullivan Theorem was first extended by the author using random walk arguments. The version presented above is both more general and more geometric. It, in turn, has been extended using quantitative approaches. He and Rodin, for example, have removed hypothesis (d) by imposing more structure on the point maps. Initially, $f : Q \longrightarrow P$ maps centers to corresponding centers. One can next define f on each interstice of Q by mapping *via* a classical conformal map to the corresponding interstice of P, with tangency points (i.e., "vertices") of one going to corresponding tangency points of the other. Finally, the identification this gives for points on the circles can be extended affinely to the circle interiors. With this extra structure on f, one can be much more quantitative (e.g., regarding quasiconformal distortion and various derivatives) than is possible with the simplicial maps that we are typically using.

These methods, too, have been eclipsed, allowing even hypothesis (c) of our theorem to be eliminated. This work was carried out by Z-X. He and Oded Schramm; here is the key result stated in their more symmetric form:

Theorem 20.6 (He–Schramm Theorem). *Let Ω and $\widetilde{\Omega}$ be Jordan domains in $\mathbb{C}$. Let p_1 be a point of Ω. For each n, let P_n be a univalent circle packing lying in Ω, let $\widetilde{P}_n$ be a combinatorially equivalent univalent circle packing lying in $\widetilde{\Omega}$, and let $f_n : P_n \longrightarrow \widetilde{P}_n$ be the associated discrete conformal mapping.*

Let δ_n be a sequence of positive numbers, $\delta_n \longrightarrow 0$. Assume that the radii of the circles of P_n are less than δ_n, that each boundary circle of P_n is within a distance δ_n of $\partial\Omega$, and that each boundary circle of $\widetilde{P}_n$ is within a distance δ_n of $\partial\widetilde{\Omega}$. Suppose that $p_1 \in \mathrm{carr}(P_n)$ and the images $f_n(p_1)$ lie in a compact subset of $\widetilde{\Omega}$. Then there exists a subsequence $\{f_{n_j}\}$ which converges uniformly on compact subsets of Ω to a conformal homeomorphism $F : \Omega \longrightarrow \widetilde{\Omega}$.

Their very clever proof uses only elementary geometry, but it is far from easy and I will not try to present it here. He and Schramm show that the maps f_n approximate Möbius transformations locally. These arguments avoid the machinery of quasiconformal mapping and thereby avoid any reliance on the Riemann Mapping Theorem. This led He and Schramm to this quite remarkable observation:

Fact: *The Koebe–Andreev–Thurston theorem provides an independent proof of the classical Riemann Mapping Theorem for plane domains.*

It is hard to imagine more dramatic evidence of the fidelity of the discrete theory to its classical roots!

21

Approximation of Analytic Functions

In view of the general discrete theory developed in Part III, it is natural to consider extending the approximation results beyond nonunivalent functions. The state of the art does not yet permit broad statements, so I can offer only a selection of examples illustrating various considerations that come into play. Being an optimist, however, I take this as my central tenet:

> **Working Tenet:** *If you can build discrete analytic functions that* **mimic** *a classical analytic function F, then by building them with successively finer circle packings you can also* **approximate** *F.*

From the several classes of functions constructed in Part III, I concentrate on finite Blaschke products in the disc and polynomials in the plane. The sphere and higher genus Riemann surfaces remain tough settings for general analytic (i.e., meromorphic) functions; significant examples exist, but I have placed them in the next chapter on conformal structures. The reader who takes our working tenet to heart might pick up the gauntlet for further development.

21.1. Approximating Blaschke Products

Recall that the classical finite Blaschke products are precisely the proper analytic self-maps of $\mathbb{D}$. We defined their discrete analogues in Section 13.3, so the next theorem illustrates our working tenet.

Theorem 21.1. *Let $B : \mathbb{D} \longrightarrow \mathbb{D}$ be an n-fold finite Blaschke product with branch set β. There exist discrete n-fold Blaschke products b_j such that $b_j \longrightarrow B$ and $b_j^{\#} \longrightarrow |B'|$ uniformly on compact subsets of $\mathbb{D}$.*

Proof. Let $\mathcal{P}_H$ be the regular hexagonal packing of the plane by circles of radius 1 with α-vertex at the origin. Define K_k, $k \geq 1$, to be the subcomplex of H formed by the vertices of generation at most k from v_1. For k sufficiently large one can choose branch structures

β_k for K_k having cardinality $n - 1$ with the property that the corresponding centers in $\mathcal{P}_{K_k}$ converge to β as $k \to \infty$, counting multiplicities. Note that since we are limited to simple branching in the hexagonal case, a branch point of B of order $j > 1$ requires a cluster of j simple branch vertices in β_k.

It is convenient to assume that $B(0) = 1$ and $B(1) = 1$. Define $b_k : \mathcal{P}_{K_k} \longrightarrow \mathcal{P}_{K_k,\beta_k}$, a discrete Blaschke product. By Lemma 13.7 we may write $b_k \equiv B_k \circ h_k$, where B_k is a classical n-fold Blaschke product and h_k is a κ_k-quasiconformal homeomorphism of $\mathbb{D}$. It is clear that no generality is lost if we assume normalizations

$$b_k = B_k \circ h_k, \quad \text{with} \quad B_k(0) = h_k(0) = 0, \quad \text{and} \quad B_k(1) = h_k(1) = 1. \tag{119}$$

In the following we will write Q_k for the domain packings $\mathcal{P}_{K_k}$ and P_k for the image packings $\mathcal{P}_{K_k,\beta_k}$, so $b_k : Q_k \longrightarrow P_k$. The first task (again) lies in showing that the boundary circles of P_k go to zero in euclidean radius. We use an extension of the argument for the Length–Area Lemma, 19.3

Begin by noting that the branch points of Q_k lie in a compact subset of $\mathbb{D}$ and by the Discrete Schwarz Lemma their images move toward the origin under b_k. The branch values therefore lie in the disc $D(0, 1 - \mathfrak{a})$ for some $\mathfrak{a} \in (0, 1)$. Fixing k, consider a boundary vertex $w \in \partial K_k$ with circle c_w in P_k. Suppose the combinatorial distance from w to the branch set β_k is greater than m. For each j, $1 \le j \le m$, let γ_j be the chain of vertices of K_k that are j generations from w. The chains γ_j, $j = 1, \cdots, m$, are disjoint and each separates w from β_k combinatorially. (These vertices are analogous to those for the chains of circles shown in Figure 19.2.)

Following the proof of the Length–Area Lemma, let S_j be the chain of circles in P_k corresponding to γ_j, $j = 1, \cdots, m$. Assume that S_j has combinatorial length n_j, euclidean radii ρ_{ji}, $1 \le i \le n_j$, and length $l_j = 2 \sum_i \rho_{ji}$. We claim that

$$l_j \ge \min\{\mathfrak{a}, R\}, \ R \text{ the euclidean radius of } c_w. \tag{120}$$

This depends on simple mapping properties of branched coverings. Fix j and let L be the simply connected subcomplex of K_k cut off by γ_j and containing w. If b_k is univalent on carr(L), then the chain S_j, which starts and ends at the unit circle, must separate c_w from at least one branch value. S_j stretches either around c_w or around a branch value in $D(0, 1 - \mathfrak{a})$, and (120) follows. On the other hand, if b_k is not univalent on carr(L), then since L does not contain any branch vertices, topological considerations imply that the chain S_j must separate at least one branch value from the unit circle, so again (120) holds.

We can now apply the argument leading to the inequality (107); however, the S_j do not have mutually disjoint interiors and instead we use the fact that P_k is n-valent. The argument is the same, giving the inequality (107) with the 4 on the right side replaced by $4n$. As a result, $l = \min\{l_1, l_2, \cdots, l_m\}$ satisfies $l^2 \le 4n(n_1^{-1} + n_2^{-1} + \cdots + n_k^{-1})^{-1}$. With hexagonal combinatorics, $n_j < 6j$, so this inequality implies

$$l \le \frac{6n}{(1 + 1/2 + \cdots + 1/m)}.$$

Given any positive $\epsilon < \mathfrak{a}$, for k sufficiently large we can choose m so large that $l \leq \epsilon$. Inequality (120) then implies radius $(c_w) \leq \epsilon$.

We are now going to bootstrap our way to the conclusion by studying the two families $\{h_k\}$ and $\{B_k\}$. Several details are left for the reader, such as extraction of sequences and normal families, but here are the main ideas.

The Family $\{h_k\}$: The packings P_k are of finite valence n and the complexes K_k have uniformly bounded degree, so we are in position to apply the Ring Lemma in the Large, Lemma 14.5. In particular, given any disc $D_r = D(0, r)$, $0 < r < 1$, the combinatorial depth in K_k of vertices whose circles in Q_k intersect D_r will grow without bound as k grows. Thus, for large k, if f is any face of carr(Q_k) intersecting D_r, then there is a uniform bound on the ratio of the radii for the circles of the corresponding face in carr(P_k). By Lemma A.9 this implies a uniform bound κ on the quasiconformal dilatation of b_k on f; since $b_k = B_k \circ h_k$ and B_k is analytic, this in turn implies that h_k is κ-quasiconformal on f. By standard results in the quasiconformal theory, h_k is κ-quasiconformal on all of D_r. This holds for any $r \in (0, 1)$, so by normal families there is a subsequence $\{h_{k_j}\}$ which converges to a κ-quasiconformal map h on all of $\mathbb{D}$. For notational convenience we may assume that the full sequence h_k converges to h (which can be shown after the fact, in any case).

The Family $\{B_k\}$: The branch sets β_k converge to $\beta = \mathrm{br}(B)$, so they lie in a compact subset of $\mathbb{D}$. The uniform κ-quasiconformality of the h_k and the normalization $h_k(0) = 0$ imply that the image sets $h_k(\beta_k)$ will lie in a compact subset of $\mathbb{D}$. But $h_k(\beta_k) = \mathrm{br}(B_k)$, and $\{B_k\}$ is a normal family (because its members are uniformly bounded). Using the normalizations in (119) and extracting a further subsequence we may assume that B_k converges to an n-fold finite Blaschke product $\widetilde{B}$ satisfying $\widetilde{B}(0) = 0$, $\widetilde{B}(1) = 1$, and $\mathrm{br}(\widetilde{B}) = \lim_k h_k(\beta_k) = h(\beta)$.

Now we can return for a closer look at the h_k. Consider any point $z \in \mathbb{D} \backslash \beta$. The image point $h(z)$ lies in $\mathbb{D} \backslash \mathrm{br}(\widetilde{B})$, so $\widetilde{B}$ is univalent on a small neighborhood $h(z)$. A classical winding number argument (Hurwitz's Theorem) shows that there is an open neighborhood of $h(z)$ in which B_k will be univalent for large k. It follows that there is an open neighborhood U of z in which $b_k = B_k \circ h_k$ is univalent for large k. Let L_k be the simply connected subcomplex associated with the circles of Q_k lying in U. Given a compact set $E \subset U$, the combinatorial depth of any circle of Q_k intersecting E will grow as k grows. By univalence of b_k on U, the circles of P_k corresponding to L_k also form a univalent packing. Applying the Hexagonal Packing Lemma, Lemma 19.4, we conclude that the quasiconformal dilatation of h_k converges to 1 in a neighborhood of z. In other words, h is conformal in $\mathbb{D} \backslash \beta$. Since we can remove isolated singularities, h is conformal on all of $\mathbb{D}$. Since the boundary circles of P_k go to zero in radius, the ranges of the functions h_k exhaust $\mathbb{D}$ and by the Carathéodory Kernel Theorem h maps onto $\mathbb{D}$. As a consequence, $h \in \mathrm{Aut}(\mathbb{D})$. The normalizations of (119) tell us that h is the identity function. In other words, $b_k \longrightarrow \widetilde{B}$, where $\widetilde{B}(0) = 0 = B(0)$, $\widetilde{B}(1) = 1 = B(1)$,

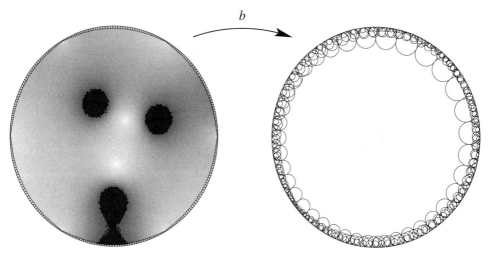

Figure 21.1. Approximating a Blaschke product.

and $\mathrm{br}(\widetilde{B}) = \beta = \mathrm{br}(B)$. These conditions imply $\widetilde{B} \equiv B$. The convergence $b_k^{\#} \longrightarrow |B'|$ follows from Proposition 20.5. □

21.1.1. An Example

Pictured in Figure 21.1 is a refinement of the discrete Blaschke product b of Figure 2.6 in Part I. The four-sheeted range packing is far too complicated to view in full, so I have again shown only the boundary horocycles, which go four times around $\mathbb{T}$, and the (extremely small!) branch circles. I have shaded the euclidean carrier in the domain to reflect the quasiconformal dilatation of b on the various faces, shades of grey indicating $\kappa \in [1, 1.1]$, black indicating higher distortion. There is certainly no difficulty in guessing the general location of the branch points in the domain, since branching increases local distortion.

21.2. Approximating Polynomials

Theorem 21.2. *Given a polynomial $F(z)$ of degree n, there exists a sequence $\{f_k\}$ of discrete polynomials of degree n so that $f_k \longrightarrow F$ and $f_k^{\#} \longrightarrow |F'|$ uniformly on compact subsets of the plane.*

Proof. One can base the discrete polynomials f_k on the scaled regular hexagonal packings $(1/k)\mathcal{P}_H$ and proceed basically as we did in the previous Blaschke product case. We may assume $F(0) = 0$, $F(1) = 1$. As noted after Definition 14.4 $f_k \equiv F_k \circ h_k$ for some polynomial F_k of degree n and homeomorphism $h_k : \mathbb{C} \longrightarrow \mathbb{C}$. We may assume without loss of generality the normalizations

$$F_k \circ h_k, \quad \text{with} \quad F_k(0) = 0 = h(0), \quad \text{and} \quad F_k(1) = 1 = h_k(1).$$

The h_k will be κ-quasiconformal for a uniform constant κ by the Ring Lemma in the Large applied to the mappings $\mathcal{P}_{H,\beta_k}$. A new feature must be used here: since the h_k fix the points 0, 1, and ∞ on the sphere, they form a normal family and one can extract a subsequence that converges to a κ-quasiconformal function h. The rest of the proof is essentially that of the Blaschke case.   □

I have used hexagonal combinatorics in both the previous constructions for convenience. Methods for more general combinatorics, as we used in the previous chapter, can likely be extended, at least to bounded degree and finite valence situations.

Example: In contemplating an example here you will notice a rather bothersome little detail: we must work with infinite packings! Classical function theory gives us an out. Polynomial behavior in the large is topologically that of finite Blaschke products:

$$F(z) = a_n z^n + a_{n-1} z^{n-1} + \cdots + a_1 z + a_0$$
$$\implies |F(z)|/|z|^n = |a_n| + o(1) \text{ as } |z| \to \infty.$$

Define $\Omega_\rho = F^{-1}(D(0, \rho))$. For ρ sufficiently large, Ω_ρ contains all branch points of F and Ω_ρ will be simply connected by the classical maximum principle. Define a conformal mapping $\phi_\rho : \mathbb{D} \longrightarrow \Omega_\rho$. The function

$$B_\rho = \left(\frac{1}{\rho}\right) F \circ \phi_\rho : \mathbb{D} \longrightarrow \mathbb{D}$$

is finite valence and has boundary values of modulus one – it is a finite Blaschke product. If one normalizes ϕ_ρ appropriately, then $\rho B_\rho \longrightarrow F$ uniformly on compacta as $\rho \to \infty$.

If you trace back through our construction of discrete polynomials f_j, you find that we use precisely the discrete analogues of these functions B_ρ. In practice, then, we can cut out the middleman f_j and approximate F directly by (modified) discrete Blaschke products. (Of course, in illustrations of discrete polynomials one exploits this by framing the packing image so that the viewer does not see that there is a large exterior circle containing the packing.)

21.3. Further Examples

For our purposes, the matter of approximating classical functions serves mainly to confirm the faithfulness of our discrete world: *If you get the geometry right, the approximation is automatic.* We can see geometry at work in different ways with the exponential and sine functions.

Exponential: Choose as domain the regular hexagonal packing of constant radius $\epsilon > 0$ and as image the Doyle spiral about the origin with parameters a, b. Let $f = f_{\epsilon,a,b}$ denote the associated discrete analytic function, normalized so that $f(0) = 1$ and $f(1) > 1$. Letting a, b converge to 1, one can choose $\epsilon = \epsilon(a, b) > 0$ such that $f_{\epsilon,a,b}$ converges to the exponential function.

The geometry is quite clear: the carrier of a Doyle spiral is a universal covering surface of $\mathbb{C}^*$, so f and the exponential share the same image surfaces. In a Doyle spiral, each circle lies in a univalent patch of circles on one "sheet" of this covering surface. This patch has hexagonal combinatorics, and as $a, b \longrightarrow 1$ the number of hexagonal generations in the patch grows. The Hexagonal Packing Lemma implies that the quasiconformal distortion of f associated with each such univalent patch goes to 1. There is another normalization needed, however, to pin down the preimages of a given patch in the domain packings so they do not, e.g., slide off to ∞ or otherwise degenerate. One possibility is to choose ϵ so that $f^\#(0) = 1$, but there are still details to handle.

An alternative is to argue via Stoïlov's Theorem. $f(z) \equiv \exp(h(z))$ for some homeomorphism $h = h_{\epsilon,a,b}$. By choosing ϵ appropriately, we can arrange that $h(0) = 0$, $h(1) = 1$. All faces of a Doyle spiral are similar, so the local quasiconformal dilatation of f, and hence the dilatation of h, is a constant $\kappa(a, b)$. An explicit computation (which I leave to the reader) shows that $\kappa(a, b) \longrightarrow 1$ as $a, b \longrightarrow 1$. With our normalizations, we conclude that h converges to the identity, and hence f converges to the exponential function.

Again, we have relied on hexagonal packings. It is certainly a worthy challenge to see if one can construct and analyze discrete exponentials based on more general combinatorics.

Sine: Recall the sine construction in Figure 14.7. The image surfaces with this construction are precisely those of the classical sine function. (Remember that the image packing is not a regular hexagonal packing, as the pictures suggest; each circle is the projection of infinitely many circles from the image packing.) One improves the fidelity of the discrete sine functions by increasing the number n of circles along the edge γ_c in the initial domain half-strip (see Figure 14.5(a), where $n = 7$). The geometry of the image surface ensures that each point aside from ± 1 has a univalent neighborhood of circles with hexagonal combinatorics and with a number of generations which grow with n. The Hexagonal Packing Lemma therefore provides local control on distortion. The fact that $\pm\pi/2$ map to ± 1 in the construction ensures nondegenerate convergence, and analyticity extends to points over ± 1 because they are isolated. In conclusion, then, these discrete sine functions converge to the classical sine function.

You might have more faith in this conclusion if you were to see how our function performs when restricted to the real axis, where it is the familiar "real" sine function. In Figure 21.2 the circles from $[-\pi, \pi]$ in the domain have been laid on the x-axis while their image circles from the interval $[-1, 1]$ are laid along the y-axis; the polygonal curve is then drawn through the points (x, y), the two coordinates being the centers of the two corresponding circles. The top graph illustrates the result at the resolution of the construction from Figure 14.7; the next does the same after two further stages of hexagonal refinement, $n = 25$.

Regarding the other discrete examples we developed in Part III, in each instance the identical constructions with ever finer circle packings will converge to their classical models, and in some circumstances, as with the Blaschke product of Figure 13.3 or the disc algebra example of Figure 13.5, these are perhaps the most effective approximation methods available for the target functions.

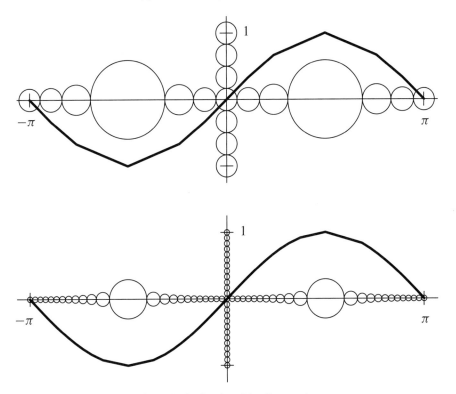

Figure 21.2. Graphs of the discrete sine.

Approximations like these that exploit geometric properties of the target functions are, in my view, the most satisfying. I will be the first to admit, however, that this keeps our roster of examples rather short. In general, one must rely on approximations that are not geometric per se. There are several classical results to use. By Runge's theorem, for example, any entire function can be approximated uniformly on compacta by classical polynomials, hence by discrete polynomials, and, therefore, as we have seen, by scaled discrete Blaschke products. There are also the methods using boundary values, as with the erf(z) construction of Section 14.4 and the disc algebra construction of Section 13.4.1, which are perhaps more practical and which also seem likely to converge under appropriate refinement to their classical targets. General formulations and proofs about such approximation remain to be carried out, however.

22

Approximation of Conformal Structures

The classical notion of conformal structure for surfaces can be quite abstract. But not always. In many circumstances a surface can be described in some constructive way which provides an atlas of explicit charts and maps. To what extent does one then "know" that surface? How can one extract specific information about it?

Here we lay the groundwork for applications in the next chapter by investigating *polyhedral* surfaces, that is, surfaces formed by pasting together euclidean polygons. Though locally modeled on familiar euclidean pieces, these have global properties which are often beyond the reach of classical methods. We begin with *equilateral* surfaces to see the central ideas, and then discuss additional subtleties for more general cases.

22.1. Polyhedral Surfaces

In technical terms, a *polyhedral surface* is an oriented topological surface S that has a locally finite cell decomposition whose faces are euclidean polygons. Thus it is a metric surface whose edges (1-cells) are (isometric to) euclidean segments and whose faces (2-cells) are (isometrically isomorphic to) euclidean polygons. The vertices (0-cells) v are called *cone points*, and each has a *cone angle* that is the sum of the euclidean angles at v in the link (i.e., the flower) of faces to which v belongs.

Examples are everywhere, from the surfaces of platonic solids, to the affine scenery in computer games, to abstract surfaces that cannot be realized in 3-space. Every polyhedral surface inherits a *natural* conformal structure, namely, one whose restriction to each face is compatible with its conformal structure as a euclidean polygon. Thus polyhedral surfaces are also Riemann surfaces, and we use the notation $\mathcal{S}$.

Let me briefly describe the conformal structure on $\mathcal{S}$. For each face f_j there exists an isometry $\phi_j : f_j \longrightarrow p_j$ to a polygon p_j in the plane; ϕ_j provides a local coordinate for the int(f_j) (the interior of f_j) and a local border coordinate for any edge of f_j lying in ∂S. An interior edge e will belong to two faces f_j and f_k; with rigid motions one can bring p_j and p_k together along the corresponding edges in the plane, and the map $\phi_e : f_j \cup f_k \longrightarrow p_j \cup p_k$ provides a local coordinate for int$(f_j \cup f_k)$, a neighborhood of int(e). In particular, note that the transition maps between these coordinates are rigid

275

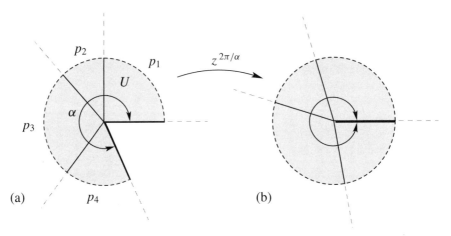

Figure 22.1. Power map for a cone point.

motions in $\mathbb{C}$, hence analytic. (Should f_j and f_k happen to be the same face, the argument works in a neighborhood of e by letting p_k be a clone of p_j.)

Things are only slightly more complicated at cone points, where power maps are used to define the local coordinates. Suppose v is an interior vertex with oriented link $\{f_1, \ldots, f_n\}$ and cone angle α. The local coordinates can be translated and rotated so their image polygons $p_1, \ldots, p_n$ are laid out in succession with the vertex corresponding to v at the origin, as in Figure 22.1(a). In a small metric disc U about v in $\mathcal{S}$ we can define the map $g_v : U \longrightarrow \mathbb{C}$ on each f_j as the restriction $\phi_j|_U$. If $\alpha = 2\pi$, then g_v will be well defined and univalent and provides a local coordinate, $\phi_v = g_v$ in U. In general, however, one must compose g_v with the power map $z \mapsto z^{2\pi/\alpha}$, which brings the edges of p_1 and p_n together, as in Figure 22.1(b). The result, $\phi_v = (g_v)^{2\pi/\alpha} : U \longrightarrow \mathbb{C}$, is univalent on U and for any j, the intersection $U \cap \text{int}(f_j)$ does not contain v, so the transition map is again analytic. A similar construction using $z \mapsto z^{\pi/\alpha}$ defines a border coordinate when v is a boundary vertex of $\mathcal{S}$. These power maps complete our atlas.

A special subclass among polyhedral surfaces is central to our work. A surface $\mathcal{S}$ is *equilateral* if its faces are (isometrically isomorphic to) euclidean equilateral triangles. The cell decomposition is a simplicial complex, and for convenience we assume henceforth that this is a complex K appropriate to circle packing. (If not, one can apply barycentric subdivision (see Section 12.8) without affecting the conformal structure of $\mathcal{S}$.)

For equilateral surfaces $\mathcal{S}$ one has, in addition to its classical conformal structure, a *discrete conformal structure* associated with K, as defined in Chap. 17. Classical conformal mappings of $\mathcal{S}$ can well have parallel discrete conformal mappings based on K, and it is the relations between these two classes of mappings that concern us here.

22.2. The Ten-Triangle-Toy

I want us to be rooted in a concrete example, so let me pick up again on the ten-triangle-toy example described in Section 17.1. The complex K is a combinatorial disc with 10 faces;

S is the associated equilateral surface, a conformal disc. S has two interior cone points, p with degree 3, hence cone angle $\pi = 3(\pi/3)$, and q with degree 9, cone angle 3π. It also has eight boundary cone points of various cone angles, four of which have been designated as corners. (The fact that S can be realized geometrically in 3-space as in Figure 17.1 is neither needed nor used.) One way to demonstrate that you "know" S would be to map it to a standard plane domain and be able to see, for example, the shapes of the faces. That is what we take as our goal.

Recall the classical theory: There exists a classical conformal mapping $F : S \longrightarrow \mathfrak{R}$ where $\mathfrak{R} \subset \mathbb{C}$ is a euclidean rectangle, and F and $\mathfrak{R}$ are essentially unique when we require that the designated corners of S be mapped to the corners of $\mathfrak{R}$. The *existence* of F would have been evident to Riemann himself 150 years ago, but it is only with the advent of circle packing methods that general mappings of this sort can be approximated numerically.

We began the process back in Section 17.1, with the in situ packing Q and the rectangular packing P of Figure 17.2 and the associated discrete conformal map $f : Q \longrightarrow P$. However, one can hardly expect 10 circles to carry all the conformal information of S, so some process of refinement is needed. Hexagonal refinement, $K \longrightarrow \delta K$ (see Section 12.8), is admirably suited to our purpose: Figure 22.2 illustrates an equilateral face, its "coarse" packing, and the $n = 3$ and $n = 4$ stage in situ hexagonal refinements. In a surface of equilateral faces, one simply refines all of them at once and in place; halving all the radii of packing Q_n at stage $\delta^n K$ automatically yields a packing label for an in situ packing Q_{n+1} for the next stage $\delta^{n+1} K$. While each stage is technically a different equilateral surface (four times as many faces), you can see as in Figure 22.2 that S has precisely the same metric and conformal structures.

Computing the rectangular packing P_n for refined complex $\delta^n K$, we define the discrete conformal mapping $f_n : Q_n \longrightarrow P_n$. Figure 22.3 illustrates the packings P_3 and P_6 with rectangles $\mathfrak{R}_3$ and $\mathfrak{R}_6$ as their carriers. The circles are in greyscale: the darker the shading, the more the radius changed during repacking from its constant value in Q_3 or Q_6, respectively. In each image I have outlined the 10 original faces of S whose shapes we are investigating.

22.3. Convergence

The discrete theory illustrated in the previous example has provable convergence. Assume S is an equilateral surface, not necessarily simply connected, whose complex K has

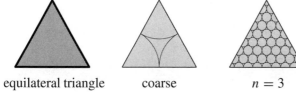

| equilateral triangle | coarse | $n = 3$ | $n = 4$ |

Figure 22.2. In situ hexagonal refinement.

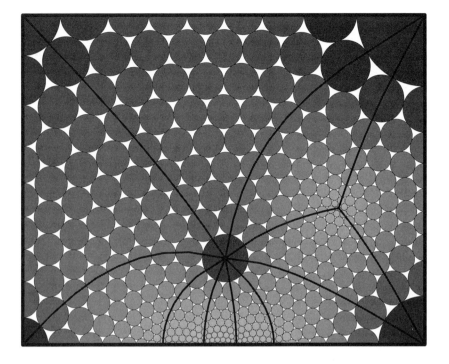

Figure 22.3. Refinement stages 3 and 6 of the ten-triangle-toy.

bounded degree. We show that the universal covering maps associated with the sequence of hexagonal refinements of K converge to the classical universal covering map of S itself. The setup is this: for $n \geq 0$ let $K_n = \delta^n K$ be the n-stage hexagonal refinement of K, and let $\widetilde{K}_n$ be its universal covering complex. For each n there is an in situ circle packing Q_n for K_n in the piecewise euclidean metric of S and we denote by f_n the discrete universal covering map $f_n : \mathcal{P}_{\widetilde{K}_n} \longrightarrow Q_n$.

Theorem 22.1. *Let S be an equilateral surface whose complex K has bounded degree. For each n let f_n be the discrete universal covering map from $\mathbb{G}$ ($\mathbb{P}$, $\mathbb{C}$, or $\mathbb{D}$, as appropriate) to S as described above. Then there exist automorphisms ϕ_n of $\mathbb{G}$ such that $f_n \circ \phi_n$ converges uniformly on compact subsets of $\mathbb{G}$ as $n \to \infty$ to a classical universal covering map $F : \mathbb{G} \longrightarrow S$.*

Proof. Fix a universal covering map $F : \mathbb{G} \longrightarrow S$. For each n, f_n is a simplicial map from $\mathrm{carr}(\mathcal{P}_{\widetilde{K}_n}) \subset \widetilde{\mathbb{G}}$ to $\mathrm{carr}(Q_n) = S$ for $\widetilde{\mathbb{G}}$ one of $\mathbb{P}$, $\mathbb{C}$, or $\mathbb{D}$. (We will resolve the identity of $\widetilde{\mathbb{G}}$ in a moment.) For any $\phi \in \mathrm{Aut}(\widetilde{\mathbb{G}})$, $f_n \circ \phi : \widetilde{\mathbb{G}} \longrightarrow S$ lifts under F^{-1} to a homeomorphism $h_n : \widetilde{\mathbb{G}} \longrightarrow \mathbb{G}$. We can choose ϕ so that $h_n(0) = 0$ and $h_n(1) > 0$. Note that S is a sphere if and only if the K_n are combinatorial spheres, in which case $\mathbb{G} = \widetilde{\mathbb{G}} = \mathbb{P}$ and we may also arrange that $h_n(\infty) = \infty$.

As to the identity of $\widetilde{\mathbb{G}}$ when S is not a sphere, if S has a border, then $\mathbb{G} = \mathbb{D}$; since K_n then also has boundary, K_n is hyperbolic and $\mathbb{G} = \widetilde{\mathbb{G}} = \mathbb{D}$. In the remaining cases, $\widetilde{\mathbb{G}}$ is either the plane or the disc. Since $\deg(K_n) = \max\{\deg(K), 6\}$, the K_n have a universal bound d on their degrees, so by the Ring Lemma, the simplicial map f_n is locally κ-quasiconformal for some $\kappa = \kappa(d)$. In particular, h_n is κ-quasiconformal (since F is conformal), so again we find $\widetilde{\mathbb{G}} = \mathbb{G}$. Summarizing:

Lemma 22.2. *If S is an equilateral surface with complex K having bounded degree, then the types of K and its hexagonal refinements are the same as the type of S.*

Let us put the cases where S is a sphere or has a border aside for the moment. In the remaining cases, the h_n are normalized κ-quasiconformal maps of $\mathbb{G}$ ($\mathbb{C}$ or $\mathbb{D}$) into itself. By quasiconformal normal families arguments, $h_n \longrightarrow h$ for some κ-quasiconformal homeomorphism h of $\mathbb{G}$. Our aim is to show that h is conformal.

If you anticipate forcing the dilatations of our maps h_n to 1 as n grows, you will be disappointed. Figure 22.4 illustrates the coordinate chart at a cone point with cone angle $4\pi/3$; notice the distortion in the faces containing the cone point itself. This represents local behavior at the cone point, independent of refinement level, so it is clear that κ is bounded away from 1.

On the other hand, this local behavior is restricted to smaller and smaller regions as one refines. If x is a point of S that is some distance $\delta > 0$ from all cone points, then in a small metric disc $D \subseteq D(x, \delta/2)$ on S, the number of generations of hexagonal combinatorics in $P_n \cap D$ will grow with n, and the Hexagonal Packing Lemma, 19.4, implies that the distortion of f_n restricted to the corresponding circles of $\mathcal{P}_{\widetilde{K}_n}$ will converge to 1. In other

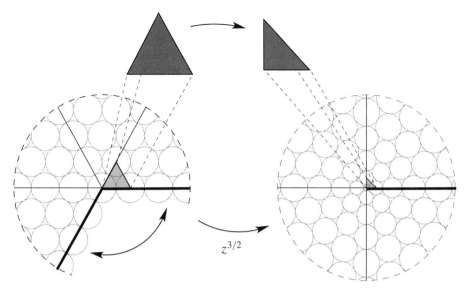

Figure 22.4. The local power map.

words, the limit function h is conformal outside the preimages of cone points of S. Since this set of preimages is countable, h is conformal.

Our last task is to pick up the spherical and bordered cases. There is no Ring Lemma in spherical geometry, but we can put some cone point at ∞, stereographically project the rest of S to $\mathbb{C}$ and proceed as above. As for bordered surfaces S, since we are concerned with uniform convergence on compact subsets and since in a hexagonal refinement scheme S is exhausted by the unions of interior faces, we do not have to concern ourselves directly with the border. Our proof is complete.   □

Observations:
(1) The automorphisms ϕ_n of the theorem are for normalization, and their use depends on the situation. Typically one might put the preimage of a designed cone point at the origin and the preimage of another on the positive real axis.

(2) When S is simply connected, the map F here is our usual conformal mapping from one of the standard geometric spaces, $\mathbb{P}$, $\mathbb{C}$, or $\mathbb{D}$.

(3) When S is compact or finitely connected (but not simply connected) then $S = \mathbb{G}/\Lambda$ for some discrete subgroup of $\text{Aut}(\mathbb{G})$. Each of the discrete universal covering maps f_n is associated with an isomorphic discrete group Λ_n, and $\mathbb{G}/\Lambda_n$ is conformally equivalent to the surface S_{K_n}, the surface supporting the maximal packing for K_n. There are various ways to address the convergence of these surfaces and discrete groups. With the *dessin* we will discuss in Section 23.1, for example, the Riemann surfaces S_{K_n} converge in the associated Teichmüller metric to S.

(4) The techniques here can be extended to equilateral surfaces punctured at one or more cone points. See the j-function and cusp examples at the end of Chap. 16.

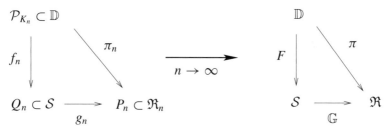

Figure 22.5. Mapping diagrams.

(5) One might question the term "conformal" for our discrete maps f_n (or, indeed, for the classical map F itself): given a cone point with degree k, say, then f_n maps a packing with angle sum 2π in $\mathbb{G}$ to one with angle sum $k\pi/3$ in $\mathcal{S}$. *How can we claim that angle is being preserved?* This is a matter of standard definitions. In a cone surface the angles at a cone point p are measured in a "market share" sense. There is no trouble in defining the signed euclidean angle, in the surface, between two smooth curves γ_1 and γ_2 through p; say that angle comes out to be β. If the cone angle at p is α, then the *market share* angle between γ_1 and γ_2 is defined as $\theta = \frac{2\pi\beta}{\alpha}$. It only makes sense, for example, that if four lines meet at p with equal angles then those should be thought of as "right" angles. Market share angles justify the term "conformal."

Let us finish with equilateral surfaces by applying this theorem to our ten-triangle-toy example. Here $\mathcal{S}$ is a conformal disc, so the universal covering map is $F : \mathbb{D} \longrightarrow \mathcal{S}$. The maps for this example are related to the maps of the theorem as in the commutative diagrams in Fig. 22.5. Here $\pi_n : \mathcal{P}_{K_n} \longrightarrow P_n$ and $\pi : \mathbb{D} \longrightarrow \mathfrak{R}$ are the Riemann mappings to rectangles, discrete and classical, respectively.

We have bounds on the dilatation in every map, so essentially everything on the discrete side – surfaces, maps, conformal structures – converges to its classical counterpart on the right. With our ten-triangle-toy, for example, one argues by extremal lengths that the image rectangles $\mathfrak{R}_n \equiv \mathrm{carr}(P_n)$ converge to the rectangle $\mathfrak{R}$ and the discrete conformal maps f_n converge to the conformal map $F : \mathcal{S} \longrightarrow \mathfrak{R}$. In particular, the original 10 faces of $\mathcal{S}$, as outlined in Figure 22.3, converge to their correct conformal shapes in $\mathfrak{R}$. It is this type of information that has been missing in the past.

22.4. More General in Situ Packings

It is easy to overlook certain subtleties in these conformal structures for more general polyhedral surfaces $\mathcal{S}$. This is a strongly *cautionary* section: we contrast two very simple cases, each with just two faces.

Situation 1: $\mathcal{S}_1$ consists of two regular euclidean pentagons attached along an edge. Packings can be used to define the euclidean structure on each face, as shown in Figure 22.6 at the coarse and $n = 3$ refinement stages. Two packings can be geometrically pasted along an edge at the coarse stage and then hexagonally refined and repacked with specified

Regular pentagon   Course packing $n = 3$

Figure 22.6. A regular pentagon and two in situ packings.

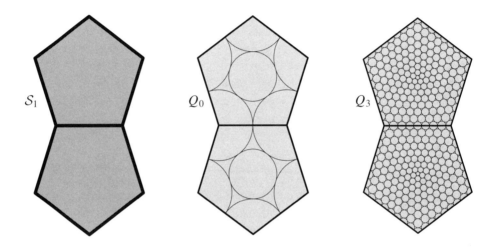

Figure 22.7. $\mathcal{S}_1$ and two in situ packings.

boundary angle sums to obtain in situ packings Q_n in $\mathcal{S}_1$, as in Figure 22.7 (or equivalently, the faces can be refined and repacked first and then pasted). As we saw in the equilateral case, these packings permit discrete conformal maps which will converge to their classical counterparts. For example, the discrete conformal maps $f_n : \mathcal{P}_{K_n} \longrightarrow Q_n$ both mimic and, as $n \to \infty$, converge to the Riemann mapping $F : \mathbb{D} \longrightarrow \mathcal{S}_1$.

Situation 2: Our second situation may seem equally straightforward: for $\mathcal{S}_2$ one of the pentagons is replaced by an equilateral triangle. Figure 22.8 may seem to parallel Figure 22.7, but in fact it is not working! The geometric pasting is fine at the coarse stage, but not after refinement. A close look at the edges of the 3-level packings in Figures 22.2 and 22.6 reveals that corresponding circles are not the same size – these edges are not eligible for "geometric" pasting.

What, then, is happening with the hexagonal refinements in Figure 22.8? There are many (uncountably many!) ways to attach two polygons edge to edge. That used for $\mathcal{S}_2$ is perhaps most natural – identifying the edges via a euclidean isometry. However, the

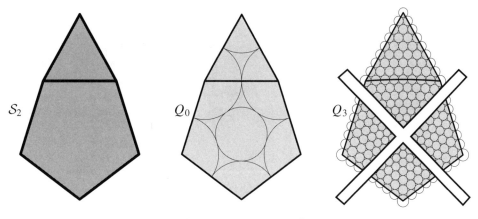

Figure 22.8. A failure in S_2.

refined packings, which arise from the "combinatorial" pasting of the refined triangle and pentagon, think they live in an equilateral surface – namely the surface in which the pentagon is replaced by five equilateral triangles. Call this equilateral surface S_3.

Riemann surfaces S_2 and S_3 are conformally equivalent (as conformal discs); however, they are conformally *distinct* when their faces are marked. You can see this in the slight curvature to the edge between the pentagon and triangle in the last image of Figure 22.8, whereas this edge is a straight line in S_2. If you reflect the triangular face across this curved edge, its top corner in S_3 would land precisely at the center of the pentagon, while the top corner in S_2 reflected in the straight edge will fall below the center.

> **Caution:** It is tempting to assume that when one has found some circle packing P lying in a Riemann surface S, then its hexagonal refinements will approximate the conformal structure of S. This is not necessarily the case! These refinements will approximate the *equilateral* structure associated with the *discrete conformal structure* – i.e., with the complex K underlying P.

So our methods, by their nature, apply only in *equilateral* situations. There are two mild generalizations that we will need in the applications of the next chapter. The conformal structure of a polyhedral surface S all of whose faces are regular n-gons for the *same n* is precisely that induced by replacing each n-gon with n equilateral triangles. This applies, for example, to our surface S_1. Constructing it from 10 equilateral triangles would give the same *conformal* structure, though a different *metric* structure (one in which we would not need any repacking to get our in situ packings). In a similar vein, the equilateral structure of a simplicial complex and its barycentric subdivision (see Section 12.8) induce the same conformal structure on the surface.

Of course, one would hope to approximate more general Riemann surfaces, and this is a topic of active research. Let me close, therefore, by briefly illustrating two methods for working around the problems encountered with the surface S_2 above.

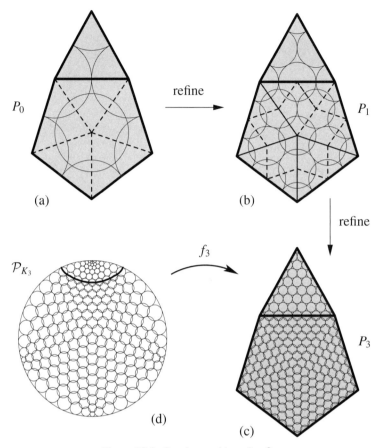

Figure 22.9. Overlap packings for $\mathcal{S}_2$.

Overlap Packings: Many of the properties we have developed in this book for *tangency* circle packings extend to configurations with prescribed overlaps among the circles. I discuss these in the broader context of *inversive distance* packings in Appendix E. Figure 22.9 demonstrates their use with $\mathcal{S}_2$.

Figure 22.9 shows $\mathcal{S}_2$ with in situ overlap packings. The coarse packing (a) has complex K identical to that of the coarse packing in Figure 22.8, but all the circles have radius $1/2$ and the dashed edges indicate circles which necessarily overlap (in this case by angle $\phi \approx 0.35242\pi$). If we break each triangle into four similar triangles half its size and center circles of radius $1/4$ at all vertices, we get (b). In fact, the patterns of overlaps for the triangles nicely replicate the originals; the dashed edges again indicate overlap angles ϕ. This is just hexagonal refinement in this setting; we have another in situ packing with no repacking computation required. Packing (c) shows two further stages of hexagonal refinement. Finally, we show the associated (overlapping) maximal packing for (c) with the edge between the faces shown for reference. The discrete Riemann mappings f_n converge on compacta of $\mathbb{D}$ to the classical Riemann mapping $F : \mathbb{D} \longrightarrow \mathcal{S}_2$, just as

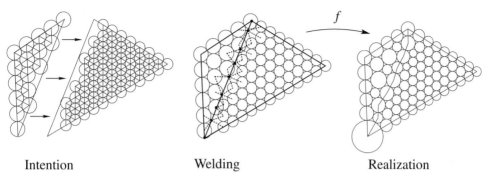

| Intention | Welding | Realization |

Figure 22.10. A combinatorial welding.

they would in the tangency case. I will have to leave the details on this for another time.

Welding: An even more general workaround employs *combinatorial welding*, a tool described in Chap. 12. In the pasting for S_2 we found that the refined faces had combinatorics and geometry which were out of step. With a mandate to respect the geometry, we might modify the combinatorics. Figure 22.10 illustrates how to paste two triangles along fairly incompatible looking edges.

Note the new vertices introduced in the pasting region: the union of the two triangles lacks an in situ packing, but it does have an in situ triangulation. Mapping that via f to its circle packing (computed here with specified boundary angle sums) involves a certain amount of quasiconformal distortion. In regions where the combinatorics were not changed that distortion is controlled in the usual way via the Ring Lemma. We do not have this control in the pasting region itself, but under typical refinement schemes this region will have area going to zero and, as with isolated cone points, will not effect the conformality that emerges in the limit.

23

Applications

Conformal structures have to do with surfaces, and surfaces are everywhere in science, engineering, graphics, and, of course, mathematics. In this section we use the methods that have been developed earlier in four applications areas. It is in the nature of applications to stretch one's thinking, so I encourage the reader to skip around, perhaps digesting some of the geometry in the images before going back for the mathematics.

23.1. *"Dessins d'Enfants"* of Grothendieck

The reader may recall the discrete Belyĭ meromorphic functions we constructed back in Section 16.3, each branching only over $\{0, 1, \infty\}$. These are not casual examples, but rather are the heart of the topic of *Dessins d'Enfants* ("children's drawings"), conceived by Grothendieck, which tightly binds combinatorics, meromorphic functions, and algebraic number fields.

Here is the cast of characters in this surprising story: (1) D is a child's *drawing* on a surface S; (2) T_D is an associated tripartite complex; (3) $\mathcal{S}_D$ is a Riemann surface; (4) $B_D : \mathcal{S}_D \longrightarrow \mathbb{P}$ is a Belyĭ meromorphic function; and (5) $\mathbb{F}_D$ is an algebraic number field, that is, a finite extension of the rational numbers $\mathbb{Q}$.

The storyline, very briefly, is this: A child casually draws a graph D on S; more precisely, D is a connected finite graph whose complement in S is a disjoint union of open discs (clever child!). T_D is a tripartite complex obtained from D as follows. Place a "•" at each node of D, a "×" in the interior of each edge of D, and a "○" interior to each 2-cell in the complement of D. Now walk around the boundary of each 2-cell and draw nonintersecting edges from the ○ of that cell to each • and × that you encounter. The result is a triangulation T_D, and every face will have one vertex of each type •, ×, and ○ either positively ordered (*unshaded* face) or negatively ordered (*shaded* face). Although T_D may technically fail to be a triangulation (two faces may intersect in more than one vertex and/or edge) one can nonetheless use it to impose an equilateral structure on S by declaring each face to be a unit, equilateral, euclidean triangle. This makes S into a Riemann surface, which we denote by $\mathcal{S}_D$. As in the discrete situation of Section 16.3, every face of $\mathcal{S}_D$ can be mapped conformally to the appropriate hemisphere of $\mathbb{P}$ ($\mathbb{H}^+$ for unshaded and $\mathbb{H}^-$ for shaded) in such a way that $\bullet$, $\times$, and $\circ$ are mapped to 0, 1, and ∞, respectively. The edges $\bullet\!\!-\!\!-\!\!\times$, $\times\!\!-\!\!-\!\!\circ$, and $\circ\!\!-\!\!-\!\!\bullet$ of each face are mapped to real intervals $[0, 1]$, $[1, \infty]$, and $[-\infty, 0]$, respectively. Consequently, when two faces of T_D share an edge, a Schwarz reflection argument implies that the maps on the faces are analytic continuations of one another across that edge. In other words, our face-by-face defined function is analytic outside the vertex set of T_D. These vertices are isolated, and hence removable, so our function extends to become a meromorphic function B_D on all of $\mathcal{S}_D$. B_D is locally univalent on faces and across edges of T_D, so it can branch only at the points $\bullet$, $\times$, and $\circ$, and so by construction its branch *values* lie in $\{0, 1, \infty\}$. The function B_D is known as a Belyǐ meromorphic function and the pair $(\mathcal{S}_D, B_D)$ is a *Belyǐ pair*. You will note that the original drawing D in S is represented by the preimage $B_D^{-1}([0, 1])$.

This process is reversible, also. If you start with a Belyǐ meromorphic function $B : S \longrightarrow \mathbb{P}$, then the preimages of the upper and lower half planes under B form a tripartite simplicial complex T on $\mathcal{S}$, the set $D = B^{-1}([0, 1])$ is a drawing on $\mathcal{S}$, and $T_D = T$. So all these objects are tightly bound to one another:

$$D \longleftrightarrow T_D \longleftrightarrow \mathcal{S}_D \longleftrightarrow B_D.$$

Enter cast member $\mathbb{F}_D$, stage right. This is the most surprising character in the story, but I need to say a word for those not familiar with the theory of compact Riemann surfaces. A *real algebraic curve* is a connected component of the solution set $\{(x, y) : p(x, y) = 0\}$ for an irreducible real polynomial p in the *complex projective plane* $P_{\mathbb{C}}^1$ (i.e., in $\mathbb{P}$). Likewise, a *complex algebraic curve* is a component of $\{(z, w) : P(z, w) = 0\}$ for an irreducible complex polynomial in the complex projective plane $P_{\mathbb{C}}^2$. This "curve" is of complex dimension 1 and inherits a conformal structure from the way it is embedded in $P_{\mathbb{C}}^2$ (based on the implicit function theorem), so it is a compact Riemann surface. One of the triumphs of nineteenth century mathematics is the realization that every compact Riemann surface is (conformally equivalent to) such a complex algebraic curve. That is, every compact Riemann surface has a *defining equation* $P(z, w) = 0$. One more definition: an *algebraic number field* is a finite algebraic extension of the rational numbers $\mathbb{Q}$; that is, it is a field extension $\mathbb{Q}[\rho]$, where ρ is the solution of a polynomial equation $p(x) = 0$ for a polynomial with coefficients in $\mathbb{Z} + i\mathbb{Z}$. $\mathbb{F}_D$ appears with the rest of the cast in this wonderful theorem.

Theorem 23.1 (Belyǐ's Theorem). *For a Riemann surface $\mathcal{S}$ of genus $g \geq 1$, the following statements are equivalent:*

(a) $\mathcal{S}$ is conformally equivalent to an equilateral surface.

(b) There exists a nonconstant meromorphic function $B : \mathcal{S} \longrightarrow \mathbb{P}$ that branches only over the points $\{0, 1, \infty\}$.

(c) There exists a defining equation for $\mathcal{S}$ whose coefficients lie in an algebraic number field $\mathbb{F}$.

Here we find a collection of Riemann surfaces characterized simultaneously by combinatoric, function theoretic, and algebraic conditions, all associated with children's drawings! I should say that number fields also enter in a related way for genus 0 *dessins*.

The algebraic features have been the main impetus for studying *dessins*, the target being what is known as the *Inverse Galois Problem*. There are books dedicated to this topic and I must leave it to the interested reader to investigate. For us, the point is this: *Despite the beautiful theory, for the generic dessin D almost nothing has been known in practice about S_D aside from its genus.*

This is where circle packing may have a role to play. Given a *dessin* D, we described above how to construct the classical Belyĭ pair $(\mathcal{S}_D, B_D)$ using its tripartite complex T_D. In Sect. 16.3 we described how that same T_D leads to a discrete Belyĭ pair, which we will call $(\mathfrak{s}_D, b_D)$. I hope the *discrete* $\longleftrightarrow$ *classical* parallels are evident. In particular, the topological situations are identical, and by Stoïlow's Theorem there is a homeomorphism $h_D : \mathfrak{s}_D \longrightarrow \mathcal{S}_D$ that respects the triangulations T_D in the two settings, so that

$$b_D \equiv B_D \circ h_D. \tag{121}$$

But is there a concrete *conformal* connection? (I will drop all the subscript Ds to save clutter.) In general h will not be conformal. Let us review the bidding: There is a compact surface S with tripartite complex T; imposing the equilateral structure defined by T makes S into a Riemann surface $\mathcal{S}$. Barycentrically subdividing T gives a complex K also triangulating S (see Figure 12.6), and one can show that the equilateral structure defined by K determines the *same* Riemann surface $\mathcal{S}$. As with the ten-triangle-toy, there is an in situ circle packing Q for K in the piecewise affine structure of $\mathcal{S}$. That is all fine, but where does $\mathfrak{s}$ come in? It is the Riemann surface $\mathcal{S}_K$ that supports the maximal packing $\mathcal{P}_K$. The map from the in situ packing Q to the corresponding *maximal* packing is quasiconformal but typically not conformal. One does not expect $\mathfrak{s}$ and $\mathcal{S}$ to be conformally equivalent. Too bad.

As you have seen before, one should not have expected much at a coarse packing stage: one needs to refine the combinatorics of K. Let $K^{(n)}$ denote the nth-stage hexagonal refinement of K, $\mathfrak{s}^{(n)}$ the Riemann surface for its maximal packing, and $b^{(n)} : \mathfrak{s}^{(n)} \longrightarrow \mathbb{P}$ the discrete Belyĭ map. Based on Theorem 22.1 one can show that the sequence of refined Belyĭ pairs $\{(\mathfrak{s}^{(n)}, b^{(n)})\}$ converges to the classical Belyĭ pair $(\mathcal{S}, B)$. We will not try to make this terribly precise (though I will say, for those readers familiar with Riemann surfaces, that in this compact case, if the genus is positive, then $\mathfrak{s}^{(n)}$ converges to $\mathcal{S}$ in the Teichmüller metric); suffice it to say that we can read off approximations of the covering transformations for the universal covering maps of the $\mathfrak{s}^{(n)}$, and, with appropriate normalizations, these converge to the covering transformations for $\mathcal{S}$. Moreover, as these domains converge in a conformal sense, their Belyĭ maps converge pointwise, $b^{(n)} \longrightarrow B$.

Time for some examples. Figure 23.1 shows the $n = 2$ and $n = 4$ refinement stages of the genus zero example whose coarse packing was shown in Fig. 16.5. The cartoon shows the associated *dessin* D; it appears as the darker curve in each packing. As n grows, you see the faces and the *dessin* itself converging to their correct conformal shapes – in this case, their shapes in $\mathbb{P}$.

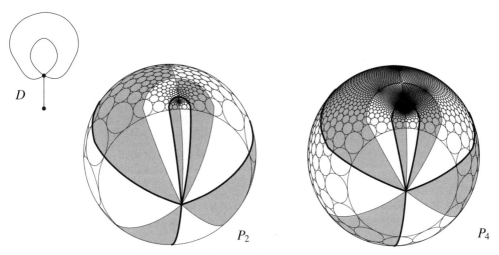

Figure 23.1. Refinement of the genus 0 example.

Figure 23.2 shows the $n = 3$ refinement stage of the universal covering map for the genus 2 example from Figure 16.5. Again you can see the shapes converging; this *dessin* D has only two vertices and you can read the side-pairings from the cartoon (identified by shared letters).

We must leave *dessins d'Enfants* now, but let me end with an intriguing fact that remains to be explored. When we discretize a *dessin* D, we expect to lose the connection with the number fields established in Belyĭ 's Theorem. However, in the case of positive genus, we gain a new connection. The covering group Λ of $\mathfrak{s}$ is a subgroup of Aut($\mathbb{C}$) ($g = 1$) or Aut($\mathbb{D}$) ($g > 1$), so its members are represented by 2×2 matrices of PSL(2, $\mathbb{C}$). According

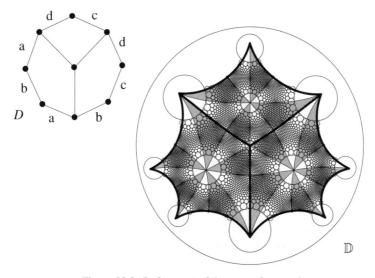

Figure 23.2. Refinement of the genus 2 example.

to a result of Gareth McCaughan, with an appropriate normalization, the matrices of the covering group Λ will have entries in some number field $\mathbb{F}$.

23.2. Conformal Tilings

"Tiling" is a topic whose artistic and mathematical roots go back millennia. By tradition one searches for the various local patterns in which a given set of geometric shapes, the *tiles*, will fit together to form an infinite pattern. We play the game in reverse, however, by starting with the pattern and asking *With what tile shapes and in which geometry can this pattern be realized?* As far as I know, this version of the topic is new and came directly out of circle packing – that story will be told in a moment.

The game plan in the topic is this. Start with a given pattern T, by which we mean a combinatorial polyhedron – a pattern of triangles, rectangles, pentagons, hexagons, etc., decomposing an open topological disc. Impose a conformal structure on the collection by declaring each polygonal face to be a unit, regular, euclidean polygon; the result is one of the *polyhedral* Riemann surfaces S we discussed in the previous chapter. According to the Uniformization Theorem, S is conformally equivalent to $\mathbb{G}$ (either $\mathbb{D}$ or $\mathbb{C}$), so there is a univalent conformal map $F : S \longrightarrow \mathbb{G}$. The images of the polygonal faces in $\mathbb{G}$ are called *conformal tiles*; they now have the correct conformal shapes to tile $\mathbb{G}$ in the pattern of T. Figure 23.3 displays four such tilings that arose from subdivision rules in a program started by Jim Cannon and being carried on with collaborators Bill Floyd and Walter Parry.

With a view to full disclosure, let me say that these images are approximations of the actual tilings – in fact, these are all made using coarse-stage circle packings. Moreover, we have as a practical matter replaced each face by a union of equilateral triangles and proceeded using the equilateral structure; by the comments at the end of the previous chapter, this is legitimate in the upper two examples, since the tiles are all pentagons, and though we are not being conformally correct in the lower images, there seems to be little visual effect. (Typically, the results will be accurate modulo a quantifiable quasiconformal distortion; see Section 22.4.)

To those in traditional tiling, where the heart of the challenge lies in the *rigidity* of the tile shapes, this probably seems like flummery – where's the fun? The challenge is indeed from quite a different direction. Stare for a time at one of the examples of Figure 23.3, perhaps letting your vision go a little out of focus. Do you sense some global pattern? perhaps even layers of pattern? maybe a bit of self-similarity? This is the theme (at least for now) in this topic:

Theme: *emergence of global geometric patterns from purely combinatorial patterns.*

I would like to illustrate the topic with the seminal "regular pentagonal" pattern. This is the tiling in the upper left of Figure 23.3 and is based on the subdivision rule of Figure 23.4(a). This will necessarily be rather terse, but it is self-contained, and I hope you can enjoy it as a bit of mathematical storytelling.

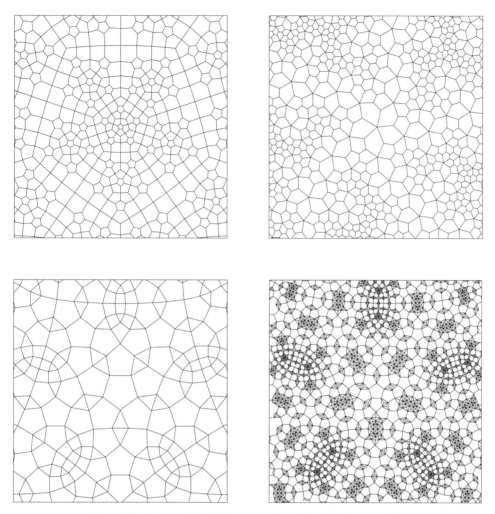

Figure 23.3. A sampling of subdivision tilings (Source: Bill Floyd).

23.2.1. Tale of a Tiling

The pentagonal rule of Figure 23.4(a) is a prescription for subdividing one pentagon into six. Successive repetitions of the rule generate increasingly complex abstract patterns with 36, then 216, then 1296 pentagons, and so forth. Our story is about these patterns and begins with Bill Floyd and his collaborators Jim Cannon and Walter Parry. They were searching for embeddings of successive subdivisions in which they could gain uniform control of the distortions of the individual pentagons – what they termed "almost round" embeddings. Bill visited Tennessee for a talk and I was immediately captivated by his wonderfully intricate patterns.

It took surprisingly long, in hindsight, but I finally suggested (what else?) a circle packing approach in which a barycenter would be added to each face to form a triangulation.

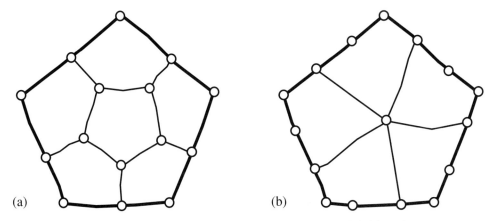

Figure 23.4. (a) Pentagonal and (b) Twisted Pentagonal subdivision rules.

(a)   (b)

Bill had software for producing reams of combinatorial data, and pushing his output through `CirclePack`, we began producing nice – even gorgeous – embeddings! We accomplished the first goal:

- *By the Ring Lemma, these circle packing embeddings are uniformly almost round.*

As so often happens, new pictures spark new insights. A few images viewed in succession suggested a recasting of the "subdivision" process into a dual "expansion" rule, and that led shortly to the infinite tiling T represented in the upper left of Figure 23.3. If you look carefully at that T, you can find all the finite patterns nested within it. Indeed, it was after staring at this image for long periods that I noted the obvious:

- *T is combinatorially invariant under the combinatorial subdivision rule.*

In other words, if every pentagon of T is simultaneously subdivided, the resulting pattern is combinatorially identical to T itself. (I have to say that I was transfixed by the pattern, and perhaps you are too. Here's a sample question. *Is the vertex at the center of this image in a combinatorially unique position?*)

Phil Bowers and I were involved in uniformizing *dessins* about this time, so we realized that T could be construed as an infinite pattern of *conformal* pentagons. By the Riemann Mapping Theorem, the conformal tiling T would fill either the plane or the hyperbolic plane, $\mathbb{G} = \mathbb{C}$ or $\mathbb{G} = \mathbb{D}$. But *Which is it?* "Refine and repack" might be the first impulse, but early experiments suggested quite a different course here. As T is invariant under subdivision, so it must be invariant under *aggregation* (un-subdividing). By putting the outlines of the first few stages of aggregation together in a *PostScript* file and doing a little fiddling to get two of them to line up, an amazing thing happened: simultaneously *all* the outlines lined up! Bowers and I came to realize that there is a *conformal* version of the pentagonal subdivision rule, which breaks a *conformal* pentagon into six *conformal* pentagons. It was a short step to prove:

- *The conformal tiling T is invariant under the conformal subdivision rule.*

Now we use the fact that, yes, the vertex at the origin of T is combinatorially unique. If ψ is a conformal automorphism of $\mathbb{G}$ mapping T to its own subdivision, then the origin must be fixed and $\psi : z \mapsto z/\lambda$. Clearly, $|\lambda| > 1$, so ψ is a contraction. But Aut($\mathbb{D}$) has no contractions. *Voilá!*

- *T tiles the euclidean plane.*

In establishing these results formally, Bowers and I introduced this "regular" pentagonal tiling as the first example of a more general notion of *conformal tiling*. We also included the experimental estimate $|\lambda| \approx 3.2$, and with the number fields of *dessins* fresh in our minds, we asked whether λ is in fact algebraic. This is where the story sat until the 1998 Barrett Lectures at the University of Tennessee, which had Jim Cannon as one of its principal speakers. Rick Kenyon had a method for encoding certain subdivision rules as branched maps. With Jim, Bill, and Walter Parry, he recognized that the pentagonal pattern was generated by *rational iteration*.

Thus the story took a new direction. Break the sphere into hemispheres, shaded and unshaded as we have done before. Let $R : \mathbb{P} \longrightarrow \mathbb{P}$ be a rational function which is real on the real line and consider preimages of the hemispheres under R – i.e., applications of R^{-1}. These preimages break the sphere into some new, finer pattern of shaded and unshaded patches. Preimages of those patches break the sphere into a yet more intricate pattern of patches. And so forth as one repeatedly applies R^{-1}. Figure 23.5 suggests a series of such iterates for a certain rational map R. (The sphere $\mathbb{P}$ has been opened along the positive real axis for display.) Do you see the pentagonal subdivision rule? To help, I have in the lower right of Figure 23.5 highlighted selected aggregates of those patches. The upshot:

- *The pentagonal subdivision rule is implemented by backward iteration of a particular rational function R.*

From the *combinatorics* alone Rick Kenyon knew the branch structure, and with the branch values being among $\{0, 1, \infty\}$ he and the others were able to deduce the precise rational function:

$$R(z) = \frac{2z(z + 9/16)^5}{27(z - 3/128)^3(z - 1)^2}. \tag{122}$$

If the patterns on the sphere seem familiar for another reason, it might be because R turns out to be a Belyĭ map! The first application of R^{-1} gives the canonical tripartite simplex T_D for this *dessin*, and further applications subdivide it (in place). In fact, this is precisely the genus 0 *dessin* example shown earlier in Figs. 16.5 and 23.1.

From (122) one deduces that $R(0) = 0$ and $|R'(0)| = 324$, making 0 a contractive fixed point for R^{-1}. Iteration at such points was studied by König and others a century ago and the behavior can be localized via a function ψ called the *Königsfunction*. In this case ψ

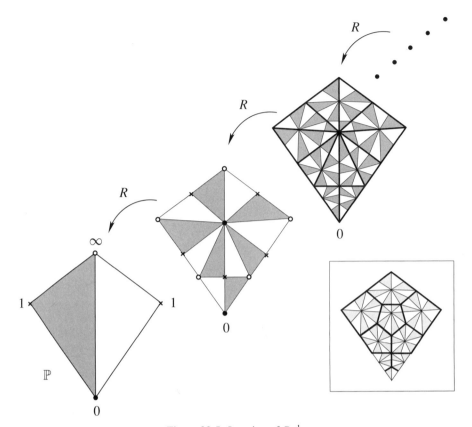

Figure 23.5. Iteration of R^{-1}.

is a meromorphic function on the plane with the property that $\lambda\psi(z) = \psi(R(z))$. If F is defined by $F(z) = \psi(z^5)$ (to ensure 5-fold symmetry at the origin), then the outlines of the tiling are a form of infinite *dessin*.

- *The tiling $\mathcal{T}$ is $F^{-1}([1, \infty])$, the preimage under the meromorphic function F of the real interval $[1, \infty]$.*

With R in hand, one can actually *compute* the contraction factor λ. At the conclusion of the Barrett Lectures, Jim Cannon presented its value to me as a gift:

$$\lambda = (-324)^{1/5}.$$

Yes, it is algebraic! The root accounts for the $\pi/5$ rotation, and its modulus is roughly 3.1777, close to the experimental estimate. I close the story with a clever observation of Cannon's:

- *The pentagonal subdivision rule is associated with the "pentagonal" number $(-324)^{1/5}$, that is, a number involving only the digits $1, 2, 3, 4$, and 5.*

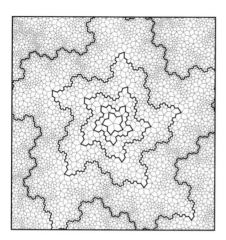 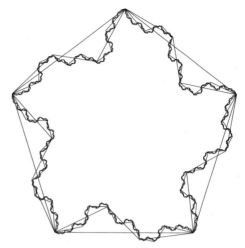

Figure 23.6. Outlined aggregates: scaled, rotated, and overlaid.

This concludes our story about this example, but there are certainly many other tales to look forward to in conformal tiling. Let me mention two others. First is the "twisted" pentagonal tiling in the upper right of Figure 23.3. The rule is shown in Fig. 23.4(b).

If you outline aggregates as described earlier for the pentagonal case, you get the image on the left of Figure 23.6. It is clear that scale invariance is out – or is it? If you scale, rotate, and juxtapose these outlines you get the image on the right in Figure 23.6. Motivated by this very experiment, Cannon et al. have proven that the vertices in each outline lie *exactly* at vertices in the next! Many of the results we saw before go through, but the final "limit" tiling is more intriguing here – it seems to be a self-similar fractal tiling, which, according to our development, supports a fractal subdivision rule.

Our final example is actually the first in Cannon's program, the *dodecahedral* subdivision rule. There are three tile types and the abstract rule tells how each is subdivided into a pattern of those same three types. Starting with a rectangle, applying three stages of subdivision, and then circle packing the result provides the tiling of Figure 23.7 (Thanks to Bill Floyd for the data; he has computed a fourth stage, but with 1,600,000 circles, the current circle packing record, it is far too intricate to reproduce here.) So many questions! Does the aspect ratio of the rectangle have a limiting value under continued subdivision? Do the various aggregate shapes stabilize? Is there an infinite tiling with perfect scaling, as in the regular pentagonal case? It is unclear that questions like these would even be contemplated were it not for the experimentation and approximation available through circle packing.

23.3. Conformal Welding

We discussed welding in an abstract combinatorial sense in Section 12.5. To set the more concrete scene, consider constructing a beach ball. You start with two rubber discs and sew (i.e., identify) them together along their boundaries. Pump the resulting sphere full of air and the "seam" becomes some rigid curve – the *welding curve*. With perfect stitching,

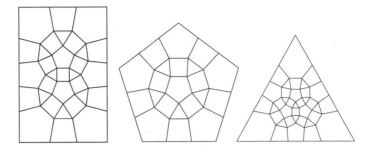

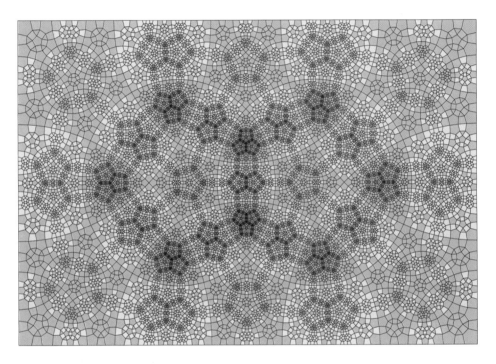

Figure 23.7. Dodecahedral tiling, color-coded by tile ancestry.

the curve would clearly be a great circle by symmetry. In general, however, as you gather/stretch opposite sides while sewing, irregularities creep in. It is hardly surprising that those irregularities are reflected in the curve's final shape.

This construction gives you the rudiments of *conformal welding*; the "conformal" part comes with the mathematical translation. Let $\omega : \partial \mathbb{D} \longrightarrow \partial \mathbb{D}$ be an orientation-reversing homeomorphism. Attach two copies D_1 and D_2 of the unit disc by identifying $z \in \partial D_1$ with $\omega(z) \in \partial D_2$. The result is, of course, a topological sphere S, which we map via a homeomorphism h to the Riemann sphere $\mathbb{P}$ – our mathematical beach ball. If ω is sufficiently well behaved, there will exist an $h : S \longrightarrow \mathbb{P}$ whose restrictions to D_1 and D_2 in S are both conformal mappings – this is where the "conformal" enters. The image of the common boundary of D_1 and D_2 is the welding curve $\Gamma = \Gamma_\omega$.

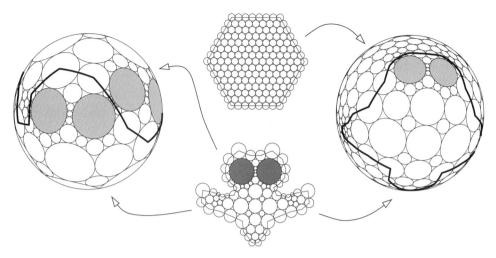

Figure 23.8. Simple weldings.

There is a deep and well-developed theory of conformal welding that has found many uses, for example, in the study of 3-manifolds and the mapping class group. The key result for us concerns the rigid link between two classes of functions: on the one hand are the *quasisymmetries* ω of $\mathbb{D}$; among them, e.g., are all orientation-reversing, continuously differentiable homeomorphisms of $\partial\mathbb{D}$. On the other hand are the curves Γ on the sphere known as *quasicircles*, curves that are images of $\partial\mathbb{D}$ under quasiconformal maps of the full sphere. It was proved by Ahlfors and Bers that every quasisymmetric welding map ω leads to an essentially unique quasicircle welding curve Γ and vice versa.

Let us move, now, to the discrete side. There is a naive formulation of discrete welding that is actually quite entertaining. Closed discs K_1 and K_2 having equal numbers of boundary vertices can be adjoined along their boundaries in reverse orientation to get a complex K which is a combinatorial sphere. The welding curve is the seam embedded in $\mathbb{P}$ as an edge-path in the carrier of the maximal packing $\mathcal{P}_K$. In Figure 23.8 I have used two copies of Owl on the left, welded wingtip-to-ear, and one copy of Owl welded to a hexagonal pattern on the right.

Fun as these simple examples can be, they are driven purely by the combinatorics, not the geometry. It is George (Brock) Williams who succeeded in transplanting the "conformal" into the discrete setting. Figure 23.9 illustrates the process. Suppose K is a combinatorial closed disc and $\omega : \partial\mathbb{D} \longrightarrow \partial\mathbb{D}$ is an orientation-reversing homeomorphism, our desired welding map. A copy of $\mathcal{P}_K$ in the northern hemisphere and an orientation-reversed copy in the southern hemisphere define locations for their boundary vertices on the equator (i.e., on $\partial\mathbb{D}$). Combinatorial welding is described in Sect. 12.5. That mechanism frees us from strict combinatorics by allowing additional boundary vertices and edges in the two complexes so that the welding on the equator can interpolate ω. The maximal packing in $\mathbb{P}$ for the welded complex gives a geometric embedding of this common boundary as a *discrete welding curve*, which we denote by Γ_ω^K.

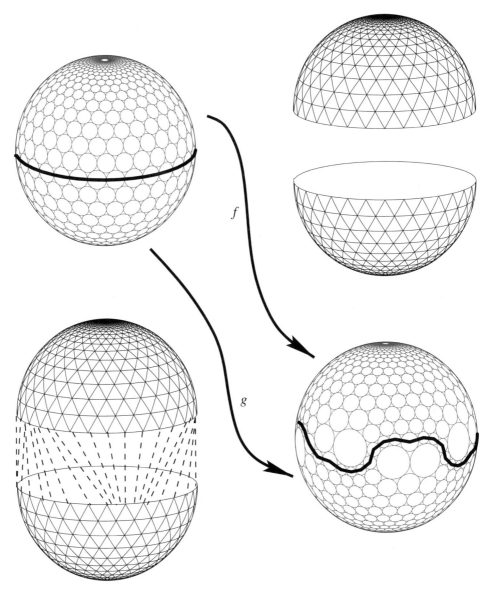

Figure 23.9. Discrete conformal welding detail.

Theorem 23.2 (Brock Williams). *Let ω be an orientation-reversing quasisymmetry of the unit circle and let Γ_ω be the associated classical conformal welding curve in $\mathbb{P}$. Given a closed disc K, the sequence $\{\Gamma_\omega^{K_n}\}_n$ of discrete welding curves associated with ω and successive hexagonal refinements K_n of K will converge, after appropriate normalization, to Γ_ω.*

The proof rests on bounding the quasiconformal dilatation associated with the weld region by using the fact that ω is a quasisymmetry and then showing that these distortions

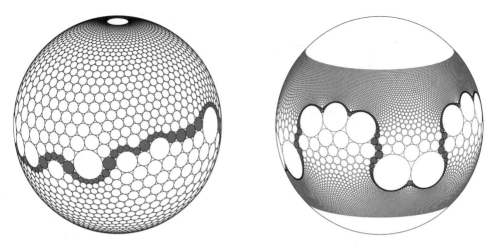

Figure 23.10. Discrete conformal weldings (source: Brock Williams).

are restricted to a region of small area close to the welding curve. Away from the welding curve, the dilatation is controlled by the Ring Lemma and goes to 1 under hexagonal refinement as in results we have seen earlier.

This is yet another topic with a theory that was lacking any practical realization until circle packing methods arrived. As you can see in the examples of Fig 23.10, provided by G. Brock Williams, these can also be beautiful curves.

23.4. Brain Flattening

Some of us study mathematics for its purity and intellectual detachment from the ordinary world, others of us precisely for its relevance. All of us, I think, are pleasantly surprised when a topic stretches between these two poles, so I want to conclude the book by showing some recent applications of circle packing being developed for real world use.

Complex analysis has traditionally played important roles in the physical sciences and engineering – topics such as electrostatics, fluid flow, airfoil design, and residue computations, to name a few. But these roles have now largely been taken over by numerical partial differential equations, symbolic packages, simulations, and brute force computations. But science is ever changing, and it is my opinion that the geometric core of complex analysis is too fundamental to go missing for long. Currently, for example, *surfaces* embedded in three-space are becoming ubiquitous – as in medical imaging, computer graphics, nanoscience constructs, and image analysis. Conformal geometry is all about just such surfaces. With new tools to (faithfully) access conformality, perhaps complex analysis has new roles to play.

Let me describe a recent collaboration among neuroscientists and mathematicians: the topic is *brain flattening*. Figure 23.11 illustrates two samples of the type of 3D human brain data that have become routinely available through noninvasive imaging techniques such as MRI (magnetic resonance imaging), fMRI (functional MRI), and PET (positron emission tomography). Mental processing in the brain occurs largely in the neurons of the

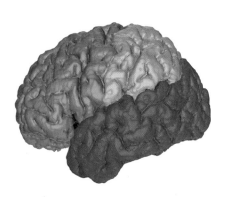

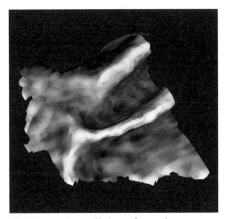

(a) Cerebral hemisphere   (b) Ventral medial prefrontal cortex

Figure 23.11. Examples of typical 3D cortical surfaces (sources: (a), Monica Hurdal, David Rottenberg, Kelly Rehm, and Lili Ju; (b), Monica Hurdal, Michael Miller, and Kelly Botteron).

cerebral cortex, the thin layer of gray matter on the brain surface. The image on the left of Figure 23.11 represents a human left cerebral hemisphere colored to reflect anatomical features termed "lobes." On the right is a smaller patch, known as the left ventral medial prefrontal cortical region, which is color coded to reflect the mean curvature of the surface as it sits in 3-space. (My thanks to Monica Hurdal (Florida State U.) for processing and analyzing the data and generating the images. Thanks to Dr. David Rottenberg and Kelly Rehm (U. Minnesota) for the MRI cortical data and to Michael Miller (Johns Hopkins U.) and Kelly Botteron (Washington U. School of Medicine) for providing the MRI prefrontal cortex data.)

Neuroscientists have become interested in applying surface-based techniques to the cortex because of its intrinsic 2D sheet topology. They wish to map regions of interest to flat domains – hence the topic of "brain flattening." Triangulations of the cortex can be obtained from the volumetric data acquired in brain images, but as you can see, the cortex is an extremely convoluted surface; it is estimated that 60%–70% of the surface is buried in the deep folds and fissures. In particular, the cortex has extreme variations in gaussian curvature, and as we have learned over two millennia in trying to flat-map the earth, it is impossible to bring familiar metric features such as surface areas and lengths through the flattening process. However, Bernhard Riemann proved 150 years ago that one can preserve *conformal* features – i.e., there do exist *conformal* flat maps. Developments since then have shown repeatedly how rich is the structure that comes along with conformality. With circle packing we finally have a chance to test these maps in practice.

A sampling of flat maps is shown on the next page. Figure 23.12(a) is the cerebral hemisphere from Figure 23.11 as it is mapped to a round sphere by one of our conformal flat maps. In our terminology, this is the maximal packing for the 3D triangulation ($\sim$145,000 vertices), with lobe colors brought along. (Curiously enough, maps to $\mathbb{P}$ count as "flat" in this context.) When the corpus callosum and ventricle (nongray matter) are removed

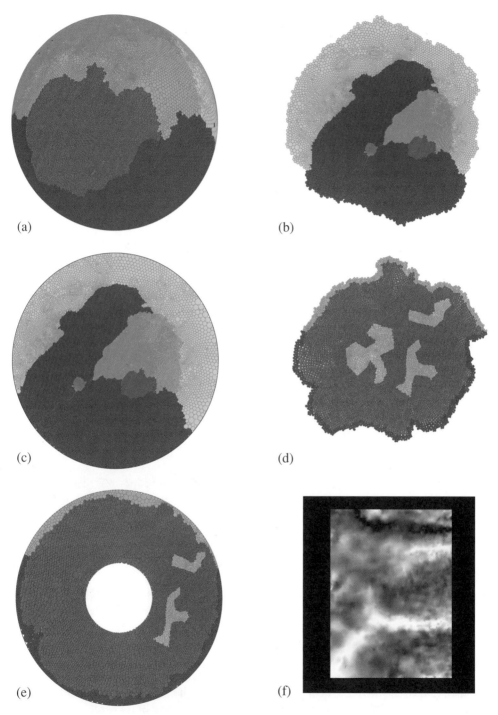

Figure 23.12. A sampler of cortical flat maps.

from the triangulation, the remainder, the cortex itself, is a topological disc; Figures 23.12(b) and (c) display the corresponding discrete conformal flat maps in euclidean and hyperbolic geometry, respectively. For the former the 3D boundary edge lengths provide the boundary label, while the latter is the maximal packing. My neuroscience colleagues have a surprising affinity for the hyperbolic maps – perhaps they are reminiscent of the typical view in a microscope, where one can bring the area of interest into the center for closer inspection (in this case using automorphisms of the disc).

Figure 23.12(d) isolates a region of the cerebral cortex known as the *occipital* lobe in a euclidean flat mapping. (This lobe contains the main areas for visual processing in the brain.) There are three islands in yellow; these are fabricated, but represent the type of information one might gather in studies of brain activity using, e.g., fMRI. When the largest island is removed, the result can be mapped as in (e) to a round annulus, giving an estimate of the conformal modulus of this cortical ring domain. In a similar demonstration of flat mapping manipulations, the prefrontal cortex shown in Figure 23.11 has been flat-mapped to the rectangle in (f) by identifying four anatomical landmarks on its boundary and prescribing these as the corners. The carrier is shown with colors brought along from the 3D surface to help in visual identification (giving the flat map an eerie 3D look). Here the map gives an estimate of the conformal modulus of the quadrilateral region on the cortex itself.

23.5. Summary

It is not our goal here to discuss the potential scientific uses for these cortical flat maps. Nor in the earlier applications to *dessins*, tilings, or conformal welding do I claim that circle packing has some predefined role it should play. It is best to regard it as perhaps a new tool in one's toolbox. The examples of this chapter do, however, suggest some points to keep in mind.

- First, it is not necessary to believe that conformality per se has any relevance to an application to exploit its amazing richness – rigidity, related notions such as extremal length, the associated function theory.

- Second, approximation of true conformality may be superfluous if its companion structures appear faithfully at coarse stages, as seems often to be the case with our circle packing examples.

- Finally, the structures themselves take precedence over technique. Circle packing and its experiments can contribute to our understanding of a topic even if other methods, say using PDEs, ultimately prevail in practice. (It would, nonetheless, be nice to hear at the end of a neurological consult "you know, we need to hire another conformal geometer.")

People are drawn to mathematics for a variety of reasons – the beauty of elementary geometry, the richly layered theory and deep questions, the joys of teaching, and not least, the satisfaction of applying the results in the real world. Many of these attractions can be found in circle packing, along with some new pleasures for those with an experimental spirit and an eye for beautiful pictures. Perhaps you can find uses for circle packing in your own favorite topic.

Practicum IV

The mysteries of circle packing seem only to deepen with industrial strength challenges. It becomes clear when packings reach 50,000+ circles, when the algorithm runs for 2 hours and yet the layout remains jumbled and broken, that the practical side of circle packing needs serious attention. It is a testament to the synergy between *practice* and *theory* that the geometry itself keeps center stage.

"Follow the curvature" was the dictum in Practicum III – packing is about the movement of curvature between circles. With large packings, that model becomes untenable – more regional, even global concepts are needed. Moreover, with large packings you quickly find that layout is no longer an afterthought. So, what to do? Your first move is, naturally, to "tune" your latest code. When that is not enough, you develop some methods for watching the repacking process. You know that curvature is the key, and when you have watched enough examples, you start catching some hints of *trends* in curvature movement. Description of trends and some ideas for exploiting them are the topics for this final Practicum. The details will be scant, but some nice mathematical issues bubble to the surface.

Let us look at label adjustments first. In Practicum III we seemed to have the basic algorithm in pretty good shape thanks to the "uniform neighbor" model. For larger packings, however, formerly minor issues such as storage, communication, and roundoff can cripple the code. Numerical analysts have a few tricks up their sleeves and Chuck Collins pulled out a *superstep* method that improved our performance by an order of magnitude. The basic idea is very simple: as you carry out the normal label adjustments, store the label changes in vector form; when two or three successive vectors are roughly parallel, change the labels with a "super" step in that general direction. This is not as cavalier as it sounds; there are standard methods for choosing the timing and size of supersteps.

However, superstepping is a *generic* trick with no notion of what the computations represent. It has been a nice surprise, then, that it not only works, but also brings out unsuspected trends in the computations. A typical large repacking, for example, will experience a period of slow improvement in the normal packing iterations, which seem to set up for a very effective superstep. Successive supersteps are less impressive, however, and the algorithm soon returns to normal iterations and slow improvement. After a period of retrenchment, the circles are ready for another big superstep. This cycle repeats. It is as though the circles "learn" their proper roles through a few practice cycles, then, like a circle packing Bose–Einstein condensate, move with one will in the optimal direction; this changes their roles, however, and the condensate breaks up. Here are some possible ways to exploit the group dynamics. (It may be good to revisit my observations in Practicum III. Think of nonzero curvature as "error." Repacking is the attempt to banish error – to get it to cancel or to push it to the boundary.)

Packing Strategies

(a) The packing algorithm is (almost) "embarrassingly parallel" since each circle's adjustments depend only on its immediate neighbors. One promising approach is suggested in Figure IV.1: decompose the packing into several regions and dedicate a computer processor to each. A master processor negotiates transfer of information among neighboring regions. Charles Collins and I, with students Joel Mejeur and George Butler, have implemented such a strategy in MPI, *Message Passing Interface*, which permits use of multiple machines on a generic computer network. This code is promising for improving speed, but perhaps even more for insight. The individual regions pack in the usual way, but then there is a flow of error *between* neighboring regions as they reconcile common boundaries. Studying this has been intriguing; more than a mere *multigrid* approach, it is like trying to capture the geometry under *coarsening* of a complex – a reversal of our familiar *refinement* process.

(b) An experienced human circle packer can make very reliable predictions about the outcomes of certain circle packing runs – for example, in estimating the effects of changes in some of a packing's boundary radii. Is there a *neural network* or *genetic algorithm* approach to circle packing? The notion that the circles – either individually or in regions – "learn" their roles might be true in some literal sense.

(c) Radius adjustments are temporary during repacking; a given change might well turn out, in hindsight, to have been in the wrong direction. Why not dampen adjustments with a *relaxation* parameter $t \in [0, 1]$; in other words, if change Δr in a label will result in zero curvature, make the relaxed change $t\Delta r$ instead? Intuition suggests that by moderating the fluctuations, reliable trends in the radii adjustments may appear more quickly.

(d) In practice, error is never totally banished; a "packing label" is simply declared when the total error is small and fairly evenly *distributed*. *Simulated annealing* might be useful in the final stages of repacking. The idea would be to make small, random perturbations in all the radii; the local error introduced would, on average, quickly cancel as repacking continued, but the jiggling might shake the process out of some rut that is slowing convergence.

Some of these ideas are already paying dividends. The most striking early results can be illustrated with the ten-triangle-toy surface S which we studied in Sect. 22.2. Recall that this equilateral surface has 10 cone points and that we used circle packings to flatten it to a plane rectangle $\mathfrak{R}$. Each cone point has an initial cone angle in S and a final cone angle in $\mathfrak{R}$. (For instance, at the interior point p, the initial cone angle is π, while the target cone angle is 2π.)

Motivated by the observation (c) above, Collins and I ran experiments in which we chose to adjust the initial label R_0 associated with S (in which all radii are identical) to the final label R_1 for $\mathfrak{R}$ (in which radii differ by four orders of magnitude) in a series of 50 incremental stages $\{R_t : t = 0.0, 0.02, 0.04, \ldots, 0.98, 1.0\}$. During each stage, the cone angles were moved another 2% of the way toward their final values.

Our hope was to somehow visualize the evolving radii. We decided to monitor the relative changes in radii given by the function

$$f_t(v) = (R_{t+0.02}(v) - R_t(v))/R_t(v), v \in R. \tag{123}$$

We plotted this using the locations of the vertices v in the final flat packing – that is, using the centers of the circles in the lower panel of Figure 22.3. f_t itself is difficult to display, so Collins had *Matlab*

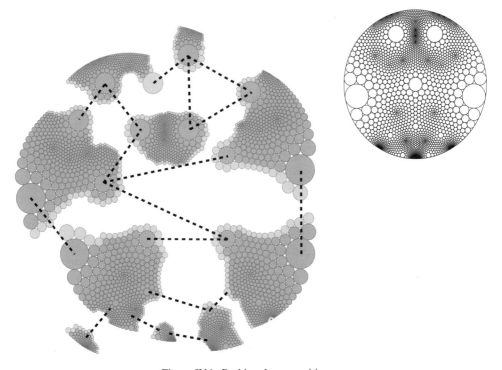

Figure IV.1. Packing decomposition.

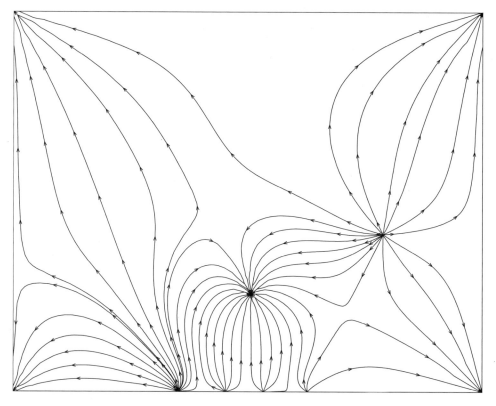

Figure IV.2. Curvature flow.

drop simulated fluid on its graph and record the pattern of flow, the flow lines for the gradient field ∇f_t. We were pleasantly surprised by the first images that we printed out, one of which is shown in Figure B. Since the packing process involves movement of "curvature," we call this a graph of *curvature flow*.

You might well ask "Which value of $t \in [0, 1]$ is represented in this image?" This was the shock as our 50 successive pictures were produced: *the gradient fields of f_t are essentially indistinguishable for $t \in [0, 1]$.* We uncovered a curvature flow in the computations that was invariant over the 50 values of t! This was totally unexpected. It suggests that the circles learned how to cooperate from the very beginning of the process and never deviated until they had packed $\mathfrak{R}$. Collins, Tobin Driscoll, and the author have since uncovered the analogous classical flow in novel forms of the Schwarz–Christoffel method and are continuing to study this quite unexpected phenomenon.

Let me move now to packing layout. Meta-code for a basic procedure was given in Practicum II. A contrived example there ((b) of Figure II.3) showed that a single erroneous radius can play havoc with a layout. In large packings, problems like this are inevitable; they might turn a presumed "packing" into a meaningless jumble of circles, or appear as localized blemishes when you focus in on a subregion. Our meta-code was too simplistic. There is a tradeoff possible between sophistication and accuracy – with care you can get *better* layouts with *less* accurate labels. The key is "even distribution" of the inevitable packing errors, and this requires that the layout process become dynamic, depending not only on combinatorics, but also on the computed labels and on developments during layout. Here are some ideas.

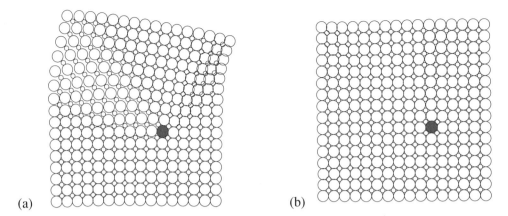

Figure IV.3. Layout repairs.

Layout Strategies

(1) In the placement of each circle c_v, average the centers computed using all available neighboring pairs rather than just one (as in the meta-code).

(2) Do not use unreliable circles in subsequent placement decisions. For example, exclude circles of small degree, say, less than 5, because their radii tend to be smaller and hence more prone to layout errors.

(3) Monitor placement *during* layout so that only well-placed circles are used for subsequent layout decisions. This requires developing a supple strategy which can adjust during the computation.

(4) Localized *blemishes* often occur in regions with extremely small circles due to roundoff errors. One can excise a small patch of the packing containing the blemish, repack and lay out this patch separately, preserving its boundary radii, and then blend the result back into the full packing.

(5) After a packing is laid out, some minor flaws can be corrected by "mollification"; that is, by passing several times through the packing, each time adjusting the center of each circle to be the average of its locations using *all* contiguous pairs of petal circles.

Let us see how two of these ideas might affect the broken packing we saw in (b) of Figure II.3. Applying (1) gives the layout Figure IV.3(a). Though largely intact, it still suffers from the erroneous radius of the shaded circle. By applying (2) after explicitly designating the shaded circle as "unreliable," the layout is fixed.

I hope the four "Practicum" sections of the book have given you some flavor of the practical side of circle packing – the problems and the opportunities. Much remains to be done.

Notes IV

Chap. 19: The seminal work by Bernhard Riemann dates from 1851 (Riemann, 1851). The Conjecture in Thurston's talk (1985) was never published. The proof by Burt Rodin and Dennis Sullivan is in (Rodin and Sullivan, 1987). Uniqueness of the hexagonal packing, now very accessible (see Chap.8), originally relied on deep results, such as Mostow rigidity (Sullivan, 1981). Considerable early study went into the constants s_n of the Hexagonal Packing Lemma (Lemma 19.4), which Rodin and Sullivan conjectured to be $o(n)$. This was proven by Z-X. He (1991) and, using different methods, by Aharonov (1990, 1994). Other work includes Doyle, He, and Rodin (1994a, 1994b) and Rodin (1987,1989), culminating in He and Schramm (1998) which proves that convergence in the Rodin–Sullivan Theorem is C^∞. Among the advantages of hexagonal combinatorics is that euclidean packing labels and their reciprocals are subharmonic, see Bárány, Füredi, and Pach (1984).

Chap. 20: The probabilistic proof of Theorem 20.1 by Stephenson (1990,1996) used tailored random walks in the hyperbolic plane. The result was extended by He and Rodin (1993). The more general He–Schramm Theorem is from (1996). Don Marshall's conformal mapping software is at (Marshall).

Chap. 21: Approximations with Blaschke products and polynomials were established by Dubejko (1995, 1997c) in the hexagonal case. See also Colin de Verdière and Mathéus (1994a). Most convergence results involve uniform convergence on compacta, but Dubejko's work in (1997a) gives convergence up to the boundary.

Chap. 22: For conformal structures on polyhedral surfaces, see, for example, Beardon (1994). The methods described here and with discrete *dessins* are developed by Bowers and Stephenson (2004).

Chap. 23: Our subdivision tilings are associated with work of Jim Cannon, Bill Floyd, and Walter Parry (Cannon, 1991, 1994; Cannon et al., 2001); "conformal tiling" was defined by Bowers and Stephenson (1997); see Cannon et al. (2003) for details of our tiling "story." Bill Floyd provided pivotal data and software for creating the tiling images used here. For classical welding see Lehto (1987); the discrete theory is due to G. Brock Williams, (2004), who provided the images of Fig. 23.10. Regarding brain mapping applications see Hurdal et al. (1999) and Hurdal and Stephenson (2004).

Practicum IV: For curvature flow in the ten-triangle-toy, see Collins, Driscoll, and Stephenson (2003).

Appendix A

Primer on Classical Complex Analysis

I will assume that the reader has a basic familiarity with complex numbers and complex functions. I provide here a broad, but very selective, overview of analytic and quasiconformal function theory which, I hope, brings out the geometric features most relevant to the book. For background, I especially recommend the complex analysis books of Ahlfors (1978), Beardon (1979), and Needham (1997) for their more geometric outlooks, Ahlfors (1973) for some spice, and Lehto and Virtanen (1973) and Lehto (1987) for quasiconformal mapping.

A.1. Analytic Mappings

The fundamental objects of study are complex functions $F : \Omega \longrightarrow \mathbb{C}$ on plane domains Ω. I will routinely use the term *mapping* in place of *function* to keep us in the proper frame of mind, and I recommend the "rubber sheet" metaphor from topology, which posits F as a physical warping of domain Ω onto range $F(\Omega)$.

A mapping $F : \Omega \longrightarrow \mathbb{C}$ is said to be *analytic* at $z_0 \in \Omega$ if it has a complex derivative defined by

$$F'(z_0) = \lim_{z \to z_0} \frac{F(z) - F(z_0)}{z - z_0}; \tag{124}$$

F is analytic *on* Ω if is analytic at each point of Ω. The usual calculus shows that familiar functions such as (complex) polynomials, rational functions, the exponential function, and trigonometric functions are analytic, and that sums, products, quotients, and compositions of analytic functions are again analytic (where defined).

Geometry is part of the bedrock of the topic, entering with the complex numbers through their identification with $\mathbb{R}^2$ and the resulting geometry of complex addition and multiplication. The latter, in particular, makes existence of the complex derivative $F'(z)$ a much stronger statement than one might at first suspect. For z near z_0, write $\Delta z = z - z_0$ for the change in z. By (124),

$$\Delta F = F(z_0 + \Delta z) - F(z_0) = F'(z_0)\Delta z + 0(|\Delta z|).$$

In other words, a small change in z causes a change in F which is approximately Δz rotated by $\arg(F'(z_0))$ and scaled by $|F'(z_0)|$. As this action is independent of the direction of Δz, a circle c of small radius r centered at z_0 is carried to a curve $F(c)$ that is roughly a circle of radius $|F'(z_0)|r$ centered at $F(z_0)$; moreover, F preserves orientation, since as z moves counterclockwise about c, $F(z)$ clearly moves counterclockwise around $F(c)$. This bit of local geometry is neatly summed up:

Adage: An analytic function is one that maps infinitesimal circles to infinitesimal circles.

The local geometric action has a second important consequence: when $F'(z_0)$ is nonzero the mapping preserves the angle (magnitude and orientation) between any two curves meeting at z_0. Thus an analytic function is said to be a *conformal mapping* at points where its derivative does not vanish. Regions Ω, $\widetilde{\Omega}$

309

are *conformally equivalent* if there is a univalent (i.e., one-to-one) analytic bijection $F : \Omega \longrightarrow \widetilde{\Omega}$. In this case, $F^{-1} : \widetilde{\Omega} \longrightarrow \Omega$ is also univalent and analytic.

Line Integrals: Leaving local considerations, we come to geometric behavior entering *via* line integrals. Let us start with the *index* of a point relative to a closed curve: if γ is a (piecewise smooth) closed curve in the plane then

$$\mathfrak{w}(\gamma; z_0) = \frac{1}{2\pi i} \int_{\gamma} \frac{1}{(z - z_0)} \, dz, \qquad z_0 \notin \gamma, \tag{125}$$

is integer-valued and counts the number of times that γ winds about z_0, also called the *winding number*. (By convention, positive orientation is always *counterclockwise*.)

Suppose now that Ω is simply connected. If F is analytic in Ω and $\gamma \subset \Omega$ is a closed curve, then the famous Cauchy Theorem tells us that $\int_{\gamma} F(z) \, dz = 0$. In conjunction with indices this yields a version of Cauchy's Integral Formula:

$$\mathfrak{w}(\gamma; z_0) \cdot F(z_0) = \frac{1}{2\pi i} \int_{\gamma} \frac{F(z)}{z - z_0} \, dz, \quad z_0 \in \Omega \backslash \gamma. \tag{126}$$

Assume for convenience that γ is a Jordan curve (simple, closed). The Cauchy Integral Formula says that the values of F *inside* γ are determined by its values *on* γ. This is an extremely strong *rigidity* statement – these are not garden variety maps from $\mathbb{R}^2$ to $\mathbb{R}^2$! Couple this machinery with change of variables and one can count the number of points $z \in \Omega$ with $F(z) = w$ (assuming $w \notin F(\gamma)$):

$$\mathrm{Card}(F^{-1}(\omega)) = \mathfrak{w}(F(\gamma); w) = \left(\frac{1}{2\pi i}\right) \int_{\gamma} \frac{F'(z)}{F(z) - w} \, dz. \tag{127}$$

Of course, this is just the number of times that $F(\gamma)$ winds around w. This result is also known as the *argument principle*: the integrand is the derivative of $\log(F(z) - w)$, so the integral computes the change in argument of $F(z) - w$ as one transits γ.

Many of our premiere tools – computational, geometric, and topological – arrive via line integrals. It is a short step from (127) to the fact that nonconstant analytic functions F are *open mappings* and hence satisfy the *maximum principle*: $|F|$ cannot have a local maximum in the interior of Ω. There is Hurwitz Theorem: if functions F_n are analytic in a neighborhood of a Jordan domain Ω (bounded by a Jordan curve) and converge uniformly to an analytic function F which is univalent in Ω, then for sufficiently large n, the F_n themselves are univalent in Ω. Decompositions and homotopies are tools which come naturally with line integrals. Cauchy's Theorem implies, for example, that if γ and σ are homotopic within a region in which F is analytic, then $\int_{\gamma} F \, dz = \int_{\sigma} F \, dz$. Also, if a curve γ is decomposed as a sum of curves, $\gamma = \gamma_1 + \cdots + \gamma_n$, then $\int_{\gamma} F \, dz = \sum_{j=1}^{n} \int_{\gamma_j} F \, dz$. Putting such decompositions and algebraic cancellations together often allows one to leverage local results into global results. It is important to note that many of these results apply to more general functions that share certain key topological properties of analytic functions, for instance, so-called *light-interior* mappings. We will make good use of this machinery in the book.

Power Series: A little calculus applied to the kernel in (126), $1/(z - z_0)$, and its geometric series proves that analytic functions are always represented locally by power series. In particular, there exists $\rho > 0$ such that if $|z - z_0| < \rho$, then

$$F(z) = \sum_{n=0}^{\infty} a_n (z - z_0)^n.$$

Among the consequences, the derivatives $F^{(n)}(z)$ exist for all n and are again analytic, so the coefficients a_n are given by $a_n = F^{(n)}(z_0)/n!$. Also, if F is not identically zero, then its zeros must be isolated,

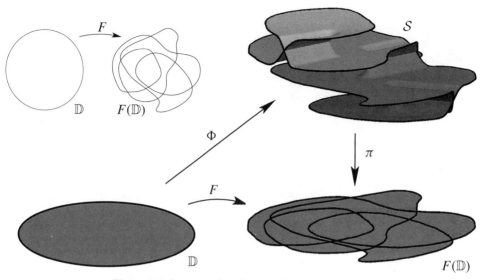

Figure A.1. Image surface for a 4-valent function on $\mathbb{D}$.

leading to another surprising rigidity feature: *If two analytic functions on a connected open set Ω agree on a sequence of points having a limit point in Ω, then they are identical throughout Ω.*

Image Surfaces: The geometric significance of the power series representation lies with its local mapping implications: $F(z) - F(z_0)$ behaves near z_0 like $a_k(z - z_0)^k$, where k is the smallest positive index for which $a_k \neq 0$. The restriction of F to $D(z_0, \delta) = \{|z - z_0| < \delta\}$ is roughly k-to-1 onto its image for small δ. When $F'(z_0) = 0$, then $k > 1$ and F is said to have a *branch point* of order $k - 1$ at z_0. Branch points must be isolated (since F' is analytic), so at a generic point F' will be nonzero and F will be locally univalent.

The ease with which we can visualize the local behaviors of F encourages me to introduce an important pictorial device, the so-called *image surface*. Figure A.1 illustrates with a 4-valent analytic function $F : \mathbb{D} \longrightarrow \mathbb{C}$. In the upper left is a typical flat sketch of F, which tells one very little about the mapping's geometry. The alternative view depicts the standard ingredients of the image surface approach: the domain disc, the multisheeted image surface $\mathcal{S}$, the univalent analytic function Φ from $\mathbb{D}$ onto $\mathcal{S}$, and the natural "projection" π from $\mathcal{S}$ to the planar image. The map F is the composition $F \equiv \pi \circ \Phi$.

Admittedly, the image surface for a generic analytic function would be horribly more complicated; nonetheless, it has essentially the same local features. In the surface $\mathcal{S}$ of the figure, the four sheets are attached *via* cross-connections over the branch points. These cross-connects cannot actually be embedded in 3-space, so I use the classical ruse of drawing the surface as though it passed through itself without really intersecting.

I am a fan of the "image surface" philosophy, irrespective of whether you can actually picture a given surface or not. It is a matter of *replacing a complicated mapping F on a simple domain $\mathbb{D}$ by a simple mapping π on a complicated domain $\mathcal{S}$.* As you read this book you may come to be a fan of image surfaces yourself. If you would like to test drive the concept, try applying the argument principle (127) on the image surface of Figure A.1 to confirm that its valence is 4.

I should say that image surfaces can be rigorously defined as portions of the Riemann surfaces associated with *germs* of inverse functions F^{-1}. The exponential function, for example, is commonly understood as a map of $\mathbb{C}$ onto the universal covering surface of $\mathbb{C}\backslash\{0\}$; that universal cover (the image surface) is, in turn, the domain for $\log(\cdot) \equiv \exp^{-1}(\cdot)$.

Differentials: The differential approach to analytic functions also has its grounding in geometry. Writing $z = x + iy$ and $w = u + iv$, a complex function $w = F(z)$ may be treated as a map from

$\mathbb{R}^2$ to $\mathbb{R}^2$,

$$w = F(z) = F(x, y) = u(x, y) + i\,v(x, y),$$

where u and v are the *real* and *imaginary parts* of F. Assuming u and v have continuous partials, the associated differentials are

$$du = u_x\,dx + u_y\,dy, \quad dv = v_x\,dx + v_y\,dy,$$
$$dw = F_x\,dx + F_y\,dy = (u_x + iv_x)dx + (u_y + iv_y)dy.$$

It is convenient to introduce complex 1-forms, $dz = dx + i\,dy$, $d\bar{z} = dx - i\,dy$, so

$$dx = \frac{dz + d\bar{z}}{2}, \qquad dy = \frac{dz - d\bar{z}}{2i}.$$

Now we may write

$$\begin{cases} F_z &= \frac{1}{2}(F_x - i\,F_y) \\ F_{\bar{z}} &= \frac{1}{2}(F_x + i\,F_y) \end{cases} \implies dw = F_z\,dz + F_{\bar{z}}\,d\bar{z}. \tag{128}$$

The existence of the complex derivative $F'(z_0)$ in (124) implies the famous *Cauchy–Riemann* (C–R) equations: $u_x = v_y$, $v_x = -u_y$. Substituting in (128) gives $F_{\bar{z}} \equiv 0$. So among differentiable complex functions, the analytic ones are those which do not depend on $\bar{z}$, and $F_z \equiv \frac{dw}{dz} \equiv F'$. The geometry of the C–R equations has to do with the level sets of $u = \Re F$ and $v = \Im F$, since they imply that these are orthogonal at points z_0 where $F'(z_0)$ is nonzero. This orthogonality is another expression of the conformality of F.

Taking further derivatives in the C–R equations implies that u and v satisfy the *Laplace Equation*,

$$\Delta u \equiv 0 \text{ and } \Delta v \equiv 0, \quad \Delta = \frac{\partial^2}{\partial x^2} + \frac{\partial^2}{\partial y^2}.$$

Solutions are termed *harmonic* and have a purely geometric characterization: *A continuous function h (real or complex) on Ω is harmonic if and only if the average of h on any disc in Ω is its value at the center.*

Function Theory: The transition from the local features of analyticity to the global properties of analytic functions is, in my view, the most fascinating aspect of function theory. The study is organized largely around the standard domains, the sphere $\mathbb{P}$, the plane $\mathbb{C}$, and the disc $\mathbb{D}$, and in each case the "local-to-global" transition is intimately bound up with geometric considerations.

In the disc, the most directly geometric functions are the univalent ones, and the most important result is the Riemann Mapping Theorem of 1851 (proof completed by Osgood): *If Ω is an open, simply connected, proper subset of the plane, then there exists a one-to-one analytic function F, unique up to automorphisms of $\mathbb{D}$, mapping $\mathbb{D}$ onto Ω.* That is, the geometry of every simply connected Ω is rigidly encoded in an analytic function. A key preliminary is the Schwarz–Pick Lemma which can be restated as the *hyperbolic contraction principle*: *Analytic self-maps of $\mathbb{D}$ are contractions in the hyperbolic metric.* Geometric notions play key roles in the broader theory through area, harmonic measure, extremal length, distortion theorems, line integrals, and so forth.

In the plane, geometry shows up most directly in various *value distribution* results. Most famous are the Liouville and Picard theorems: *A nonconstant analytic function on $\mathbb{C}$ cannot be bounded; in fact, it must assume every complex value with at most one exception.* Much of the study then turns on valence, rates of growth, asymptotic behavior, and so forth; to some extend these may be viewed as studies of the geometry of the image surfaces. Among the elementary observations are the facts that an analytic function $F : \mathbb{C} \longrightarrow \mathbb{C}$ is finite valence $\iff$ F is proper $\iff$ F is a polynomial.

Finally, the sphere is the most rigid of the standard domains. Analytic functions there are necessarily rational, so in addition to geometry, they also bring *algebraic* properties, with results which can be quite stunning.

The local-to-global motif becomes bedrock in the theory of Riemann surfaces, where a global object is defined entirely by compatibility conditions within an *atlas* of local pieces. Geometry enters in a surprising and beautiful way through covering groups. The conformal automorphisms, Aut($\mathbb{G}$), of $\mathbb{G}$ ($\mathbb{P}$, $\mathbb{C}$, or $\mathbb{D}$) consist of Möbius transformations (see Section 3.2); by the Uniformization Theorem, every Riemann surface S is conformally equivalent to one of these spaces $\mathbb{G}$ modulo a group $\Lambda \subset$ Aut($\mathbb{G}$). The geometry of S is thus reflected both in the algebraic structure of Λ and in the geometry of its individual automorphisms. This is particularly propitious for our purposes, since Möbius transformations are precisely the (orientation preserving) homeomorphisms of $\mathbb{P}$, $\mathbb{C}$, or $\mathbb{D}$ which map circles to circles.

A.2. Quasiconformal Mappings

The geometric rigidity at the heart of analytic function theory has its down side in practice: it is notoriously difficult to construct analytic functions from geometric descriptions. Fortunately, analysts have loosened the straightjacket with controlled loss of rigidity – *quasiconformal* distortion. The local definition is very natural, making the wealth of global consequences quite a pleasant surprise. In circle packing, we not only exploit quasiconformal machinery as a bridge to analyticity, but also see in it a continuous model for local-to-global transitions that circle packings themselves manifest.

A.2.1. Local Behavior

Let F be a continuous mapping from a domain $\Omega \subset \mathbb{C}$ into the plane and fix $z_0 \in \Omega$, $w_0 = F(z_0)$. For $\epsilon > 0$, define L_ϵ and l_ϵ by

$$L_\epsilon = \max\{|F(z) - w_0| : |z - z_0| = \epsilon\}, \qquad l_\epsilon = \min\{|F(z) - w_0| : |z - z_0| = \epsilon\}.$$

A typical situation is illustrated in Figure A.2. The ratio of L_ϵ and l_ϵ measures the distortion the circle $\{|z - z_0| = \epsilon\}$ suffers under F. To pin it down precisely at z_0, define the *dilatation* by

$$D_F = D_F(z_0) = \limsup_{\epsilon \to 0} \frac{L_\epsilon}{l_\epsilon}.$$

If F is differentiable at z, say $w = F(z) = F(x, y) = u(x, y) + iv(x, y)$, then the Jacobian (determinant) of the map is

$$J = u_x v_y - u_y v_x = |F_z|^2 - |F_{\bar{z}}|^2,$$

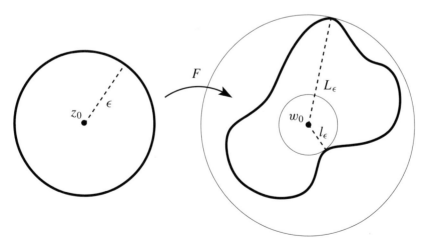

Figure A.2. Quasiconformal dilatation.

using our previously defined differential notation. We are interested only in the sense-preserving mappings, so J must be positive, meaning that

$$|F_{\bar{z}}| < |F_z| \implies (|F_z| - |F_{\bar{z}}|)|dz| \leq |dw| \leq (|F_z| + |F_{\bar{z}}|)|dz|. \tag{129}$$

This says that to first order F maps infinitesimal circles at z to infinitesimal ellipses at $w = F(z)$, and implies that

$$D_F = \frac{|F_z| + |F_{\bar{z}}|}{|F_z| - |F_{\bar{z}}|} \geq 1. \tag{130}$$

These infinitesimal ellipses are infinitesimal *circles* $\iff D_F = 1 \iff u$ and v satisfy the Cauchy–Riemann equations at $z \iff |F_{\bar{z}}| = 0 \iff dw = F_z(z)\,dz = F'(z)\,dz$. In other words, conformality at a point is a limiting case of quasiconformality – it represents the absence of quasiconformal distortion.

A.2.2. Global Behavior

Maps having bounded dilatation are said, informally, to have *bounded distortion*. Of course these maps also distort angles, hence the name "quasiconformal."

Definition A.1. An orientation-preserving homeomorphism between plane domains, $F : \Omega \longrightarrow \Omega'$, is said to be **quasiconformal** if its dilatation D_F is bounded above on Ω. In particular, F is κ-**quasiconformal** for $\kappa \in [1, \infty)$ if $D_F(z) \leq \kappa, z \in \Omega$.

The univalence condition here is actually no impediment; one can define related multivalent mappings, termed *quasiregular*, but they in fact are precisely the functions of the form $G \circ F$ where G is analytic and F is quasiconformal.

The weakening of conformal geometry represented by the prefix "quasi" may seem *ad hoc* and only local in character, but in fact quasiconformal mappings display surprisingly rich global properties. As a teaser, here is the Liouville result:

Theorem A.2. *There exists no quasiconformal mapping from $\mathbb{C}$ to the unit disc.*

One approach to the global properties is *via* derivatives: F is quasiconformal on Ω if and only if it solves a *Beltrami differential equation*

$$F_{\bar{z}} = \mu F_z. \tag{131}$$

The *Beltrami coefficient* $\mu = \mu(z)$ represents a "complex" dilatation which we will not be using in the book. Nonetheless, we will use numerous properties which come out of this representation. For example, one can prove that the inverse function $F^{-1} : \Omega' \longrightarrow \Omega$ for a κ-quasiconformal mapping is also κ-quasiconformal and that the composition of a κ_1-quasiconformal map with a κ_2-quasiconformal map is $\kappa_1 \kappa_2$-quasiconformal. Analyticity fits in naturally.

Proposition A.3. *A one-to-one conformal mapping on a domain $\Omega \subset \mathbb{C}$ is necessarily 1-quasiconformal and a 1-quasiconformal mapping is necessarily analytic. Therefore, the following three classes of one-to-one, orientation-preserving, complex-valued mappings on Ω are identical:*

$$conformal \iff \textbf{\textit{1-quasiconformal}} \iff analytic.$$

The mappings in this book associated with circle packings are largely piecewise affine, hence piecewise quasiconformal; we use the following result for stitching the pieces together.

Proposition A.4. *Let $F : \Omega \longrightarrow \Omega'$ be an orientation-preserving homeomorphism. Suppose $E \subset \Omega$ is a countable union of analytic arcs and suppose that the restriction of F to $\Omega \backslash E$ is κ-quasiconformal. Then F is κ-quasiconformal on all of Ω.*

One can get some geometric sense of global distortion in a κ-quasiconformal mapping $F : \Omega \longrightarrow \Omega'$ by studying certain standard domains in the theory, namely, rectangles and *ring domains* (annuli). For a euclidean rectangle $\mathfrak{R}$ with two designated "ends," the *modulus* is $\mathrm{Mod}(\mathfrak{R}) = $ height/length (the heights of the two ends divided by their distance apart). If A is a round annulus, meaning $A = \{r < |z| < R\}$, some $0 \leq r < R \leq \infty$, then its *modulus* is $\mathrm{Mod}(A) = 2\pi / \log(R/r)$.

Suppose $F : \Omega \longrightarrow \Omega'$ is κ-quasiconformal. If Ω and Ω' are both rectangles and F maps the designated ends of one to those of the other, or if Ω and Ω' are both round annuli, then modulus is a "quasi-invariant":

$$\frac{1}{\kappa} \mathrm{Mod}(\Omega) \leq \mathrm{Mod}(\Omega') \leq \kappa \, \mathrm{Mod}(\Omega). \tag{132}$$

The same considerations apply to less regular domains. A *conformal rectangle* is a conformal disc $\mathcal{R}$ in which disjoint boundary arcs α, β have been designated as "ends"; we write $\langle \mathcal{R}; \alpha, \beta \rangle$. If we require that ends be mapped to ends, then $\mathcal{R}$ is conformally equivalent to an essentially unique euclidean rectangle $\mathfrak{R}$. One can define $\mathrm{Mod}(\mathcal{R}) = \mathrm{Mod}(\mathfrak{R})$, and by (132) (with $\kappa = 1$) this is unambiguous. Likewise, every ring domain $\mathcal{A} \subset \mathbb{C}$ is conformally equivalent to some round annulus A and one defines $\mathrm{Mod}(\mathcal{A}) = \mathrm{Mod}(A)$, again without ambiguity. If F is κ-quasiconformal between two conformal rectangles (or two ring domains), then since pre- or post-composition by conformal maps will not change its dilatation, the inequalities of (132) remain valid.

A.2.3. Convergence

The terminology and computations all suggest that quasiconformal maps are somehow "more conformal" as their dilatations approach 1. This is true and is the key to our use of the quasiconformal theory with sequences of mappings.

Proposition A.5. *Let $\{F_n\}$ be a sequence of κ-quasiconformal mappings defined in Ω and assume that the sequence converges to a limit function F. Then F is either a constant (possibly infinity), a mapping of Ω to two points, or a κ-quasiconformal mapping of Ω.*

For an example of the middle outcome, let $F_n : z \mapsto nz$: the limit function fixes 0 and maps everything else to ∞. This behavior cannot occur when the functions' images are uniformly bounded, as is generally the case in our applications: this leaves just the two alternatives, constant or κ-quasiconformal.

Crucial for finding convergent sequences are the rich normal-family results in the theory. Recall that a family $\mathcal{F}$ of complex-valued functions on Ω is a *normal family* if every sequence $\{F_j\} \subset \mathcal{F}$ has a subsequence $\{F_{j_n}\}$ that converges uniformly on compact subsets of Ω to a function F (not necessarily in $\mathcal{F}$) or to a constant (possibly diverging uniformly on compacta to infinity).

Proposition A.6. *Let $\Omega \subseteq \mathbb{C}$ and $\kappa \geq 1$ be given and let w_1, w_2 be two distinct points in $\mathbb{C}$. Let $\mathcal{F}$ be the family of κ-quasiconformal mappings $F : \Omega \longrightarrow \mathbb{C}$ that omit the values w_1, w_2. Then $\mathcal{F}$ is a normal family.*

Typically a limit F of functions in a normal family need not be back in the family. However, we are fortunate here – if F is nonconstant, we know it is κ-quasiconformal and we can even say something about its range through a result of Carathéodory.

Theorem A.7 (Carathéodory Kernel Theorem). *Assume Ω is a plane region with at least two boundary points and suppose z_0 is a point of Ω. Let $\{F_j\}$ be a sequence of κ-quasiconformal mappings $F_j : \Omega_j \longrightarrow \mathbb{C}$ for open sets $\Omega_j \uparrow \Omega$ and suppose the F_j converge to a homeomorphism $F : \Omega \longrightarrow \mathbb{C}$, with $w_0 = \lim_{j \to \infty} F_j(z_0)$. Let Γ be the kernel of the ranges, defined by*

$$\Gamma = \bigcup_{n=1}^{\infty} \bigcap_{j=n}^{\infty} F_j(\Omega_j). \tag{133}$$

Then F is a κ-quasiconformal mapping from Ω onto Σ, where Σ is the connected component of Γ containing w_0.

A point lies in Γ if and only if it has a neighborhood contained in $F_j(\Omega_j)$ for all sufficiently large j. To paraphrase the conclusion: *The F_j converge to F and the ranges of the F_j converge to the range of F.*

We will be working in this book with sequences of mappings which are quasiconformal, but have decreasing distortion. Here is a formulation of standard results suitable for our use.

Proposition A.8. *Let Ω and $\{F_j\}$ be as defined in the previous proposition with limit function F : $\Omega \longrightarrow \mathbb{C}$. If for any $\epsilon > 0$ the maps F_j are $(1 + \epsilon)$-quasiconformal for all sufficiently large j, then the limit function F is a conformal mapping and $F(\Omega)$ is a component of the kernel of the sets $F_j(\Omega_j)$.*

A.2.4. Computation

You may be disappointed to find how few concrete computations we need. Our maps are generally simplicial – piecewise affine – maps between collections of euclidean triangles. We need bounds on dilatations and conditions that drive those bounds to 1, but we need few explicit values.

Suppose $T = \langle a, b, c \rangle$ and $\widetilde{T} = \langle \tilde{a}, \tilde{b}, \tilde{c} \rangle$ are (nondegenerate, positively oriented) euclidean triangles. Every point $z \in T$ has unique *barycentric coordinates* $\lambda_1, \lambda_2, \lambda_3$; that is, $z = \lambda_1 a + \lambda_2 b + \lambda_3 c$, where the λ_j are nonnegative with $\lambda_1 + \lambda_2 + \lambda_3 = 1$. Our simplicial maps are *barycentric*, meaning $f : T \longrightarrow \widetilde{T}$ is defined by

$$f(\lambda_1 a + \lambda_2 b + \lambda_3 c) = \lambda_1 \tilde{a} + \lambda_2 \tilde{b} + \lambda_3 \tilde{c}.$$

The mapping f identifies corresponding vertices and edges of T and $\widetilde{T}$ and is *real affine*; that is,

$$f(x, y) = \begin{pmatrix} \tilde{x} \\ \tilde{y} \end{pmatrix} = \begin{pmatrix} a_{11} & a_{12} \\ a_{21} & a_{22} \end{pmatrix} \begin{pmatrix} x \\ y \end{pmatrix} + \begin{pmatrix} s \\ t \end{pmatrix}.$$

In particular, the determinant $a_{11}a_{22} - a_{12}a_{21}$ is positive when the triangles are nondegenerate and share orientations; the dilatation of f can be computed from (130),

$$D_f(z) \equiv Q + \sqrt{Q^2 - 1}, \quad \text{where} \quad Q = \frac{a_{11}^2 + a_{12}^2 + a_{21}^2 + a_{22}^2}{2(a_{11}a_{22} - a_{12}a_{21})}. \tag{134}$$

Since dilatation is independent of euclidean scaling, one can normalize the domain and range triangles as in Figure A.3 so that f maps 0, 1, and z to 0, 1, and z', respectively, with z, z' in the open upper half plane $\mathbb{H}^+$. In this setting we can formulate quantitative statements about dilatation that will be used frequently in parts III and IV. The reader can verify the following distortion lemma.

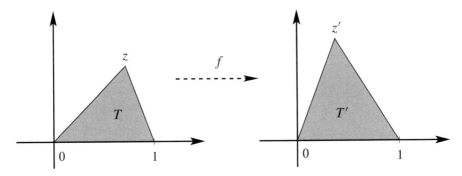

Figure A.3. Affine map between normalized triangles

Lemma A.9. *Given E a compact subset of* $\mathbb{H}^+$, *there exist constants* $\kappa = \kappa(E) \geq 1$ *and* $\mathfrak{c} = \mathfrak{c}(E) > 0$ *such that the following hold, where* f *is as depicted in Figure A.3:*

(a) If z and z′ belong to E, then $D_f \leq \kappa$.

(b) If z and z′ belong to E, then $D_f \leq 1 + \mathfrak{c}|z - z'|$.

The concrete situations we face in the book involve affine maps f between triangles formed by triples of mutually tangent euclidean circles. The intuition is clear: the closer the triangles are in "shape", the closer D_f is to 1 and vice-versa. We will frequently quantify this as follows. Given $\mathfrak{C} > 1$, we say a triple $\{r_1, r_2, r_3\}$ of euclidean radii is $\mathfrak{C}$-*bound* if $r_i/r_j \leq \mathfrak{C}$, $i, j = 1, 2, 3$. It is elementary to show the existence of a compact set $E = E_{\mathfrak{C}} \subset \mathbb{H}^+$ so that the normalized triangle for any $\mathfrak{C}$-bound triple will have its vertex z in $E_{\mathfrak{C}}$. By part (a) of the lemma, if triples $\{r_1, r_2, r_3\}$ and $\{r'_1, r'_2, r'_3\}$ are each $\mathfrak{C}$-bound, then the affine map f between their triangles will have dilatation bounded by $\kappa(E)$. You might note that part (a) of the lemma follows from part (b), but we split it out because the bound will be all that is needed in several situations.

Part (b) has to do with pushing the bound on D_f down to 1 as z and z' approach one another in E. In the situations of interest we will be facing $\mathfrak{C}$-bound triples for which the three ratios $r_1/r_2, r_1/r_3, r_2/r_3$ approach the corresponding ratios $r'_1/r'_2, r'_1/r'_3, r'_2/r'_3$. Elementary trigonometry shows that convergence of these ratios implies that $|z - z'|$ goes to zero uniformly for $z, z' \in E$. Part (b) of the lemma then gives bounds on D_f which converge to 1. That is, we will be able to conclude that the affine maps f are approaching conformality.

References: Ahlfors (1966, 1973, 1978), Beardon (1979), Garnett (1981), Lehto and Virtanen (1973), Lehto (1987), Needham (1997).

Appendix B

The Ring Lemma

The Ring Lemma of Rodin and Sullivan, Lemma 8.2, tells us that in a univalent flower with k petals, the petal circles cannot be too small compared to the center circle – there is a lower bound $\mathfrak{c}(k)$ on the ratio of their radii. Originally a utilitarian observation used in proving Thurston's Conjecture, this is, on closer inspection, a real geometric gem. We work in the euclidean setting – see the note at the end regarding the hyperbolic case.

Existence. The Rodin and Sullivan argument is basically this. Suppose $\{c_0; c_1, \cdots, c_k\}$ is a (closed) univalent k-flower, some $k \geq 3$. By scaling we may assume that the central circle c_0 has radius 1. At least one petal, say c_1, must be as large as the k petals would be if they were all the same size. Consider now what we can say about c_2. The smaller a circle tangent to c_0 and c_1 is, the deeper it must lie in the cravasse formed by c_1 and c_0. And if c_2 lies deep in that cravasse, univalence implies that the next petal, c_3, is also drawn deeply into the cravasse, as in Figure B.1. Note here the importance of *univalence* – c_3 must be small only because it is not allowed to overlap c_1.

Succeeding petals would likewise be drawn into the cravasse one after the other. In particular, if c_2 were too small we would see that the chain of k petals could not possibly reach out of the cravasse and around c_0 to close up the flower. In other words, there exists some lower bound on the radius of c_2. Now repeat this reasoning, with c_3 lying in the cravasse formed by c_0 and c_2 to get a lower bound on the radius of c_3, then with c_4 and so forth for succeeding indices. We conclude that there is a positive lower bound $\mathfrak{c}(k)$ which applies to *all* k petal radii, and existence is established.

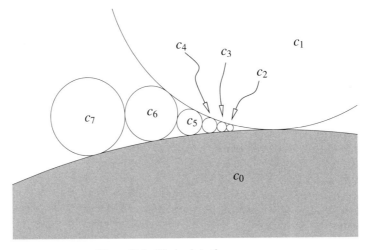

Figure B.1. Circles into the crevasse.

Extremal Flowers. Would you like more detailed information about $c(k)$? It is natural to start by determining the extremal flowers. The key lies with these two monotonicity results, easy consequences of the monotonicity of angle sums in Lemma 6.3. We use the term *bearing* (as in "ball bearing") to denote the largest circle that can be placed inside the interstice of a triple of tangent circles:

(a) *For a flower to remain intact with the central circle unchanged, one petal can be made smaller only by making one or more of the other petals larger.*

(b) *The radius of the bearing for a triple is monotone increasing in the radii of the circles of that triple.*

The central circle c_0 will be of unit radius in what follows. Consider first the case $k = 3$. It is clear from (a) that a given petal can be made as small as possible only by making the remaining two petals as large as possible. In the limit, then, these two petals become straight lines (infinite radius circles), which, because their discs are nonoverlapping, must be parallel. They are distance 2 apart because they are both tangent to c_0, so the third petal has the same radius as c_0, and this third petal cannot get any smaller by (a). We conclude that $c(3) = 1$ and denote this extremal flower by $\mathcal{F}_3$.

How are we to apply this philosophy for general k? Let us describe the flowers $\mathcal{F}_k$, $k = 4, 5, \ldots$, inductively first and then come back to observe that they are extremal. For notation, each $\mathcal{F}_k$ will have petals p_j, $j = 1, \ldots, k$; two (and only two because of univalence) will be halfplanes, denote these by p_1 and p_2. Petal p_k will always denote the smallest petal, that is, the one of radius $c(k)$. (Note that beyond $k = 3$ the indices will no longer correspond to the tangential *order* of the petals.)

Given that the $\mathcal{F}_n$ have been defined, $n = 3, \ldots, k$, one can create a $k + 1$-flower by simply adding the bearing to one of the interstices of $\mathcal{F}_k$ bordering c_0. By (b), the smallest of these bearings must be that for the triple $\langle c_0, p_{k-1}, p_k \rangle$, because p_k is the smallest petal of $\mathcal{F}_k$ and p_{k-1}, being the smallest of $\mathcal{F}_{k-1}$, is the second smallest of $\mathcal{F}_k$. Label this bearing as petal p_{k+1} and the resulting flower as $\mathcal{F}_{k+1}$. It is clear that a bearing is smaller than each circle in its triple, so p_{k+1} is the smallest petal of $\mathcal{F}_{k+1}$. That completes the inductive step in the construction. Figure B.2 illustrates the successive placement of petals.

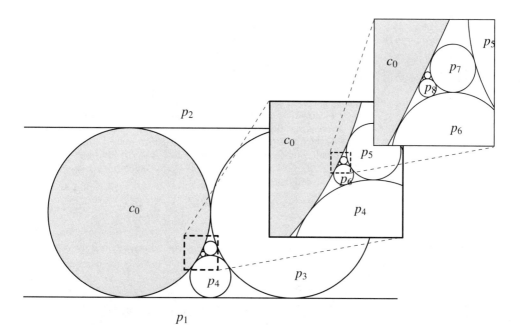

Figure B.2. Adding successive bearings.

We use induction again to prove extremality. We know $\mathcal{F}_3$ is extremal. Assume $\mathcal{F}_n$, $n = 3, \ldots, k$, is extremal and consider a general $(k + 1)$-flower $\mathcal{F}$. Suppose for convenience that the petals of $\mathcal{F}$ are again indexed by size (not necessarily tangency), c_1 being the largest, c_{k+1} the smallest. Modify $\mathcal{F}$ in stages, at the nth stage increasing petal c_n as much as possible, while compensating to maintain a univalent flower by decreasing some or all of the petals $c_{n+1}, \ldots, c_{k+1}$. Clearly, c_1 can be increased until it becomes a halfplane; then c_2 grows to become a parallel halfplane; and c_3 grows to radius 1. We cannot be sure of the fates of c_4, c_5, etc.; these depend on their locations vis-à-vis earlier petals. We can, however, observe the following: each petal c_n in turn grows until it is hemmed in by two among the petals $c_1, \ldots, c_{n-1}$. That is, for every $n \leq k + 1$, $c_1, \ldots, c_n$ must be the petals of a univalent flower about c_0. In particular, by the induction hypothesis, circle c_n has radius bounded below by $\mathfrak{c}(n)$ for each n, $3 \leq n \leq k$.

Of course, the last petal c_{k+1} has done nothing except possibly shrink in the adjustment processes so far. Let us check its size. Suppose c_i and c_j are its two neighboring petals in $\mathcal{F}$ (after all the adjustments), and suppose $1 \leq i < j \leq k$. Note that c_i, c_j must be tangent to one another as petals in the k-flower formed by $c_1, \ldots, c_k$, so c_{k+1} is the bearing for the triple $\langle c_0, c_i, c_j \rangle$. Since $i < j \leq k$, radius $(c_j) \geq \mathfrak{c}(k)$ and radius $(c_i) \geq \mathfrak{c}(k - 1)$. Compare to $\mathcal{F}_{k+1}$. The last petal of $\mathcal{F}_{k+1}$ is the bearing for neighbors with radii *equal* to $\mathfrak{c}(k)$ and $\mathfrak{c}(k - 1)$, so by (b), the radius $(c_{k+1}) \geq \mathfrak{c}(k + 1)$. In other words, $\mathcal{F}_{k+1}$ is extremal among univalent $(k + 1)$-flowers.

Explicit Constants. Knowing the geometry of the extremal flowers, we can compute the constants $\mathfrak{c}(k)$ using the *Descartes Circle Theorem* (DCT). (The most pleasant reference I know for this theorem is the 1936 poem "The Kiss Precise" by F. Soddy.) A *quad* is a configuration of four mutually tangent circles in the plane bounding discs with mutually disjoint interiors. The "bend" of a circle in the plane is the reciprocal of its eucidean radius, with the conventions that the bend of a straight line is zero and the bend of an "improper" circle (one whose "disc" is its *exterior*) is the negative of its reciprocal radius.

Theorem B.1 (DCT). *Let b_1, b_2, b_3, and b_4 be the bends associated with a quad. Then*

$$\left(\sum_{j=1}^{4} b_j \right)^2 = 2 \sum_{j=1}^{4} b_j^2. \tag{135}$$

Note that a quad is just a mutually tangent triple with a bearing and our extremal flowers consist of successive bearings added to $\mathcal{F}_3$. Write $\langle\!\langle b_1, b_2, b_3, b_4 \rangle\!\rangle$ for the bends of a quad. If you know the bends b_1, b_2, b_3 of a triple, you can solve (135) for the bend b_4 of its bearing *via* the quadratic formula:

$$b_4 = (b_1 + b_2 + b_3) \pm 2\sqrt{b_1 b_2 + b_1 b_3 + b_2 b_3}.$$

A triple has two interstices on the sphere, hence two bearings (of which one may be improper).

Let us compute. Let β_0 denote the bend of c_0 and β_j the bends of successive extremal petals p_j, $j = 1, 2, \ldots$. For $k \geq 3$, petal p_k is the bearing for triple $\langle c_0, p_{k-2}, p_{k-1} \rangle$, so the quads of interest have bends $\langle\!\langle \beta_0, \beta_{k-1}, \beta_k, \beta_{k+1} \rangle\!\rangle$. We know the first three: $\beta_0 = 1$, and p_1, p_2 are halfplanes, so $\beta_1 = \beta_2 = 0$. From (135) we compute $\beta_3 = 1$, confirming that the radius of p_3 is $1 = 1/\beta_3$. Shifting to the quad $\langle\!\langle \beta_0, \beta_2, \beta_3, \beta_4 \rangle\!\rangle$ we can now compute $\beta_4 = 4$, giving $\mathfrak{c}(4) = 1/4$. It is clear that we can recursively compute the bends and constants; here are the first few:

$$\mathfrak{c}(3) = 1, \quad \mathfrak{c}(4) = 1/4, \quad \mathfrak{c}(5) = 1/12, \quad \mathfrak{c}(6) = 1/33,$$
$$\mathfrak{c}(7) = 1/88, \quad \mathfrak{c}(8) = 1/232, \quad \mathfrak{c}(9) = 1/609, \quad \mathfrak{c}(10) = 1/1596.$$

What is happening here? Where do all these integer bends come from? Look again at the process. In our computation, successive quads

$$\cdots \longrightarrow \langle\!\langle \beta_0, \beta_{k-2}, \beta_{k-1}, \beta_k \rangle\!\rangle \longrightarrow \langle\!\langle \beta_0, \beta_{k-1}, \beta_k, \beta_{k+1} \rangle\!\rangle \longrightarrow \cdots$$

share the triple $\langle c_0, p_{k-1}, p_k \rangle$. As the two solutions of (135), β_{k+1} and β_{k-2} satisfy

$$\beta_{k+1} - \beta_{k+2} = 2(\beta_0 + \beta_{k-1} + \beta_k).$$

Thus the β_j satisfy the recursion

$$\beta_0 = 1, \quad \beta_1 = \beta_2 = 0, \quad \beta_3 = 1, \quad \text{and} \quad \beta_k = 2(\beta_0 + \beta_{k-2} + \beta_{k-1}) - \beta_{k-3}, \ k \geq 4.$$

In particular, *all the bends are integers, so all the radii are reciprocal integers!* I leave it to the reader to deduce the closed form expression for $c(k)$ given in (14), which follows from difference equation methods. You might also come across the *golden ratio* φ and Fibonacci numbers lurking here. For example, formula (14) may be restated to give

$$c(k) = \frac{1}{a_{k-2}^2 + a_{k-1}^2 - 1}, \quad k \geq 3,$$

where the a_j are the Fibonacci numbers, defined recursively by $a_1 = a_2 = 1$, and $a_{j+1} = a_j + a_{j-1}$. The sequence of blowups in Fig. B.2 strongly suggests the asymptotic geometric behavior. The ratios of radii of successive new petals converges to a limit which by the DCT must be

$$\frac{\text{radius}\,(p_{k+1})}{\text{radius}\,(p_k)} \longrightarrow \frac{1}{\varphi^2} = \frac{2}{3 + \sqrt{5}} \approx 0.38196601.$$

If you like the number theory that comes along with the geometry here, you might investigate recent papers on the Apollonian packing which we saw back in Figure 16.8 where similar ideas are in play. See citations below.

Note. The hyperbolic result follows the euclidean. The further a circle in $\mathbb{D}$ is from the origin, the greater the ratio between its hyperbolic and euclidean radii. After centering c_0 at the origin, then, the ratio between a petal radius and the radius of c_0 is greater in the hyperbolic than in the euclidean metric, meaning that the lower bound $c(k)$ established in the euclidean case applies. On the other hand, shrinking the flowers gives radii ratios which are roughly euclidean, meaning that no smaller constant would work. There is no ring lemma in spherical geometry.

References. The original Ring Lemma is in Rodin and Sullivan (1987). Extremal flowers and a recursion formula for the $c(k)$ were described by Hansen (1988); explicit constants are due to Aharonov (1997); see also Aharonov and Stephenson (1997). For number theory connections, see Graham et al. (2003), Lagarias, Mallows, and Wilks (2002), and Soddy (Jan. 9, 1937); for Soddy's poems (Soddy, June 20, 1936; Dec. 5, 1936); for Japanese temple geometry, Rothman (1998).

Appendix C
Doyle Spirals

A challenge to the reader: *Show that for any positive real numbers a and b a chain of six circles of successive radii $\{a, b, b/a, 1/a, 1/b, a/b\}$ will precisely close up around a central circle of radius 1 to form a hexagonal flower.* This observation by Peter Doyle led to the two-parameter family $P_{a,b}$ of spirals we have already used to generate discrete exponential functions; see Figures 1.2(b) and 2.7.

In this appendix we pick up on a simple "Growth Rule" which is playing out behind the scenes. It is illustrated in Figure C.1: given a triple with radii, $\{r_1, r_2, r_3\}$, add a fourth circle as shown whose radius r_4 is determined by the condition $r_4 = r_2r_3/r_1$, or more symmetrically, $r_1r_4 = r_2r_3$. You might try this out first by growing the Doyle flower in the figure.

There is nothing wrong with our efficient construction of the spirals $P_{a,b}$ in Section 14.3, but in light of this Growth Rule it is more interesting and instructive to let the Doyle spirals "self-assemble" (no batteries required). Start with a packing having a single triple of circles. Create a succession of packings by continually adding new circles. In particular, whenever a triple has two of its circles in the boundary of the current packing, attach a new circle as specified by the Growth Rule. The added circles tangent to any single circle c_v will have radii fitting the Doyle pattern (after scaling by the radius of c_v), so these circles are petals for a hexagonal flower around c_v and we will identify them modulo 6. Let the growth continue ad infinitum, generating an infinite circle packing P. You can check that P is simply connected, has no boundary (otherwise the construction would keep going), and that every vertex is degree 6. If the original triple's radii were $\{r_1, r_2, r_3\}$, then P is Möbius equivalent to the packing $P_{a,b}$, where $a = r_2/r_1$, $b = r_3/r_1$.

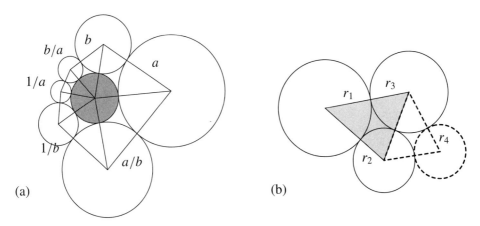

(a) (b)

Figure C.1. (a) Sample Doyle flower; (b) the Growth Rule.

322

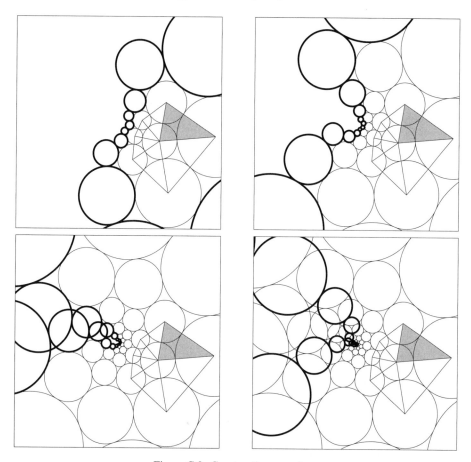

Figure C.2. Construction snapshots.

It is important to demonstrate this construction, though we are again restricted to static pictures when a movie is really called for. The snapshots in Figure C.2 show how a spiral experiment develops. The shaded triangle marks the original triple. We first generate the Doyle flower around one of its circles; then in each snapshot I show the fringe of circles added to the previous packing during one generation of new growth.

When you watch such a picture emerge for the first time there are some surprises. Aside from the cases $a = 1$ and/or $b = 1$, one always sees spiral paths of circles emerge in the pattern. Moreover, there is always a unique point z_0 somewhere in the plane which the circles never cover – a point where those spirals accumulate. In the live experiments you get a definite sense that you are watching a familiar surface emerge, and that is correct: carr(P) is a universal covering surface of $\mathbb{C} \backslash \{z_0\}$. As the growing packing first reaches around z_0, however, you generally find that the circles begin overlapping, as in the third snapshot here, and the image typically degrades from then on. Indeed, for generic a and b the centers of the circles of $P_{a,b}$ form a dense set in the plane, frustrating any hopes of seeing the global picture.

In a real experiment, of course, it is natural to twiddle the parameters. If you are persistent, you can arrange that as the growing pattern reaches around the spiral point z_0, some new circle will line up precisely with one already in place. In fact, there is another surprise in store here, for if you manage this coincidence with *one* circle, all the rest automatically line up as well – that is, any two circles will either be identical or have mutually disjoint interiors. You will have found, in that case, what is known as a *coherent* Doyle spiral; several are shown in Figure C.3.

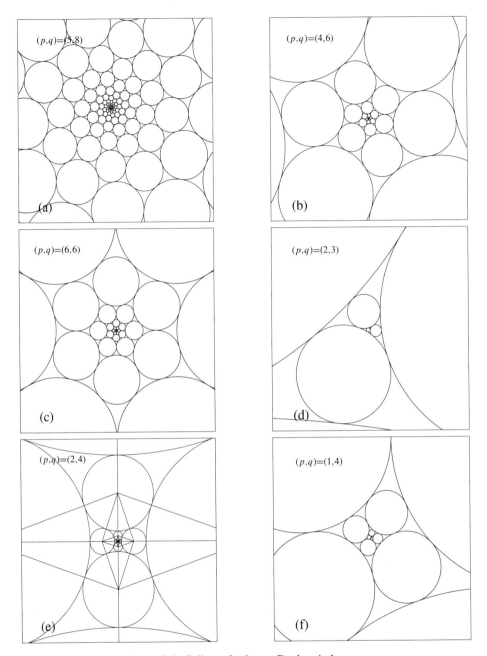

Figure C.3. Gallery of coherent Doyle spirals.

These coherent spirals are in fact *combinatoric* entities: each projects, when you identify coincident circles, to a univalent circle packing $Q_{a,b}$ of $\mathbb{C}\backslash\{z_0\}$ whose complex is an open annulus. Placing the spiral point z_0 at the origin, the self-similarities evident in the packing are associated with a discrete group Λ generated by $\phi : z \mapsto \lambda z$ and $\psi : z \mapsto \sigma z$. The coefficients satisfy $a = |\lambda|$, $b = |\sigma|$, and $a^p = b^q$ for some positive integers p and q. These integers can be discovered in the combinatorics of the spiral: starting at any circle c, stepping along p circles in one of the three hexagonal axes, then turning and stepping along q circles in another hexagonal axis will return you to c. There are only countably many coherent Doyle spirals (up to conformal transformations) and they are in one-to-one correspondence with the pairs (p, q) satisfying $0 < p \le q \le 2p$ (and omitting cases $(1, 1)$ and $(1, 2)$). Try your hand with the spirals of the Gallery, where the pairs are indicated.

Of course, in practice one would prefer to start with the parameters a, b. What are some parameter pairs which lead to *coherent* spirals? Most have been found by experimental trial and error, as with (a) and (b) in the Gallery. Degenerate situations such as (c), where $a = 1$ and one spiral direction is closed, are easy to figure out. As far as I know, there are only three other spirals explicitly known, namely, (d), (e), and (f). Figure C.3(d) is the *Coxeter spiral*. Coxeter arrived at his spiral differently; it is a pattern of *quads*, mutually tangent quadruples of circles. Using Descartes' Circle Theorem (see the previous Appendix), Coxeter computed the parameters $a = t^2$ and $b = t^3$, where $t = \varphi + \sqrt{\varphi}$, with φ the golden ratio. In (e) I have added the edges and perhaps you can see some internal self-similarities; these lead to the parameters $a = \varphi + \sqrt{\varphi}$, $b = a^2$. The parameters for the "Brooks spiral" in (f) will be found in the next Appendix.

It is no accident that the known parameter pairs are all algebraic numbers. The reasoning requires a deeper look at the Doyle flowers. Pick a, b (not both 1) and form a Doyle flower. The associated spiral (whether coherent or not) will omit some point z_0; translate, rotate, and scale the flower so its central circle is centered at 1 and the spiral point is placed at the origin. The centers of the flower's petal circles

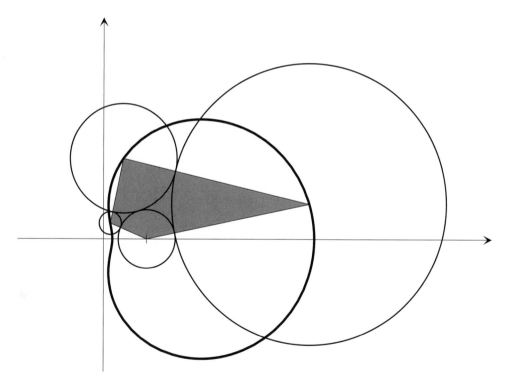

Figure C.4. An "Oval of Descartes."

can be shown to lie on a common level curve of the function $k(z) = \frac{1+|z|}{|1-z|}$. An example (with three of the petals) is displayed in Figure C.4; the level curve is the thicker curve and it is known classically as one of the *ovals of Descartes*. (It is also, curiously enough, the image of a circle under $z \mapsto z^2$.)

The spiral $P_{a,b}$ is associated with maps $\phi(z) = \lambda z$ and $\psi(z) = \sigma z$, as noted earlier. Although the group they generate will in general fail to be discrete, nonetheless,

$$\frac{1+|\lambda|}{|1-\lambda|} = \frac{1+|\sigma|}{|1-\sigma|} = \frac{1+|\lambda\sigma|}{|1-\lambda\sigma|}. \tag{136}$$

In other words, λ and σ are algebraically related. In the case that a, b generate a *coherent* spiral, there also exist the integers p and q so that $|\lambda|^p = |\sigma|^q$. This second relationship implies that $a = |\lambda|$ and $b = |\sigma|$ are algebraic. (My thanks to David Wright and Dov Aharonov for this observation.)

Let me close with the claim that you have probably seen coherent Doyle spirals in nature. *Phyllotaxis* is the study of arrangements of leaves or seeds in plant growth, as in the seed head of a sunflower. Doyle spirals provide one type of mathematical model for such spontaneous patterns. Those occurring in nature almost invariably have parameters p and q which are successive numbers in the Fibonacci sequence, as in the spiral (a) of our Gallery, where one can count five spiral arms in one direction and eight in the other. Try this counting with a pine cone on your next hike.

References. The principal reference is Beardon, Dubejko, and Stephenson (1994); related spirals occur in Mumford, Series, and Wright (2002). For Coxeter's spiral, see Coxeter (1968); relations to the Apollonian packing, Aharonov and Stephenson (1997); references to phyllotaxis, Rothen and Koch (1989).

Appendix D
The Brooks Parameter

Our insistence on *triangulations* in packing combinatorics is associated with the *rigidity* of triangular interstices, the ultimate source for the uniqueness we have seen again and again. Play with four coins on a tabletop and you immediately see that a *quadrilateral* interstice can wobble like a bookcase without cross bracing. Robert Brooks exploited this nonuniqueness by developing an associated parameter. We must forego his sophisticated application (parameterizing Schottky groups), but the geometry of this parameter is fascinating in its own right.

The features of our various examples are illustrated in Fig. D.1. There will be four shaded circles representing an initial *quadruple*, that is, a chain of four successively tangent circles with mutually disjoint interiors which are labeled T,R,B, and L (for *top*, *right*, *bottom*, and *left*). Brooks introduced the following protocol for adding a new circle to their quadrilateral interstice.

Brooks Protocol. *Start a small circle in the crevasse between T and L. Let it grow, remaining tangent to T and L, until it first becomes tangent to either R or B; in the former case label the circle T and in the latter case label it L.*

Typically, the added circle will leave a new quadrilateral interstice bounded by circles labeled T,R,B,L, so the protocol can be repeated. We keep track of successive outcomes with a sequence of H's and V's, H for *horizontal* (tangent to T, L, and R) and V for *vertical* (tangent to T, L, and B). From a sequence

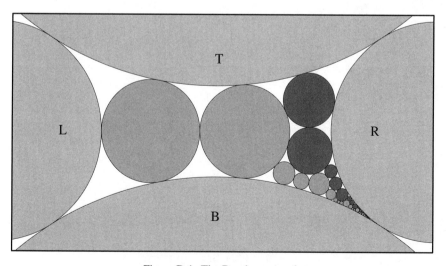

Figure D.1. The Brooks protocol.

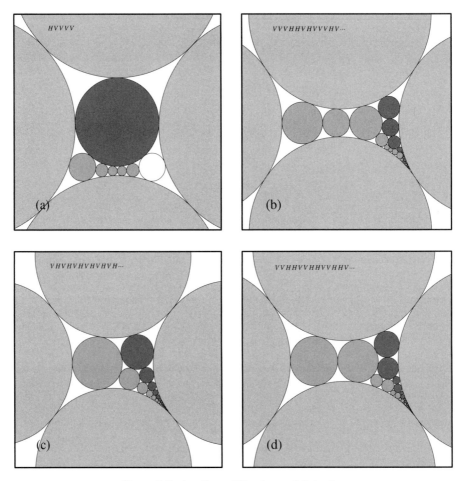

Figure D.2. A gallery of Brooks quadrilaterals.

of V's and H's one can read a sequence of integers $\{n_j\}_{j=1}^{\infty}$, where n_1 is the number of initial V's, n_2 is then the number of H's, n_3 the number of V's, n_4 the number of H's, and so forth. Define the *Brooks parameter* $t > 0$ for the quadrilateral as the continued fraction

$$t = n_1 + \cfrac{1}{n_2 + \cfrac{1}{n_3 + \cfrac{1}{n_4 + \cdots}}} = [n_1, n_2, n_3, n_4, \ldots].$$

For Fig. D.1, the two sequences are

$$\text{V H H V V V H H V H V V V V H H} \ldots \text{ and } t = [1, 2, 3, 2, 1, 1, 4, \ldots].$$

Brooks developed this parameter with a view to its Möbius invariance. If a quadruple $\langle c_1, c_2, c_3, c_4 \rangle$ has Brooks parameter t and if $\phi \in \mathrm{Aut}(\mathbb{P})$, then the Brooks parameter for $\langle \phi(c_1), \phi(c_2), \phi(c_3), \phi(c_4) \rangle$ is also t. The converse is a little more slippery: an initial quadruple generically defines *two* quadrilateral interstices on the sphere; two quadruples are Möbius images of one another if and only if the Brooks parameters t, t' for *both* interstices agree.

Figure D.2 is a gallery of examples, each with its sequence of V's and H's. Note that in (a) the process has ended at a finite stage. This happens if and only if the last circle is simultaneously tangent to all four circles bounding its interstice if and only if the Brooks parameter t is rational. Rational Brooks parameters t connect directly with circle packings, since the initial quadrilateral interstice is "packed" with a finite number of circles. Circle packings in turn provide a means for approximating arbitrary Brooks quadrilaterals. Namely, given t, truncation of its continued fraction determines a finite complex K with four boundary vertices. If one specify four boundary labels, one can compute and lay out the packing to create an associated interstice. Brooks established continuity for his parameter, so these interstices for truncated expansions will converge to an interstice for t itself. The choice of boundary radii involves one degree of freedom (modulo $\text{Aut}(\mathbb{P})$); if you are given companion parameters, t, t', and truncate both, you are led to combinatorial spheres with no degrees of freedom and you can approximate both interstices simultaneously.

The *golden ratio* φ comes to mind whenever the topic is continued fractions since its expansion is $\varphi = [1, 1, 1, 1, 1, \ldots]$. The associated (approximate) quadrilateral is Figure D.2(c). I must say that I was disappointed with this image. I expected more from this famous number, especially with its vaunted self-similarity. We can pump up the drama by exploiting Möbius invariance to give an alternate construction for precisely the *same* interstice.

For any quadruple $\langle T, R, B, L \rangle$ there exists a Möbius transformation that interchanges both T/B and L/R. A vertical circle (tangent to T and B) can be placed tangent to L (as the protocol dictates) or equally well tangent to R, without changing the geometry; the quadrilateral interstices remaining would be Möbius equivalent. Likewise, a horizontal circle can be placed tangent to B or to T. For an irrational Brooks parameter t, then, there are infinitely many circle patterns for *the same interstice*. Here is a new protocol that provides better balance and avoids high degrees in the emerging packings.

Balanced Protocol. *Alternate the placement of vertical circles, left to right; alternate the placement of horizontal circles, up to down.*

Applying this protocol for the golden ratio φ gives the circle pattern of Figure D.3. Note that the original interstice is precisely that of Figure D.2(c), despite the distinct combinatorics. I have zoomed in on the interior of this packing to bring out the apparent self-similarity. As one goes deeper into the pattern and the influences of the ad hoc radii for the initial quadruple fade, symmetry tells us that asymptotically each new circle radius will be a constant multiple λ of that of the previous circle. Now, you might ask,

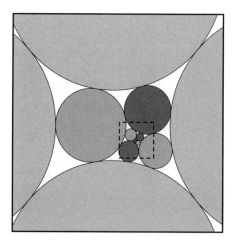
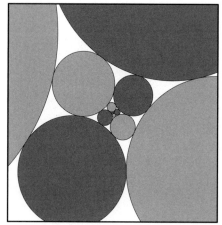

Figure D.3. The Brooks Spiral.

"What is λ?" Well, if you look carefully at any circle c_0 inside this pattern you will find that it is the center of a 6-flower, and that the ratios of petal radii will be, in order, $\{\lambda, \lambda^4, \lambda^3, \lambda^{-1}, \lambda^{-4}, \lambda^{-3}\}$.

Aha! This is one of the Doyle Spirals of the previous Appendix, namely that associated with parameters $p = 1$ and $q = 4$ in Figure C.3(f) (hence the name *Brooks Spiral*). The associated spiral parameters are $a = \lambda$, $b = \lambda^4$. CirclePack gives $\lambda \sim 1.623$, but David Wright has computed the precise (algebraic) value: $\lambda = |\sigma|^2 \approx 1.623275178$, where σ satisfies $\sigma + 1/\sigma = 1 - (1 - i)/(1 + \sqrt{3})$. You might note that the Coxeter Spiral of Figure C.3(d) is the "triangular interstice" analogue to our construction of the Brooks Spiral. Is some quadrilateral version of the Descartes Circle Theorem lurking here?

The asymptotic Doyle Spiral of Fig. D.4 is yet another that arises using our balanced protocol. This is associated with the Brooks parameter $t = 1 + \sqrt{2} = [2, 2, 2, 2, \ldots]$. We computed the factor in Appendix C as $\lambda = \varphi + \sqrt{\varphi} \sim 2.89005363\ldots$, one of the two parameters occurring with the Coxeter Spiral. If you start the Brooks construction with the four boundary circles having their proper relationships, the balanced construction is *exactly* the Doyle construction.

References. The principal reference is Brooks (1985). See also Brooks (1992) and Coxeter (1968).

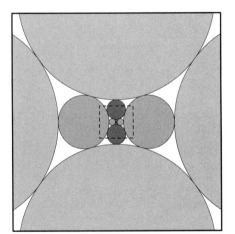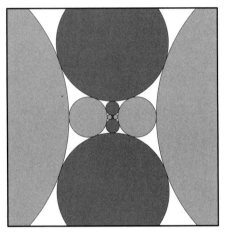

Figure D.4. Another Brooks/Doyle spiral.

Appendix E

Inversive Distance Packings

We have concentrated almost exclusively on *tangency* circle packings in this book. However, the topic began with a broader definition, and here we extend that even further, bringing new twists to the geometry and opening further questions.

The classical notion of *inversive distance* describes the relationships between pairs of spheres in euclidean space in a Möbius invariant way – in our setting, between pairs of circles in $\mathbb{P}$, $\mathbb{C}$, or $\mathbb{D}$. The distance $\sigma(c_1, c_2)$ between two circles is independent of the geometry, but in the case of euclidean centers z_1, z_2 and radii r_1, r_2, the formula is

$$\sigma(c_1, c_2) = \frac{|z_1 - z_2|^2 - (r_1^2 + r_2^2)}{2r_1 r_2}. \tag{137}$$

Figure E.1 illustrates a spectrum of inversive distances between C, in the center, and circles c_σ of half its size.

In the *overlap* and *deep overlap* zones for σ, the circles C and c_σ intersect one another with *overlap angle* ϕ satisfying $\sigma = \cos(\phi)$; for instance, c_{-1} is internally tangent, $c_{\frac{1}{2}}$ is orthogonal, and c_1 is externally tangent to C. In the $\sigma > 1$ zone, the circles are *separated*.

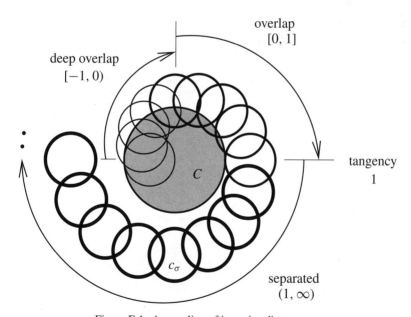

Figure E.1. A sampling of inversive distances.

We speak as usual about (oriented) triples $\langle c_1, c_2, c_3 \rangle$, but now this involves, in addition to the radii r_1, r_2, r_3, the three inversive distances

$$\sigma_{12} = \sigma(c_1, c_2), \qquad \sigma_{23} = \sigma(c_2, c_3), \qquad \sigma_{13} = \sigma(c_1, c_3),$$

which are naturally identified with edges, as indicated in Fig. E.2(a). Note that the "face" (with orientation) of a triple still makes sense, though "interstice" may not. It is a nice exercise to show, for instance, that in an *overlap* triple such as (b), the interstice vanishes if and only if $[\arccos(\sigma_{12}) + \arccos(\sigma_{23}) + \arccos(\sigma_{13})] \geq \pi$.

There is a certain comfort level with tangency packings that we are going to miss now. We have taken for granted that *any* three radii r_1, r_2, r_3 will define a tangency triple. The same holds for preassigned *overlaps* – that is when $\sigma_{12}, \sigma_{23}, \sigma_{13}$ belong to [0, 1]. However, a prescription involving one or more *deep overlaps* and/or *separations* will be *incompatible* with certain triples of radii. The triple of Figure E.2(c) is close to being incompatible: as you decrease radius r, the face degenerates to a line and any further decrease is incompatible with the given inversive distances.

Moving on to the "packing" level, we can extend our traditional notions. Given a complex K, let $\mathfrak{E}$ denote its set of edges and let $\mathcal{I} : \mathfrak{E} \longrightarrow [-1, \infty)$ be an *inversive distance prescription* for K.

Definition E.1. An **inversive distance** circle packing P for $(K, \mathcal{I})$ is a configuration $P = \{c_v : v \in K\}$ of circles with the usual properties except that $\sigma(c_v, c_u) = \mathcal{I}(e)$ whenever $e = \langle u, v \rangle$ is an edge of K. When $\mathcal{I} \equiv 1$, P is a normal "tangency" packing, and when $\mathcal{I}(\mathfrak{E}) \subset [0, 1]$ (i.e., between orthogonality and tangency), P is called an **overlap packing**.

Of course, the natural question is "Do results for tangency packings extend to inversive distance packings?" First, the good news: overlap packings have been around from the beginning. Here is Thurston's version of Theorem 7.1 (through I have phrased it terms of inversive distances $\sigma = \cos(\phi)$ instead of overlap angles ϕ themselves):

Theorem E.2 (Andreev–Thurston Theorem). *Let K be a topological sphere which is not simplicially equivalent to a tetrahedron and let $\mathcal{I} : \mathfrak{E} \longrightarrow [0, 1]$ be given. Suppose the following conditions hold:*

(A) If $e_1, e_2, e_3 \in \mathfrak{E}$ form a closed loop in K with $\sum_{i=1}^{3} \arccos(\mathcal{I}(e_i)) \geq \pi$, then (e_1, e_2, e_3) are the three edges of a face of K.

(B) If $e_1, e_2, e_3, e_4 \in \mathfrak{E}$ form a closed loop in K with $\sum_{i=1}^{4} \arccos(\mathcal{I}(e_i)) = 2\pi$, then (e_1, e_2, e_3, e_4) form the boundary of the union of two adjacent faces of K.

Then there is an essentially unique realization of K as a geodesic triangulation of $\mathbb{P}$ and a family P of circles centered at its vertices so that the two circles centered at the vertices of edge $e \in \mathfrak{E}$ meet at angle $\arccos(\mathcal{I}(e))$.

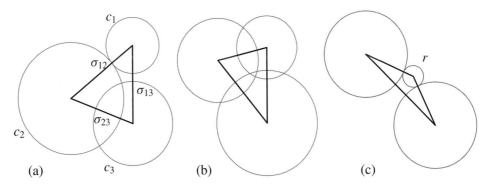

Figure E.2. Inversive distance triples.

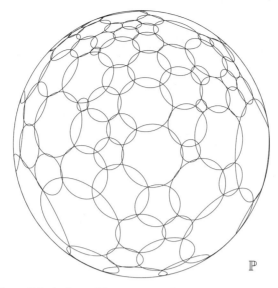

Figure E.3. Andreev–Thurston example with random overlaps.

In other words, under the given hypotheses there is a univalent overlap packing P associated with $(K, \mathcal{I})$. It is natural to call P the *maximal* packing for K *and* $\mathcal{I}$. Since the circles do not necessarily have disjoint interiors here, we reinterpret the term "univalent" to mean that $\mathrm{carr}(P)$ is univalent – the faces have mutually disjoint interiors. See Fig. E.3.

The restriction to overlap packings is necessary if one is to avoid incompatibilities and the associated theory largely mimics that of tangency packings. The ambitious reader may want to revisit Part II and work out the extensions of definitions, theorems, and proofs. The keys remain with the monotonicity properties and various bounds – though there are some interesting wrinkles to work out, for example, in the Ring Lemma and in the packing algorithm. Figure E.4 puts Owl though some paces with a randomized prescription of overlaps $\mathcal{I} \subset [0, 1]$.

The "bad" news is perhaps good – if you enjoy challenges. Any inversive distance prescription $\mathcal{I}$ having one or more values in $[-1, 0) \cup (1, \infty)$ can potentially have incompatibilities. We might classify these into "hard" and "soft." For a hard type, suppose K is a 3-flower $K = \{v_0; v_1, v_2, v_3\}$ and we set

$$\mathcal{I}(c_1, c_2) = 7, \text{ and } \mathcal{I}(c_0, c_1) = \mathcal{I}(c_0, c_2) = \mathcal{I}(c_1, c_3) = \mathcal{I}(c_2, c_3) = 1.$$

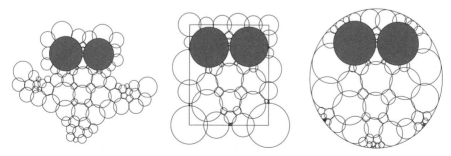

Figure E.4. Overlap Owl.

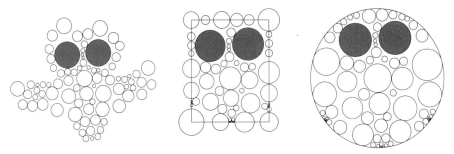

Figure E.5. Separated Owl.

In an associated packing P, the pair c_1, c_2 is separated; c_0 and c_3, being held tangent to each of them, are necessarily held somewhat close to one another – one can check that $\mathcal{I}(c_0, c_3) \leq 7$. In other words, any prescription $\mathcal{I}(c_0, c_3)$ larger than 7 can *never* be realized in a packing for K. On the other hand, a soft incompatibility might involve a prescription $\mathcal{I}$ which can be realized only under certain conditions – perhaps, e.g., there can exist no packing in $\mathbb{D}$ whose boundary circles are horocycles.

Igor Rivin has extended Theorem E.2 to allow deep overlaps by identifying an additional geometric hypothesis which is necessary and sufficient to avoid incompatibilities. No such conditions are known, however for separated circles. Figure E.5 puts the Owl through further contortions using a randomized prescription of separations $\mathcal{I} \subset [1, 3]$.

Open Question E.3. *Given complex K and inversive distance prescription $\mathcal{I} : \mathfrak{E} \longleftrightarrow [0, \infty)$, what can be said regarding existence, uniqueness, and properties of circle packings for $(K, \mathcal{I})$?*

Almost everything is open here, and progress is going to require new ideas. But there is a ray of hope: as it happens, `CirclePack` will often compute and display these packings with nary a complaint, as it has done in the last figure. That at least gives us the opportunity to probe the landscape experimentally.

References. For the Andreev–Thurston Theorem, see (Morgan, 1984; van Eeuwen, 1994); branching was introduced in Bowers and Stephenson (1996). Deep overlaps were studied by Rivin (Hodgson and Rivin, 1993; Rivin, 1986; Rivin, 1994). Inversive distances were introduced experimentally by the author; see Bowers and Hurdal (2003).

Appendix F

Graph Embedding

Circle packing methods have found direct application in various parts of graph theory, and have perhaps influenced work in other areas such as *percolation theory* and self-avoiding random walks. These topics are beyond our scope, but we would be remiss to overlook more accessible uses, particularly since the results are often so scientifically and aesthetically pleasing. Let me start with a few general observations.

Given a planar graph G, the basic strategy is to augment G to form a complex K, create a univalent circle packing P for K, embed K as carr(P), and then disregard everything but the edges belonging to G. In addition to ease of construction and choice of geometry, here are some features that tend to make these *circle packing embeddings* visually pleasing.

- *Straight-line:* The embeddings realize Fary's Theorem that every *planar* graph has an embedding with straight edges.

- *Resolution:* The faces tend to be well-proportioned. The *(angular) resolution* of a straight-line embedding of G in the plane is the minimum of the angles formed by incident edges. The *resolution* of G, Res(G), is the maximum of the resolutions among all such embeddings. Suppose deg$(G) = n < \infty$. By the Ring Lemma (see Appendix B) one can show that

$$\text{Res}(G) \geq 2 \, \arctan\left(\frac{1}{a_{n-2}(a_{n-1} + a_{n-3})}\right),$$

where $\{a_j\}$ is the Fibonacci sequence. In particular, Res$(G) \sim \text{O}(0.381966^n)$ as $n \longrightarrow \infty$. The ability to embed with known resolution is important in, e.g., numerical PDEs (partial differential equations).

- *Symmetry:* The uniqueness properties of packings mean that symmetries in the combinatorics (and in any side conditions) will result in corresponding symmetries in the embedding.

- *Dual Graph:* When G is a triangulation, then along with the embedding of G itself one gets an orthogonal embedding of the *dual graph* G^*. (Dual graphs are an integral part of the discrete integrable systems approach to circle packing; see Appendix G.)

- *Separators:* Miller, Teng, and Vavasis used circle packing to show that every planar graph G with n vertices is $\text{O}(\sqrt{n})$-*separable*, meaning that there is a subset C of vertices, card$(C) \approx \sqrt{n}$, so that $G \backslash C$ can be partitioned into sets A, B of roughly comparable size and having no edges between them. This is valuable information, for example, in analyzing "divide-and-conquer" algorithms in computer science.

Figure F.1 illustrates several uses of circle packing embeddings which I now discuss in turn.

Knot Embeddings: Mathematical *knots* are simple closed curves embedded in $\mathbb{R}^3$. Their study, which goes back centuries, remains very active and has some surprising applications; for example, in the study of DNA. Classification of knots is a major issue and circle packing plays a role through an interactive library of knots called `Knotscape`, create by Jim Hoste and Morwen Thistlethwaite. Their catalog stores two-dimensional knot projections combinatorially as patterns of over/under crossings. Thistlethwaite developed methods to convert an encoding to a simple graph, embed the graph with

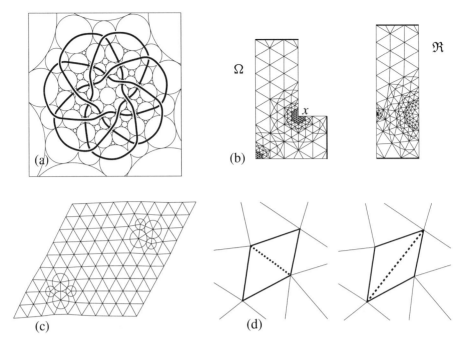

Figure F.1. Applications of circle packing embedding.

circle packing, and then display the result. Knotscape can almost instantaneously pop up a beautiful picture for any of the roughly 2 million knots currently catalogued. As in many applications, the original knot combinatorics must be augmented with vertices and edges to get a triangulation graph; after being packed, extraneous information is discarded. Figure F.1(a) illustrates with a (3, 7) torus knot. The knot is associated with a certain edge-path in the carrier, but for aesthetic reasons this edge-path is approximated by a smooth spline for the final image. Normally, the circles are not part of the picture, but, I have left them in for reference. The code is very compact, very fast, and the images have good visual balance – for example, more complicated portions of the knot do not get squeezed down in the images. The images have been important not only for display, but have aided in discovering hidden symmetries.

Grid Generation: Grid generation is a key to many numerical applications, for example, in finite element methods for numerical PDEs. I have been asked many times whether circle packing embeddings might be useful in this regard, and though I have no concrete results, let me give a (greatly simplified) example suggesting some optimism.

Suppose we face a numerical PDE on the region Ω of Figure F.1(b). A common strategy with many PDEs is to compute the conformal mapping $F : \Omega \longrightarrow \mathfrak{R}$ to a rectangle $\mathfrak{R}$, create a grid for solving the problem in $\mathfrak{R}$, then bring the answer back via F^{-1}. Suppose, however, that accuracy is a problem at interior corners of Ω, such as x. How does one create a grid in $\mathfrak{R}$ which respects the geometry back in Ω?

In the figure I suggest a circle packing solution. I have constructed a grid of the type we would like by using a circle packing P which has carrier Ω and has been locally refined near x. We then transport this grid to a rectangle $\mathfrak{r}$ by using the circle packing map $f : P \longrightarrow Q$, where $\mathfrak{r} = \text{carr}(Q)$, solve the problem in $\mathfrak{r}$, then carry the solution back to via f^{-1}. The mechanics here could be quite flexible and efficient. However, my optimism rests more on the geometry: all our experience suggests that f is *discrete conformal*, so the transplanted grid in $\mathfrak{r}$ should nicely capture the local geometry at x.

There is only one problem. The packings behind these grids are not tangency packings, but rather have overlapping circles, *à la* the previous Appendix. You would have to look even more carefully to note that certain of its overlaps have inversive distances just barely negative – they are "deep" overlaps in our terminology. In practice that seems to present no problem, as `CirclePack` seems to compute such packings without complaint. However, currently we have no *theory* that guarantees the existence and uniqueness of the rectangular packing. In other words, there is work to be done!

Lattice Dislocations: Consider Figure F.1(c). In physics, two-dimensional particle systems are often represented by "nearest neighbor" graphs, and in many situations hexagonal lattices are presumed to be lowest energy configurations. There are two "flaws" in (c), each involving two 5/7 pairs, known as *quadrupoles*. Circle packings similar to this were used by physics colleagues in simulations of so-called *hexatic phases* in 2D quenching. The circles per se had no relevance, but their embeddings were so physically natural that they provided the crucial starting locations for subsequent molecular dynamics simulations.

The quadrupoles of Figure F.1(c) were trivial to create using two *Whitehead moves* on a hexagonal complex. As illustrated in Figure F.1(d), a Whitehead move involves replacing one interior edge by another in the union of two contiguous faces. This represents a combinatoric change, but no change in topology, no change in the counts of faces, edges, or vertices, and for infinite complexes, no change in type. I personally find Whitehead moves so fascinating that I would like to end this appendix by introducing them to the reader.

Whitehead Moves: Complexes K and K' are *Whitehead equivalent* if one can be converted into the other via Whitehead moves. This may bring to mind a notion called "edge flipping" in triangulations, and indeed some of the same questions are relevant here. However, whereas edge flips are done in the context of fixed geometric vertices, combinatoric changes here lead to geometric changes under repacking. These moves are easy to implement in practice and great fun, and they may provide some insight into the connections between combinatorics and geometry.

To hook you into this, let me offer a quick challenge: *Create the "cat's eye" pattern of Figure F.2(a) from a hexagonal packing.* (I stumbled onto this accidentally, and in fact have difficulty recreating it myself.) By way of contrast, prove that Figure F.2(b) *cannot* be so obtained.

Recall that the degree of a vertex is a weak indication of its geometry: ≤ 5 positive curvature, 6 flat, ≥ 7 negative curvature. In some sense, then, Whitehead moves are redistributing the curvature. Nonetheless, deliberate sequences of Whitehead moves to reach some preassigned goal seem very difficult to plan.

Consider tori, for example. A simple calculation with Euler characteristics shows that the *average* degree for any triangulation of a torus is precisely 6, so a torus is, on average, flat. It is also known that any two tori with, say, n vertices are Whitehead equivalent. It can be proven that *every torus with $n \geq 9$*

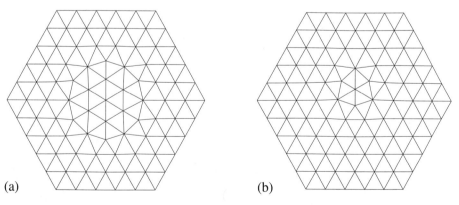

(a)   (b)

Figure F.2. Cat's eyes.

vertices is Whitehead equivalent to one with constant degree 6. You can create an example by adding two face barycenters to the torus $\mathfrak{p}_4$ of Figure 16.8(e), which already has 7 vertices. What Whitehead moves could you now use to convert this to hexagonal?

One final question which I have not been able to answer: *Does there exist a triangulation of a torus whose vertices are all of degree 6 with the exception of one vertex of degree 5 and one of degree 7?*

References. For embedding, see, e.g., Agarwal and Pach (1995), Brightwell and Scheinerman (1993), and He (1994); separators and resolution, Aharonov and Stephenson (1997) and Miller, Teng, and Vavasis (1991); knot embedding, Hoste and Thistlethwaite; lattice quenching, Somer et al. (1997). For related new developments, see, e.g., Schramm (2000).

Appendix G

Square Grid Packings

Hexagonal combinatorics have been a mainstay in many parts of this book, and the reader might yearn for an alternative more suited to our typically "rectangular" habits. The *ball-bearing* pattern, first seen in Figure 1.1(a) of the Menagerie, is a likely candidate, but the bearing circles present a certain unsettling asymmetry. We can fix that! In Figure G.1 the ball-bearing packing is morphed through several stages. First, we increase the bearing circles (gray) to the size of the lattice circles, allowing them to overlap their lattice neighbors by $\pi/2$; the result is known as a *square grid* packing and is an example of an "overlap" packing according to the terminology of Appendix E.

We have the option of treating these square grid packings much as we have treated other packings throughout the book. In Figure G.2, for example, I show two packings modeled on the *error function*, erf(z), along the lines investigated with hexagonal packings in Section 14.4; namely, by setting boundary labels based on |erf$'(z)$|. On the left I began with a square grid packing in its usual orientation, that of Figure G.1; on the right I did a preliminary rotation of the square grid by $\pi/4$. I will comment on these examples in a moment.

Returning to Figure G.1, the next morph involves removing the edges emanating from the gray bearing circles. The cells in this decomposition are now quadrilaterals. The triangular rigidity we are accustomed to is replaced by a new rigidity, namely, for each of these cells there exists a circle simultaneously orthogonal to its four vertex circles.

Such quadrilateral cell decompositions are central to a topic known as *discrete integrable systems*, which ostensibly has no connection with circle packing. This topic instead involves (among many other things) cross ratios: a cell with corners $\{z_1, z_2, z_3, z_4\}$ is associated with a *cross ratio*, the complex number defined by

$$[z_1, z_2, z_3, z_4] = \frac{(z_1 - z_3)(z_2 - z_4)}{(z_1 - z_4)(z_2 - z_3)}. \tag{138}$$

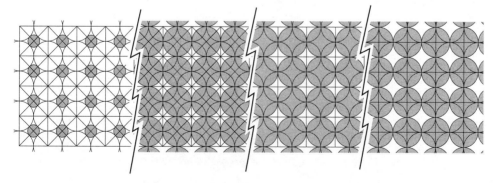

Figure G.1. Morphing the ball-bearing packing.

339

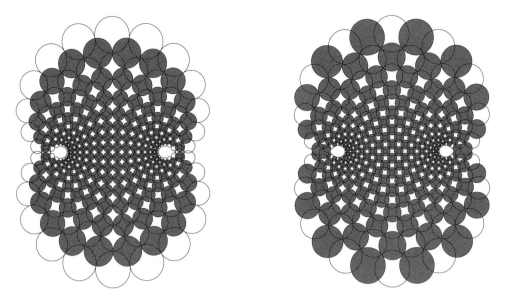

Figure G.2. Toward a discrete error function?

The key property of the cross ratio is its association with Möbius transformations: $[z_1, z_2, z_3, z_4] = [w_1, w_2, w_3, w_4]$ if and only if there exists $\phi \in \mathrm{Aut}(\mathbb{P})$ so that $w_j = \phi(z_j)$, $j = 1, \cdots, 4$. Two-dimensional discrete integrable systems involves the study of patterns of real cross ratios in quad (rilateral) graphs, and one theme revolves around the search for connections to analyticity. In work of Bobenko and Agofanov, for example, the existence of "discrete" versions of z^γ and $\log(z)$ on quad lattices are established within this theory.

As it happens, cross ratios are also connected with circles: *the cross ratio of four points is real if and only if the four points lie on a common circle*. With the publication of a paper on "square grid" packings by Oded Schramm, the topics of circle packing and discrete integrable systems began to converge. Before I discuss Schramm's setup, note the last morph in Figure G.1, which brings out a delicious ambiguity in the ball-bearing pattern: is it the gray discs that were the bearings or the others? The combinatorial "duality" between the original lattice and the lattice of gray centers is the key to the analysis of these patterns both in the integrable systems and the circle packing approaches.

Let me describe the approach of Oded Schramm for a square grid packing P. This involves two collections, τ_P and σ_P, of positive cross ratios associated with intersection points of the circles in the square grid. Figure G.3 shows the local configuration of points used in the definitions.

The τ-invariant associated with the shaded circle in the figure and the σ-invariant associated with the black vertex in the figure are defined by cross ratios as indicated. (Here "invariance" means with respect to Möbius transformations.). The invariants for all circles and for all vertices give collections τ_P and σ_P, respectively. Schramm proved that two such collections are associated with a square grid packing on the sphere if and only if they are positive and satisfy a certain discrete system of equations which he likens to nonlinear *Cauchy–Riemann equations*. I will not write these out, but the point for us is the existence of a rigid connection between circle patterns and cross ratio patterns. Schramm displayed square grid packings which parallel, among other functions, $\exp(z)$ and $\mathrm{erf}(z)$. This brings us back to Figure G.2. The packing on the right is part of an infinite pattern for Schramm's $\mathrm{erf}(z)$ example. On the other hand, Schramm's work suggests that it may be impossible for the packing on the left to be part of an infinite packing. This injects a new consideration into our work: Is the hexagonal construction of $\mathrm{erf}(z)$ sought in Section 14.4 doomed because of combinatorial bias? Would it work if the domain packing were rotated, say by $\pi/3$?

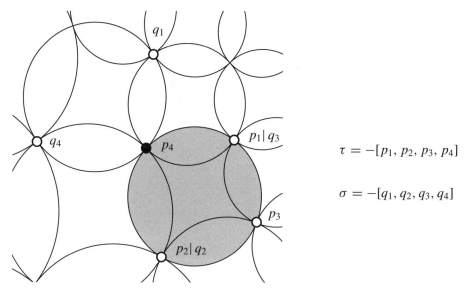

$$\tau = -[p_1, p_2, p_3, p_4]$$

$$\sigma = -[q_1, q_2, q_3, q_4]$$

Figure G.3. Square grid cross ratios τ and σ.

Solving the τ–σ systems associated with regular graphs has quite a different feel from our circle packing – it is more akin to classical discrete potential theory on lattices. However, as these topics have found one another they have begun to merge. Consider the circle packing on the left in Figure G.4. Dashed circles are added, one for each interstice of the packing, and on the right the edges of K and the edges of its dual graph are displayed. Note that the union of the two graphs defines a quad graph with kite-shaped cells. The integrable system approach begins with that quad graph and generates both the circle packing and its dual.

The blending of the integrable systems and circle packing approaches has opened many new directions for research. I will leave you with the wonderful images by Bobenko and Springborn of the discrete

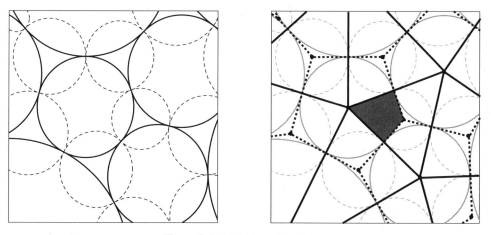

Figure G.4. A quad graph packing.

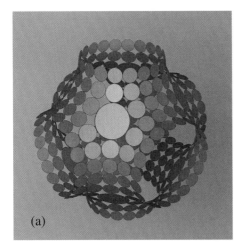
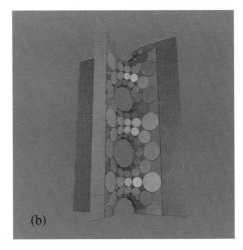

(a)   (b)

Figure G.5. Discrete (a) Schwarz and (b) Scherk minimal surfaces.

versions of two classical minimal surfaces, the Schwarz and Scherk surfaces, which are created using packings derived from the square grid approach (see Fig. G.5).

References. As an *entrée* into this alternative approach to circle packing, I highly recommend Bobenko and Springborn (2003), and references therein. For integrable systems start with Agafonov and Bobenko (2000). Square grid packings were introduced by Schramm (1997). For the minimal surfaces, see Bobenko, Hoffmann, and Springborn (2002). See also, e.g., Agafonov and Bobenko (2000), Bobenko and Hoffman (2001a, 2001b), and Bobenko, Hoffman, and Suris (2002).

Appendix H

Schwarz and Buckyballs

We construct a circle packing P of the sphere having 12 branch points. This time the circles just go along for the ride – all the work is done by some classical spherical triangles.

Our complex K is obtained from $K^{[5]}$, the constant 5-degree complex, by applying one hexagonal refinement operation. (This pattern is easily confused (I admit to this myself) with the familiar soccer ball pattern, a.k.a., buckminsterfullerene.) Of its 42 vertices, 12 have degree 5, the remainder, degree 6. We have seen this maximal packing several times in the book, but I repeat it here with some decorations. The 5-degree circles in Figure H.1 have been shaded and a pattern of geodesic segments has been superimposed. The darker segments outline the 12 faces of a regular dodecahedron, the lighter lines further decompose these into 120 spherical triangles. Caution: these triangles are *not* faces of K; indeed, using the various rotational symmetries in the pattern you can check that they are all congruent to a $(2, 3, 5)$-*Schwarz triangle* t, that is, a spherical triangle with angles $\pi/2, \pi/3, \pi/5$.

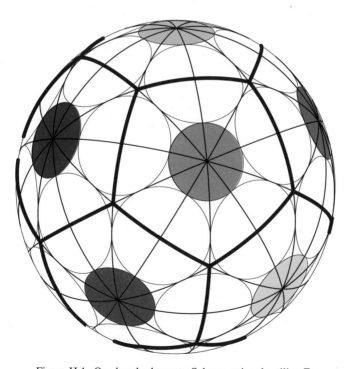

Figure H.1. One hundred twenty Schwarz triangles tiling $\mathbb{P}$.

343

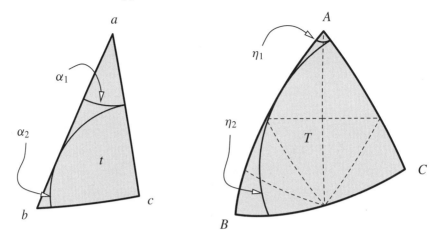

Figure H.2. Schwarz triangles t and T.

A copy of t is isolated in Figure H.2 and has been marked with arcs α_j from spherical circles centered at two of its vertices. If you begin with one copy of t on a sphere and repeatedly apply reflections – that is, copy it to a new location by *Schwarz reflection* through an edge – you generate 120 copies which close up to precisely tile the sphere. It is the circular arcs carried along on these copies which piece together to produce the maximal packing $\mathcal{P}_K$ of Figure H.1.

To construct a *branched* circle packing, we apply the same procedure, but with a new triangle, namely, with the $(2, 3, 5/2)$-Schwarz triangle T of Fig. H.2. Schwarz reflection can be carried out with T just as with t. In particular, six reflected copies of T are required about vertex B and four about vertex C to complete their neighborhoods. However, it takes 10 reflections about vertex A to close up – that is, it is not until the 10th successive copy is reflected around A that the final edge will match up to the original edge and further reflections become repeats. This represents simple branching since these ten faces wrap twice around A while the 10 copies of t wrap once about a ($10(2\pi/5) = 4\pi$ and $10(\pi/5) = 2\pi$, respectively). If you carry repeated reflections of T to completion you get 120 copies of T on $\mathbb{P}$.

The branched packing P emerges when we carry along the circular arcs η_j drawn on T. In fact, one can identify the arcs α_j on t to arcs η_j on T, $j = 1, 2$, and propagate that identification to all 120 triangles in domain and range *via* reflection. This creates the 42 circles of a branched packing P, and in fact gives a tangency preserving map of $\mathcal{P}_K$ to P. In particular, $f : \mathcal{P}_K \longrightarrow P$ is a discrete rational function.

Each branch circle of P (colored in Figure H.3) has order 1 and is associated with a vertex of degree 5 in K, so the set of 5-degree vertices of K is the branch set, $\mathrm{br}(P) = \mathrm{br}(f)$. By the Riemann–Hurwitz formula, f has valence 7. You can see this directly, also: the dashed lines on T in Figure H.2 show how it is tiled by seven copies of t, so 120 copies of T cover $\mathbb{P}$ seven layers deep. By a normalization one can place antipodal centers of $\mathcal{P}_K$ at 0 and ∞ and normalize so that these will be fixed by f. As noted in Sect. 15.2, f then interpolates the rational function $F(z) = z^2(3z^5 - 1)/(z^5 + 3)$.

You may find the combinatorics here challenging. The map f permutes the 12 vertices of a regular dodecahedron, leaving the north and south poles fixed. I will let you contemplate how this shuffling happens; the circles are color-coded in the figure so you can identify circles between domain and range. Enjoy!

References. The principal reference here is Bowers and Stephenson (1996). For the geometry of Schwarz triangles, see Coxeter (1991).

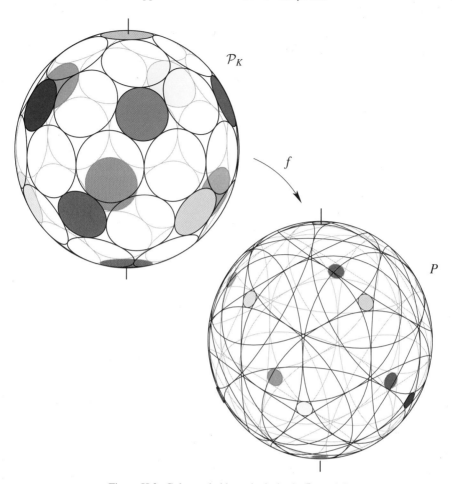

Figure H.3. Color-coded branch circles in $\mathcal{P}_K$ and P.

Appendix I

CirclePack

Virtually all the work of creating, displaying, and experimenting with circle packings for this book was carried out using CirclePack, a comprehensive software package that I have developed over several years. The latest packing algorithms are joint work with my colleague Charles R. Collins. My thanks to him and also to several students who have contributed in other ways to the software, especially George Butler, Woodrow Johnson, and Brock Williams. I further wish to thank the National Science Foundation and the Tennessee Science Alliance for their support.

I cannot say enough (if you have read the book, you may disagree) about the excitement of open-ended, live experiments with circle packing. I encourage the reader to download the software and give it a try. (The copyrighted software is freely available for personal, research, and educational purposes from my Web site (Stephenson, 1992–2004); please read the "LICENSE" file. I anticipate a Java version in the near future.)

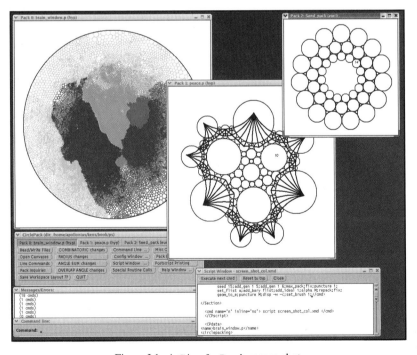

Figure I.1. A CirclePack screen shot.

Bibliography

Agafonov, Sergey I., and Alexander I. Bobenko, "Discrete z^γ and Painlevé Equations," *Internat. Math. Res. Notices*, no. 4 (2000): 165–193.

Agarwal, Pankaj K., and János Pach, *Combinatorial Geometry*, Part I, Section 8, on circle packing. New York: Wiley–Interscience, 1995.

Aharonov, Dov, "The Hexagonal Packing Lemma and Discrete Potential Theory," *Can. Math. Bull.* 33 (1990): 247–252.

"The Descartes Circle Theorem and Applications to Function Theory," (Technion Preprint, 1991).

"The Hexagonal Packing Lemma and the Rodin Sullivan Conjecture," *Trans. Amer. Math. Soc.* 343 (1994): 157–167.

"The Sharp Constant in the Ring Lemma," *Complex Variables Theory Appl.* 33 (1997): 27–31.

Aharonov, Dov, and Kenneth Stephenson, "Geometric Sequences of Discs in the Apollonian Packing," *Algebra i Analiz* [dedicated to Goluzin] 9, no. 3 (1997): 104–140.

Ahlfors L., *Lectures on Quasiconformal Mappings*. Princeton, NJ: Van Nostrand, 1966.

Conformal Invariants: Topics in Geometric Function Theory, Series in Higher Mathematics. New York/Düsseldorf/Johannesburg: McGraw–Hill, 1973.

Complex Analysis, International Series in Pure and Applied Mathematics. New York: McGraw–Hill, 1978.

Ahlfors, L. V., and L. Sario, *Riemann Surfaces*. Princeton, NJ: Princeton University Press, 1960.

Andreev, E. M., "Convex Polyhedra in Lobacevskii Space," *Mat. Sbornik* 81, No. 123 (1970a). [Russian].

"Convex Polyhedra in Lobacevskii Space," *Math. USSR Sbornik* 10 (1970b): 413–440 [English].

"Convex Polyhedra of Finite Volume in Lobacevskii Space," *Mat. Sbornik* 83 (1970c): 256–260 [Russian].

"Convex Polyhedra of Finite Volume in Lobacevskii Space," *Math. USSR Sbornik* 12 (1970d): 255–259 [English].

Bárány, Imre, Zoltán Füredi, and János Pach, "Discrete Convex Functions and Proof of the Six Circle Conjecture of Fejes Tóth," *Can. J. Math.* 36 (1984): 569–576.

Barnard, Roger, and George Brock Williams, "Combinatorial Excursions in Moduli Space," *Pacific J. Math.* 205 (2002): 3–30.

Beardon, A. F., *Complex Analysis*. Chichester/New York/Brisbane/Toronto: Wiley, 1979.

A Primer on Riemann Surfaces, London Mathematical Society Lecture Note Series, vol. 78. Cambridge: Cambridge Univ. Press, 1984.

Beardon, Alan F, *The Geometry of Discrete Groups*, Graduate Texts in Mathematics, Vol. 91. New York/Heidelberg/Berlin: Springer-Verlag, 1983.

Beardon, Alan F., Tomasz Dubejko, and Kenneth Stephenson, "Spiral Hexagonal Circle Packings in the Plane," *Geometriae Dedicata* 49 (1994): 39–70.

Beardon, Alan F., and Kenneth Stephenson, "The Uniformization Theorem for Circle Packings," *Indiana Univ. Math. J.* 39 (1990): 1383–1425.

"Circle Packings in Different Geometries," *Tohoku Math. J.* 43 (1991d): 27–36.

"The Schwarz–Pick Lemma for Circle Packings," *Illinois J. Math.* 35 (1991b): 577–606.

Benjamini, I., and O. Schramm, "Harmonic Functions on Planar and Almost Planar Graphs and Manifolds, via Circle Packings," *Invent. Math.* 126 (1996): 565–587.

Berger, Marcel, "Les placements de cercles," *Pour la Science* [French *Scientific American*] 176 (1992).

Bobenko, A. I., and T. Hoffman, "Conformally Symmetric Circle Packings: A Generalization of Doyle's Spirals," *Experiment. Math.* 10 (2001a): 141–150.

"Hexagonal circle patterns and integrable systems: Patterns with constant angles," *Duke Math. J.* 116 (2003), no. 3, 525–566.

Bobenko, A. I., T. Hoffmann, and B. Springborn, "Minimal Surfaces from Circle Patterns: Geometry from Combinatorics," (Preprint, 2002).

Bobenko, A. I., T. Hoffman, and Yu. B. Suris, "Hexagonal Circle Patterns and Integrable Systems: Patterns with the Multi-ratio Property and Lax Equations on the Regular Triangular Lattice," *Int. Math. Res. Not.* (2002): 111–164.

Bobenko, Alexander I., and Boris A. Springborn, "Variational Principles for Circle Patterns and Koebe's Theorem," *Trans. Amer. Math. Soc.* 356 (2003): 659–689.

Bowers, Philip L., "The Upper Perron Method for Labelled Complexes with Applications to Circle Packings," *Proc. Cambridge Phil. Soc.* 114 (1993): 321–345.

"Negatively Curved Graph and Planar Metrics with Applications to Type," *Michigan Math. J.* 45 (1998): 31–53.

Bowers, Philip L., and Monica K. Hurdal, "Planar Conformal Mapping of Piecewise Flat Surfaces." In *Visualization and Mathematics III,* pp. 3–34. (Berlin: Springer-Verlag, 2003).

Bowers, Philip L., and Kenneth Stephenson, "Uniformizing Dessins and Belyǐ Maps via Circle Packing," *Memoirs Amer. Math. Soc.* vol. 170, no. 805 (2004).

"The set of Circle Packing Points in the Teichmüller Space of a Surface of Finite Conformal Type is Dense," *Math. Proc. Cambridge Phil. Soc.* 111 (1992): 487–513.

"Circle Packings in Surfaces of Finite Type: An in Situ Approach with Applications to Moduli," *Topology* 32 (1993): 157–183.

"A Branched Andreev–Thurston Theorem for Circle Packings of the Sphere," *Proc. London Math. Soc. (3)* 73 (1996): 185–215.

"A 'Regular' Pentagonal Tiling of the Plane," *Conform. Geom. Dyn.* 1 (1997): 58–86.

Brägger, Walter, "Kreispackungen und Triangulierungen," *L'Enseignement Math.* 38 (1992): 201–217.

"Disk Packings and Volume of Hyperbolic Polyhedra" (Preprint, 1993).

Brightwell, Graham R., and Edward R. Scheinerman, "Representations of Planar Graphs," *SIAM J. Discrete Math.* 6, no. 2 (1993): 214–229.

Brooks, Robert, "On the Deformation Theory of Classical Schottky Groups," *Duke Math. J.* 52 (1985): 1009–1024.

"Circle Packings and Co-compact Extensions of Kleinian Groups," *Invent. Math.* 86 (1986): 461–469.

"The Continued Fraction Parameter in the Deformation Theory of Classical Schottky Groups," *Contemp. Math.* 136 (1992): 41–54.

Callahan, Kevin, "The Application of Ahlfors' Value Distribution Theory of Covering Surfaces to Circle Packing" (Ph.D. thesis, University of California, San Diego, 1993).

Callahan, Kevin, and Burt Rodin, "Circle Packing Immersions from Regularly Exhaustible Surfaces," *Complex Variables Theory Appl.* 21 (1993): 171–177.

Cannon, James W., "The Theory of Negatively Curved Spaces and Groups." In *Ergodic Theory, Symbolic Dynamics and Hyperbolic Spaces,* edited by Tim Bedford, Michael Keane, and Caroline Series, pp. 315–369. Oxford/NewYork/Tokyo: Oxford Science Publications, 1991.

"The Combinatorial Riemann Mapping Theorem," *Acta Mathematica* 173 (1994): 155–234.

Cannon, James W., W. J. Floyd, and Walter Parry, "Finite Subdivision Rules," *Conform. Geom. Dyn.* 5 (2001): 153–196.

Cannon, James W., William J. Floyd, Richard Kenyon, and Walter R. Parry, "Constructing Rational Maps from Subdivision Rules," *Conform. Geom. Dyn.* 7 (2003): 76–102.

Carter, Ithiel, "Circle Packing and Conformal Mapping (Ph.D. thesis, University of California, San Diego, 1989).

Carter, Ithiel, and Burt Rodin, "An Inverse Problem for Circle Packing and Conformal Mapping," *Trans. Amer. Math. Soc.* 334 (1992): 861–875.

Chow, Bennett, and Feng Luo, "Combinatorial Ricci Flows on Surfaces," (Preprint, 2002).

Colin de Verdière, Yves, "Empilements de cercles: Convergence d'une méthode de point fixe," *Forum Math.* 1 (1989): 395–402.

"Une principe variationnel pour les empilements de cercles," *Invent. Math.* 104 (1991): 655–669.

Colin de Verdière, Yves, and Frédéric Mathéus, "Empilements de cercles et approximations conformes." In *Comptes rendus de la Table Ronde de Geometrie Riemannienne en l'honneur de Marcel Berger*, Seminaires et Congres (S.M.F., 1994a).

"Empilements de cercles et approximations conformes." In *Les Actes de la Table Ronde de Géométrie Riemannienne en l'honneur de Marcel Berger (Lumigny, 13–17 juillet 1992)*, Seminaires et Congres, No. 1 (S.M.F., 1994b).

Collins, Charles R., Tobin A. Driscoll, and Kenneth Stephenson, "Curvature Flow in Conformal Mapping," *Comput. Methods Function Theory* (2003) 3, no. 1, 325–347.

Collins, Charles R., and Kenneth Stephenson, "A Circle Packing Algorithm," *Comput. Geom. Theory Appli.* 25 (2003): 233–256.

Conway, J. H., and N. J. A. Sloane, *Sphere Packings, Lattices and Codes*, third ed., with additional contributions by Bannai, Borcherds, Leech, Norton, Odlyzko, Parker, Queen, and Venkov (New York: Springer-Verlag, 1999).

Coxeter, H. S. M., "Loxodromic Sequences of Tangent Spheres," *Aeq. Math.* 1 (1968): 104–121.

Regular Complex Polytopes (Cambridge/New York: Cambridge Univ. Press, 1991).

Davis, Burgess, "Brownian Motion and Analytic Functions," *Ann. Probab.* 7 (1979): 913–932.

Doyle, P., Zheng-Xu He, and Burt Rodin, *"The Asymptotical Value of the Circle Packing Constants s_n,"* *Discrete Comput. Geom.* 12 (1994a): 105–116.

"Second Derivatives and Estimates for Hexagonal Circle Packings," *Discrete Comput. Geom.* 11 (1994b): 35–49.

Doyle, P. G., and J. L. Snell, *Random Walks and Electric Networks*, Carus Mathematics Monographs, No. 22 (Math Assoc. Amer., 1984).

Driscoll, Tobin A., "A Matlab Toolbox for Schwarz–Christoffel Mapping," *ACM Trans. Math. Software* 22 (1996): 168–186.

Driscoll, Tobin A., and Lloyd N. Trefethen, "Schwarz–Christoffel Mapping in the Computer Era." In *Proceedings of the International Congress of Mathematicians Berlin, Aug. 1998,* Documenta Mathematica, vol. III, pp. 532–542. (Bielefeld: Bielefeld, 1998).

Dubejko, Tomasz, "Branched Circle Packings, Discrete Complex Polynomials, and the Approximation of Analytic Functions," (Ph.D. thesis, University of Tennessee, 1993).

"Branched Circle Packings and Discrete Blaschke Products, *Trans. Amer. Math. Soc.* 347, no. 10 (1995): 4073–4103.

"Circle-Packing Connections with Random Walks and a Finite Volume Method," *Sém. Théor. Spectr. Géom.* 15 (1996–1997): 153–161.

"Approximation of Analytic Functions with Prescribed Boundary Conditions by Circle Packing Maps," *Discrete Comput. Geom.* 17 (1997a): 67–77.

"Circle Packings in the Unit Disc," *Complex Variables Theory Appl.* 32 (1997b): 29–50.

"Infinite Branched Packings and Discrete Complex Polynomials," *J. London Math. Soc.* 56 (1997c): 347–368.

"Random Walks on Circle Packings," *Contemp. Math.* 211 (1997d): 169–182.

"Recurrent Random Walks, Liouville's Theorem, and Circle Packings," *Math. Proc. Cambridge Philo. Soc.* 121, no. 3 (1997e): 531–546.

"Discrete Solutions of Dirichlet Problems, Finite Volumes, and Circle Packings," *Discrete Comput. Geom.* 22 (1999): 19–39.

Dubejko, Tomasz, and Kenneth Stephenson, "The Branched Schwarz Lemma: A Classical Result via Circle Packing," *Michigan Math. J.* 42 (1995a): 211–234.

"Circle Packing: Experiments in Discrete Analytic Function Theory," *Experiment. Math.* 4, no. 4 (1995b): 307–348.

Duffin, R. J., "Discrete Potential Theory," *Duke Math. J.* 20 (1953): 233–251.

"Extremal Length of a Network," *J. Math. Anal. Appl.* 5 (1962): 200–215.

Duren, Peter L., *Univalent Functions*, vol. 259 (New York: Springer-Verlag, 1983).

Durrett, Richard, *Brownian Motion and Martingales in Analysis* (Belmont, CA: Wadsworth, 1984).

Farkas, Hershel M., and Irwin Kra, *Riemann Surfaces*, Graduate Texts in Mathematics, vol. 71 (New York: Springer-Verlag, 1980).

Fricke, Robert, and Felix Klein, *Vorlesungen über die Theorie der automorphen Functionen*, vols. I and II (Teubner, 1897–1912).

Garnett, John B., *Bounded Analytic Functions* (New York: Academic Press, 1981).

Garrett, B. T., "Euclidean and Hyperbolic Polyhedral Surfaces Determined by Circle Packings," (Ph.D. thesis, University of California, San Diego, 1991).

    "Circle Packings and Polyhedral Surfaces," *Disc. Comput. Geom.* 8 (1992): 429–440.

Gosset, Thorold, "The Hexlet," (*Nature* Feb. 6, 1937): 251–252.

Graham, Ronald L., Jeffery C. Lagarias, Colin L. Mallows, Allan R. Wilks, and Catherine H. Yan, "Apollonian Circle Packings: Number Theory," *J. Number Theory* 100 (2003): 1–45.

Hansen, Lowell J., "On the Rodin and Sullivan Ring Lemma," *Complex Variables Theory Appl.* 10 (1988): 23–30.

He, Zheng-Xu, "Solving Beltrami Equations by Circle Packing," *Trans. Amer. Math. Soc.* 322 (1990): 657–670.

    "An Estimate for Hexagonal Circle Packings," *J. Differential Geometry* 33 (1991): 395–412.

    "Coarsely Quasi-homogeneous Circle Packings," *Michigan Math. J.* 41 (1994): 175–180.

    "Rigidity of Infinite Disk Patterns," *Ann. Math.* 149 (1999): 1–33.

He, Zheng-Xu, and Burt Rodin, "Convergence of Circle Packings of Finite Valence to Riemann Mappings," *Commun. Anal. Geom.* 1 (1993): 31–41.

He, Zheng-Xu, and Oded Schramm, "Fixed Points, Koebe Uniformization and Circle Packings," *Ann. Math.* 137 (1993): 369–406.

    "Rigidity of Circle Domains Whose Boundary Has σ-Finite Linear Measure," *Invent. Math.* 115 (1994): 297–310.

    "Hyperbolic and Parabolic Packings," *Discrete Comput. Geom.* 14 (1995a): 123–149.

    "The Inverse Riemann Mapping Theorem for Relative Circle Domains," *Pac. J. Math.* 171 (1995b): 157–165.

    "Koebe Uniformization for Almost Circle Domains," *Amer. J. Math.* 117 (1995c): 653–667.

    "On the Convergence of Circle Packings to the Riemann Map," *Invent. Math.* 125 (1996): 285–305.

    "On the Distortion of Relative Circle Domain Isomorphisms," *J. Anal. Math.* 73 (1997): 115–131.

    "The C^{∞}-Convergence of Hexagonal Disk Packings to the Riemann Map," *Acta Math.* 180 (1998): 219–245.

Helms, Lester L., *Introduction to Potential Theory* (Huntington, NY: Krieger, 1975).

Hodgson, Craig D., "Deduction of Andreev's Theorem from Rivin's Characterization of Convex Hyperbolic Polyhedra," In *Topology 90*, Proceedings of Research Semester in Low Dimensional Topology at O. S. U. (New York: de Gruyter, 1992).

Hodgson, Craig D., and Igor Rivin, "A Characterization of Compact Convex Polyhedra in Hyperbolic 3-Space," *Invent. Math.* 111 (1993): 77–111.

Hoste, Jim, and Morwen Thistlethwaite, Knotscape, [software package]. Available at http://www.math.utk.edu/~morwen.

Hurdal, M. K., P. L. Bowers, K. Stephenson, D. W. L. Sumners, K. Rehm, K. Schaper, and D. A. Rottenberg, "Quasi-conformally Flat Mapping the Human Cerebellum," In *Medical Image Computing and Computer-Assisted Intervention – MICCAI'99* edited by C. Taylor and A. Colchester, Lecture Notes in Computer Science, vol. 1679, pp. 279–286 (Berlin, Springer-Verlag, 1999).

Hurdal, M. K., and Ken Stephenson, "Cortical Cartography Using the Discrete Conformal Approach of Circle Packings" *NeuroImage* 23 (2004): S119–S128.

Hurdal, Monica K., De Witt L. Sumners, Kelly Rehm, Kirt Schaper, Philip L. Bowers, Ken Stephenson, and David A. Rottenberg, "A Quasi-conformal Map of the Cerebellar Cortex," *NeuroImage* 9 (1999): S194.

Hurdal, Monica K., De Witt L. Sumners, Ken Stephenson, Philip L. Bowers, and David A. Rottenberg, "CirclePack Software for Creating Quasi-conformal Flat Maps of the Human Brain," *NeuroImage* 9 (1999a) S250.

    "Generating Conformal Flat Maps of the Cortical Surface via Circle Packing," *NeuroImage* 9 (1999b) S195.

Jonasson, Johan, and Oded Schramm, "On the Cover Time of Planar Graphs," *Electron. Commun-Probab.* 5 (2000): 85–90 (paper 10).

Jones, G., and D. Singerman, *Complex Functions: An Algebraic and Geometric Viewpoint* (Cambridge / New York: Cambridge Univ. Press, 1987).

Kakutani, S., "Two-dimensional Brownian motion and the type problem of Riemann surfaces," *Proc. Japan Acad.* 21 (1945): 138–140.

Koebe, P., "Kontaktprobleme *der Konformen Abbildung*," *Ber. Sächs. Akad. Wiss. Leipzig Math.-Phys. Kl.* 88 (1936): 141–164.

Kojima, Sadayoshi, Shigeru Mizushima, and Ser Peow Tan, "Circle Packings on Surfaces with Projective Structures," *J. Differential Geom.* 63, no. 3 (2003): 349–397.

Lagarias, J. C., C. L. Mallows, and A. Wilks, "Beyond the Descartes Circle Theorem," *Amer. Math. Monthly* 109 (2002): 338–361.

Lehto, O., and K. I. Virtanen, *Quasiconformal Mappings in the Plane*, second ed., (New York: Springer-Verlag, 1973).

Lehto, Olli, *Univalent Functions and Teichmüller Spaces*, (New York: Springer-Verlag, 1987).

Malitz, S., and A. Papakostas, "On the Angular Resolution of Planar Graphs," *STOC* 24 (1992): 527–538.

Malitz, S., and A. Papakostas, "On the Angular Resolution of Planar Graphs," *SIAM J. Discrete Math.* 7 (1994): 172–183.

Marden, Al, and Burt Rodin, "On Thurston's Formulation and Proof of Andreev's Theorem," In *Computational Methods and Function Theory, Proceeding, Valparaiso 1989*, Lecture Notes in Mathematics, Vol. 1435, pp. 103–115. (Berlin/Heidelberg/New York/Tokyo: Springer-Verlag, 1990).

Marshall, Don, *Zipper, Software for Computation and Display of Conformal Mapping*, available at http://math.washington.edu/~marshall/personal.html.

Marx, Morris L., "Extensions of Normal Immersions of S^1 into $\mathbf{R}^2$," *Trans. Amer. Math. Soc.* 187 (1974): 309–326.

Massey, William S., *Algebraic Topology: An Introduction*, Graduate Texts in Mathematics, Vol. 56 (New York/Heidelberg/Berlin: Springer-Verlag, 1997).

Mathéus, Frédéric, "Rigidité des empilements infinis immerges proprement dans le plan et dans le disque," *Sémi. Théor. Spectr. Géom.* 12 (1993–1994), 69–85.

"Empilements de cercles: Rigidité, discretisation d'applications conformes" (Ph.D. thesis, Institut Fourier (Grenoble), 1994, These de Doctorat de l'Universite Grenoble I).

"Empilements de cercles et représentations conformes: Une nouvelle preuve du théoréme de Rodin-Sullivan," *Enseign. Math.* 42 (1996): 125–152.

McCaughan, G. J., "Some Topics in Circle Packing" (Ph.D. thesis, University of Cambridge, 1996).

McCaughan, Gareth, "A Recurrence/Transience Result for Circle Packings," *Proc. Amer. Math. Soc.* 126 (1998): 3647–3656.

Miller, Gary L., Shang-Hua Teng, and Stephen A. Vavasis, "A Unified Geometric Approach to Graph Separators," In *Proc. 32nd Symp. on Found. of Comput. Sci.* pp. 538–547 (1991).

Miller, Gary L., and William Thurston, "Separators in Two and Three Dimensions." In *Proceedings of the 22nd Annual ACM Symposium on Theory of Computing, Baltimore, May 1990* pp. 300–309 (ACM).

Minda, David, and Burt Rodin, "Circle Packing and Riemann Surfaces," *J. Analy. Math.* 57 (1991): 221–249.

Mizushima, Shigeru, "Circle Packings on Complex Affine Tori," *Osaka J. Math.* 37 (2000): 873–881.

Mohar, Bojan, "A Polynomial Time Circle Packing Algorithm," *Discrete Math.* 117 (1993): 257–263.

Morgan, John W., "On Thurston's Uniformization Theorem for Three-Dimensional Manifolds." In *The Smith Conjecture*, edited by Hyman Bass and John W. Morgan, pp. 37–126 (New York: Academic Press, 1984).

Morley, Frank, "The Hexlet," *Nature* (Jan. 9, 1937): 72–73.

Mumford, David, Caroline Series, and David Wright, *Indra's Pearls* (Cambridge: Cambridge Univ. Press, 2002).

Needham, Tristan, *Visual Complex Analysis* (Oxford/New York: Oxford Univ. Press, 1997).

Nehari, Zeev, "A Generalization of Schwarz' Lemma," *Duke Math. J.* 14 (1947): 1035–1049.

Nevanlinna, Rolf, *Analytic Functions* (New York/Heidelberg/Berlin: Springer-Verlag, 1970).

Nimersheim, Barbara, "Isometry Classes of Flat 2-Tori Appearing as Cusps of Hyperbolic 3-Manifolds Are Dense in the Moduli Space of the Torus." In *Low-Dimensional Topology* edited by K. Johannson (Boston/Knoxville, International Press, 1992).

Riemann, Bernhard, *Grundlagen für eine allgemeine Theorie der Functionen einer veränderlichen complexen Grösse, Inauguraldissertation, Göttingen, 1851*, Gesammelte mathematische Werke und wissenschaftlicher Nachlass, 2nd ed., pp. 3–48 (Leipzig, Teubner, 1876).

Rivin, I., "On Geometry of Convex Polyhedra in Hyperbolic 3-Space" (Ph.D. thesis, Princeton, 1986).

Rivin, Igor, "Euclidean Structures on Simplicial Surfaces and Hyperbolic Volume," *Ann. Math.* 139 (1994): 553–580.

"A Characterization of Ideal Polyhedra in Hyperbolic 3-Space," *Ann. Math.* 143 (1996): 51–70.

Rodin, Burt, "Schwarz's Lemma for Circle Packings," *Invent. Math.* 89 (1987): 271–289.

"Schwarz's Lemma for Circle Packings II," *J. Differential Geometry* 30 (1989): 539–554.

"On a Problem of A. Beardon and K. Stephenson," *Indiana Univ. Math. J.* 40 (1991): 271–275.

Rodin, Burt, and Dennis Sullivan, "The Convergence of Circle Packings to the Riemann Mapping," *J. Differential Geometry* 26 (1987): 349–360.

Rothen, F., and A.-J. Koch, "*Phyllotaxis or the Properties of Spiral Lattices. II. Packing of Circles along Logarithmic Spirals,*" *J. Phys. Fr.* 50 (1989): 1603–1621.

Rothman, Tony, "Japanese Temple Geometry," *Scie. Amer.* (1998).

Rudin, Walter, *Real and Complex Analysis*, 3rd ed., (New York: McGraw–Hill, 1986).

Sachs, Horst, "Coin Graphs, Polyhedra, and Conformal Mapping," *Discrete Math.* 134 (1994): 133–138.

Schramm, Oded, "Packing Two-Dimensional Bodies with Prescribed Combinatorics and Applications to the Construction of Conformal and Quasiconformal Mappings" (Ph.D. thesis, Princeton, 1990).

"Existence and Uniqueness of Packings with Specified Combinatorics," *Israel J. Math.* 73 (1991): 321–341.

"Rigidity of Infinite (Circle) Packings," *J. Amer. Math. Soc.* 4 (1991): 127–149.

"Square Tilings with Prescribed Combinatorics," *Israel J. Math.* 84 (1993): 97–118.

"Conformal Uniformization and Packings," *Israel J. Math.* 93 (1996): 399–428.

"Circle Patterns with the Combinatorics of the Square Grid," *Duke Math. J.* 86 (1997): 347–389.

"Scaling Limits of Loop-Erased Random Walks and Uniform Spanning Trees," *Israel J. Math.* 118 (2000): 221–288.

Siders, Ryan, "Layered Circlepackings and the Type Problem," *Proc. Amer. Math. Soc.* 126, no. 10 (1998): 3071–3074.

Singer, I. M., and J. A. Thorpe, *Lecture Notes on Elementary Topology and Geometry*, Undergraduate Texts in Mathematics (New York/Heidelberg: Springer-Verlag, 1967).

Sodd, F., "The Hexlet," *Nature* (Dec. 5, 1936): 958, [poem].

"The Kiss Precise," *Nature* (June 20, 1936): 1021, [poem].

"The Bowl of Integers and the Hexlet," *Nature* (Jan. 9, 1937): 77–79.

Somer, Frank L., Jr., G. S. Canright, Theodore Kaplan, Kun Chen, and Mark Mostoller, "Inherent Structures and Two-Stage Melting in Two Dimensions," *Phys. Rev. Lett.* 79 (1997): 3431.

Stephenson, Kenneth, "Construction of an Inner Function in the Little Bloch Space," *Trans. Amer. Math. Soc.* 308 (1988): 713–720.

"The Geometry of Image Surfaces of Analytic Functions," In *Bounded Mean Oscillation in Complex Analysis* edited by Ilpo Laine and Eero Posti, Univ. of Joensuu Publications in Sciences, No. 14, pp. 101–120 (1989).

"Circle Packings in the Approximation of Conformal Mappings," *Bull. Amer. Math. Soc. (Research Announcements)* 23, no. 2 (1990): 407–415.

CirclePack *software*, available at http://www.math.utk.edu/~kens (1992–2004).

"A Probabilistic Proof of Thurston's Conjecture on Circle Packings," *Rendi. Semi. Mat. Fisi. Milano* 65 (1996): 201–291.

"Approximation of Conformal Structures via Circle Packing," In *Computational Methods and Function Theory 1997, Proceedings of the Third CMFT Conference*, edited by N. Papamichael, St. Ruscheweyh, and E. B. Saff, vol. 11, pp. 551–582 (Singapore, World Scientific, 1999).

"Circle Packing and Discrete Analytic Function Theory," *Handbook of Complex Analysis*, Vol. 1, Geometric Function Theory, edited by R. Kühnau (Amsterdam, Elsevier, 2002).

Stillwell, John, *Geometry of Surfaces*, (New York: Springer-Verlag, 1992).

Sullivan, Dennis, "On the Ergodic Theory at Infinity of an Arbitrary Discrete Group of Hyperbolic Motions," In *Riemann Surfaces and Related Topics: Proceedings of the 1978 Stony Brook Conference*, Ann. of Math. Stud., vol. 97, pp. 465–496 (Princeton NJ: Princeton Univ. Press, 1981).

Thurston, William, *The Geometry and Topology of 3-Manifolds*, (Princeton University Notes, Preprint).

"The Finite Riemann Mapping Theorem," invited talk (An International Symposium at Purdue University in Celebration of de Branges' Proof of the Bieberbach Conjecture, March 1985).

Toth, L. Fejes, "Research Problems," *Period. Math. Hung.* 8 (1977): 103–104 (submitted question).

Van Eeuwen, Jeff, "The Discrete Schwarz–Pick Lemma for Overlapping Circles," *Prac. Amer. Math. Soc.* 121 (1994): 1087–1091.

Wagner, Richard, "Ein kontaktproblem der konformen abbildung," *J. Reine Angew. Math.* 196 (1956): 99–132.

Weyl, H., *The Concept of a Riemann Surface* (Reading, MA: Addison–Wesley, 1955).

Williams, George Brock, "Discrete Conformal Welding," *Indiana Univ. Math. J.* 53 (2004): 765–804.

"Approximation of Quasisymmetries Using Circle Packings," *Discrete Comput. Geom.* 25 (2001a): 103–124.

"Earthquakes and Circle Packings," *J. Anal. Math.* 85 (2001b): 371–396.

"Noncompact Surfaces are Packable," *J. Anal. Math.* 90 (2003), 243–255.

Woess, Wolfgang, *Random Walks on Infinite Graphs and Groups*, (Cambridge: Cambridge Univ. Press, 2000).

Index

354